Relativistic Magnetrons

Relativistic Magnetrons

Igor Vintizenko

CISP

CRC Press
Taylor & Francis Group
Boca Raton London New York

CRC Press is an imprint of the
Taylor & Francis Group, an **informa** business

Translated from Russian by V.E. Riecansky

CRC Press
Taylor & Francis Group
6000 Broken Sound Parkway NW, Suite 300
Boca Raton, FL 33487-2742

First issued in paperback 2020

ISBN 13: 978-0-367-57111-5 (pbk)
ISBN 13: 978-1-138-59532-3 (hbk)

Visit the Taylor & Francis Web site at
http://www.taylorandfrancis.com

and the CRC Press Web site at
http://www.crcpress.com

Contents

Introduction

The intensive development during the last decades of research in the field of high-current electronics is associated with a wide range of practical applications of high-current relativistic electron beams with a pulsed power of up to 10^{12} W at an electron energy of 10^5–10^7 eV, a current of up to 10^6 A, and a duration of 10^{-9}–10^{-6} s. Such parameters of the beams make it possible to exceed by several orders of the level of the power of electromagnetic radiation existing in traditional microwave electronics, which opens up radically new possibilities for using radiation in various fields of science and technology. The first successful experiment was carried out by a team of researchers of the Institute of Applied Physics (Gorky, USSR) and the Institute of General Physics (Moscow, USSR) under the guidance of Academician A.V. Gaponov, where pulses with a power of more than 300 MW at a frequency of 10 GHz with an efficiency of 15% were obtained, showing the possibilities of relativistic high-frequency electronics. High-current electron accelerators (HCEA) have been used to study many relativistic analogues of microwave electronics and relativistic generators of a new type. By now, the values of the pulsed power to 10^{10} W have been obtained in relativistic microwave devices. Relativistic high-frequency electronics has become one of the fastest growing areas of scientific research.

At the Tomsk Polytechnic University (Tomsk, USSR), thanks to the creation in 1974 of the Tonus-1 high-current electron accelerator with parameters of 1 MV, 40 kA and a pulse duration of 80 ns, it became possible to experimentally study the interaction of high-current relativistic electron beams with the electromagnetic fields of microwave systems. These studies were initiated on the initiative of A.N. Didenko and for many years were conducted under his leadership. The experimental base of these studies has been continuously expanded: new accelerators of nano- and microsecond duration with a pulsed current of up to 100 kA and linear induction accelerators (LIA) have been created, which make it possible to operate in a periodic mode with a repetition rate of hundreds of hertz. This made it possible to carry out investigations of a wide range of microwave devices, as are known in classical electronics (backward-wave tubes, Cherenkov tubes, klystrons, magnetrons) and based on new principles (free-electron laser, vircators, magnetrons with virtual cathode).

During the research, significant difficulties were found in the realization of the energy capabilities of the accelerators. The pulse energy of the electron beam can reach a megajoule, while the energy in the microwave pulse is only hundreds of joules. The low efficiency of energy conversion is caused by the motion of electrode plasmas, microwave breakdowns, excitation of 'parasitic' modes of oscillation, etc. The practical application of the generated microwave radiation is also difficult, which is connected with the large weight and dimension parameters of the installations and, as a rule, with the monopulse mode of their operation, the low quality of the output radiation. Finally, any relativistic microwave generator has a set of inherent specific properties that determine the requirements for the quality of the electron beam, the magnetic system, the parameters of the accelerator (internal resistance, voltage range, shape and pulse width), etc. All this imposes restrictions on the possibility of using any of the relativistic generators for the whole range of applications, which, in turn, causes that many different instruments must be used in the investigations.

The well-known advantages of devices with crossed fields predetermined interest in them from the relativistic high-frequency electronics. The frequency and phase stability of the generated radiation, high efficiency, small weight and dimensions and cost, low level of side oscillations and harmonics became the basis for intensive research of relativistic magnetrons. The first experiment in the study of the relativistic magnetron (RM) was the device developed by G.P. Fomenko and A.S. Sulakshin in 1976 in the Tomsk Polytechnic University (Fig. 1). The anode block was radically different from the classical magnetron, it had 20 rectangular resonators of different depths (a different resonator configuration was realized). This form of resonators was chosen because of the convenience of manufacturing. To increase the volume of the interaction space, the inner diameter of the anode block was 20 cm. A comparatively low power level (~100 MW) and an efficiency of less than 1%, recorded in the experiment at a voltage of 1 MV, served as an incentive for searching for the

Fig. 1. Photograph of the anode block.

optimal design and creating the device, taking into account known recommendations for devices with crossed fields.

The first experiments with relativistic magnetrons, which yielded notable results, in the USA – the Massachusetts Institute of Technology and the USSR – the Tomsk Polytechnic University – allowed power levels from hundreds of megawatts to several gigawatts with an efficiency of 10–30% .

This monograph is devoted to theoretical and experimental studies of relativistic magnetrons. Since the mid-1970s and up to the present, these devices have been studied in various scientific centres and find a number of practical applications in experiments on the scattering of an electromagnetic wave by a high-current electron beam, in tests of radioelectronic equipment, for preliminary excitation and phase synchronization of vircators, pumping of gas lasers, radars, including non-linear ones, for heat treatment and modification of semiconductor materials and metal surfaces.

Chapter 1 describes the design of relativistic magnetrons and their differences from classical instruments, presents methods for calculating the parameters of the resonator system and field theory, and describes the ways to control the output parameters of the microwave radiation.

The study of the processes of interaction of electrons with microwave waves of the anode block of planar and cylindrical geometry is the subject of Chapter 2. The use of high-voltage power supplies of RM leads to the need to take into account relativistic factors. It is known that the conditions for neglecting relativistic corrections are reduced to the requirement that the stresses of the electric fields measured in units of mc^2/e be small, and that the geometric dimensions of the magnetron are determined in units of $n\lambda/2\pi$ (m is the rest mass of the electron, c is the speed of light, e is the electron charge, n is the number of the mode of oscillation, and λ is the wavelength). In spite of the fact that in magnetron-type devices there is a transformation not of the kinetic energy of electrons, but potential energy, and, therefore, relativism is not so rigidly connected with the cathode–anode voltage, nevertheless, for a number of devices with a small magnitude of the slowing of the electromagnetic wave it is important to take the relativistic factor into account. Indeed, theoretical studies performed in the late 1970s show the influence of the relativistic dependence of the electron mass on its velocity on the microwave generation processes. On the basis of the averaging method proposed independently by P.L. Kapitsa and V.E. Nechaev for solving non-relativistic equations of motion of electrons in the interaction space of the device, an elementary theory of the relativistic magnetron of planar geometry was constructed. However, there is a significant difference between the processes of interaction of electrons with electromagnetic fields in the cylindrical geometry of the magnetron. This difference is due to the fact that in a cylindrical magnetron the condition of synchronism of the electromagnetic wave and the electron beam is exactly satisfied only at a certain radius, while for a planar magnetron the electrons are in phase with the wave along the entire height of the spoke. The impossibility of exact synchronism in the entire interaction space leads to the appearance of an additional azimuthal drift of the electrons and to distortion of the trajectories of the motion of electrons in the spokes. Therefore, the theoretical study of processes in the relativistic magnetrons of cylindrical geometry is given considerable attention in the monograph.

The sources of power for relativistic magnetrons initially included high-current electron accelerators, containing pulsed voltage generators discharged through the forming line to a magnetron diode. In spite of the fact that the voltage pulses had a nanosecond duration, the shortcomings of the RM, associated with the destruction of the anode blocks within several hundred pulses, were detected. This process is caused by several factors: 1) the high energy of the electrons of the anode current, which leads to the development of processes of evaporation, erosion, mechanical deformation of elements under the action of thermal shock; 2) inconsistency of the internal resistance of the accelerator forming line (2–24 ohms) with the impedance of the relativistic magnetron (40–100 Ohm), the appearance of repeated pulses and additional energy release in the magnetron diode. When a relativistic magnetron operates in the frequency regime, when a large number of pulses are recruited during short time intervals, it becomes especially necessary to perform thermal calculations on the surface of the resonator system to determine the parameters of the electron beams that allow the device to last for a long time. One of the paragraphs of Chapter 2 is devoted to thermal processes on the surface of the anode blocks of the RM.

Chapter 3 describes the power sources of the RM. Accelerators that are well known, as well as linear induction accelerators developed in the Tomsk Polytechnic University are briefly presented. The element base of the HCEA based on the pulsed voltage generators allows them to work exclusively in a single mode or, in extreme cases, with a frequency of several hertz. The first experiments with a relativistic magnetron in the pulse–periodic mode (a packet of three pulses with a repetition rate of 160 Hz was generated) in the TPU stimulated the development of research in this direction. The LIA has been used in these experiments and until now. Moreover, the LIAs developed at the TPU favourably differ from analogues (Compact LIA – Physics International Company and SNOMAD-Science Research Laboratory – USA) by an original layout scheme due to placement in a single case of a low-impedance strip forming line, an induction system, multichannel spark dischargers, and non-linear saturation chokes. In this way, small weight and dimensions of the installations are achieved. Chapter 3 also describes the process of modelling the power source and the non-linear load – the relativistic magnetron. Since the output parameters of the power source exert a significant influence on the processes of microwave generation by RMs, it is necessary to simulate a similar system characterized by strong feedback. The computer model for calculating the output parameters is represented by equivalent circuits, and the non-linear elements of the equivalent relativistic magnetron circuit are determined in accordance with the theory of averaged motion. A concrete realization of the model is considered with the example of a linear induction accelerator on magnetic elements intended for working with relativistic magnetrons with a high pulse repetition rate. The computer model makes it possible to calculate the processes occurring in a linear induction accelerator and enables the operative selection of the parameters of the power source and the magnetron generator for tuning to the extrema by the output power, the total and electronic efficiency.

It is the pulse–periodic mode of operation of the RM (Chapter 4) that seems to be the most promising for the practical application of devices, provided that compact radiation complexes are created with high output parameters in terms of pulse power, efficiency and pulse repetition rate. The following trends are singled out in which the use of microwave sources with a high average power is actual: for microwave power systems of linear resonant electron accelerators with a high rate of acceleration; in radar, including non-linear; in studies of the electromagnetic compatibility of radio electronic equipment; for sterilization.

The appearance of consumers of high-power microwave radiation determined the list of requirements for such complexes:

- high reproducibility of output pulses in amplitude and shape;
- range of voltages not exceeding 500 kV to ensure satisfactory radiation protection at low material costs;
- small weight and dimensions;
- 'user-friendly' type of radiated wave;
- possibility of frequency tuning;
- low level of side fluctuations and harmonics;
- durability of elements;
- maximum efficiency of conversion of primary storage energy into electromagnetic radiation;
- high pulse repetition rate;
- the duration of the microwave pulse should be close to the duration of the power pulse.

The last three parameters determine the average power of the installation, which is the most important characteristic: $P_{av} = P_{imp} \cdot \tau$, where P_{imp} is the pulse power of the RM; f is the repetition rate of pulses; τ is the pulse duration.

To obtain a high average power of electromagnetic radiation, it is necessary to realize a high frequency of repetition of pulses. This task is of a complex nature and is related to the development of: 1) power sources in terms of their parameters (voltage, current, pulse duration) most fully meeting the requirements of the RM; 2) magnetic and vacuum systems, permitting long-term operation of the RM with a high repetition rate; 3) directly pulse-periodic RM with increased durability of the elements.

The monograph also discusses relativistic magnetrons with a microsecond voltage pulse duration (Chapter 5). In the earliest experiments, effects were found that limit the duration of the microwave radiation in comparison with the duration of the voltage pulse, which is associated with the accelerated radial motion of the cathode plasma under the influence of the powerful electromagnetic fields of the anode block and the violation of the synchronism conditions between the electron beam and the microwave wave velocities. Investigations of these processes were important for determining the further development of relativistic magnetrons, which are different from the traditional configuration for choosing the most suitable one. Thus, inverted relativistic magnetrons (IRMs) appeared, including with coaxial resonators, as well as relativistic magnetrons with external injection of an electron beam into the interaction space.

In these devices, a centrifugal instability mechanism does not develop in the cathode plasma, and one can expect an increase in both the voltage pulse duration and microwave radiation. The problem of fast destruction of anode blocks is also solved. The monograph discusses some of the results of studies of such devices.

A feature of the RM is the mode instability, which manifests itself in spontaneous changes in the amplitude, time, and frequency parameters of the generated microwave pulses. This is due to a number of factors. First, the relativistic magnetron, like its classical analog, is a multifrequency device. Its oscillatory system has a distributed electrodynamic structure and is multimodal. Secondly, significant changes in the power supply fields occur in the generation mode during the pulse, which leads to the fulfillment of excitation conditions for various modes of oscillation. Another factor of frequency instability is caused by the formation of a cathode plasma, which, with its radial expansion, decreases the interelectrode gap and leads to changes in the resonant properties of the oscillatory system of the magnetron. However, direct use for the RM of traditional ways of stabilizing the generation process (the use of straps, high-Q resonators) is limited by the high level of generated power and the short duration of the microwave pulse.

The problem of increasing the power output from the resonance system and the formation of directional radiation is also topical for the relativistic microwave generating devices. For the RMs, as for most relativistic devices, this problem is due to power limitations due to high-frequency breakdowns in the output waveguide outputs.

Therefore, the final chapter 6 discusses the non-traditional possibilities of controlling the oscillating process in the RM, its spectrum and fluxes of oscillatory energy. The main method in this case is the introduction into the resonant system of the magnetron of control external links and a purposeful influence with their help on the oscillatory dynamic processes. The chapter gives a step-by-step description of the experimental investigation of the influence of the external communication channel of the resonators of the anode block of a relativistic magnetron on the conditions of existence and stability of oscillations, the spectral and energy characteristics of the generation process. An analysis of the experimental results is carried out, as well as an assessment of the control capabilities of the microwave radiation parameters of the RM. Variants of the modified circuit of the magnetron generator are discussed, differing in that a system of load-radiators is introduced into the channel for external coupling of resonators. Depending on the configuration (symmetry) of the communication channel, the modes can be realized in the system with summation or subtraction of the powers of the resonators in the total load. Details are presented of experimental results demonstrating the possibility of selective stabilization of the selected (operating) mode of RM oscillations, including with a given oscillation profile on the radiator system for spatial radiation formation.

Abbreviations

EEE – explosive electron emission
DFL – double forming line
EFS – electronic frequency shift
FEL – free electron laser
FPC – ferroelectric plasma cathode
HCEA – high-current electron accelerators
ICDMI – inverted coaxial diodes with magnetic insulation
ICM – inverted coaxial magnetron
IRM – inverted relativistic magnetrons
LIA – linear induction accelerator
MPG – magnetic pulse generator
ODP – oil-vapour diffusion pump
PVG – pulse voltage generator
RM – relativistic magnetrons
RMI – relativistic magnetron with external injection
SFL – single forming line
TMP – turbomolecular pump
TPU – Tomsk Polytechnic University
VC – virtual cathode

1

Relativistic Magnetrons and Their Difference from Classical Magnetrons

The magnetron was invented by A. Hall [1] in 1921 as a power converter and power regulation device. In 1928, two Japanese professors, Yagi and Okabe, found that a magnetron, in which the anode was cut into two or more segments, could generate high frequencies. The multiresonator magnetron was created in 1936 by Soviet engineers N.F. Alekseev and D.E. Malyarov and was described in 1940 in the press [2]. Around the same time, H. Booth and D. Randall developed a similar design in the UK. The successful use of these devices during the Second World War for radar stations caused a rapid development of theoretical and experimental research, as well as work on the practical use of magnetrons [3–7]. Such advantages as high efficiency, amplitude and frequency stability, durability, small weight dimensions and cost, characteristic for classical magnetrons and devices based on them, predetermined interest to them from relativistic high-frequency electronics [8].

The relativistic magnetron (RM) is one of the few devices of relativistic high-frequency electronics which has already found practical application: in experiments on the scattering of an electromagnetic wave on a high-current electron beam; in tests of electronic equipment; for preliminary excitation and phase synchronization of vircators; pumping of gas lasers; radar, including non-linear radiolocation; modification of semiconductor materials and metal surfaces.

The relativistic magnetrons are in fact a high-current version of the classical magnetron. Relativistic voltages are needed to excite explosive electron emission [9] and to obtain high currents. The principle of operation is identical to the classical device. However, determining the range of the operating conditions and calculating the output parameters require taking into account the relativistic factors in the corresponding formulas. For a long time, the analysis of relativistic effects in magnetron systems was not given serious significance. Only the conditions under which these effects can be neglected were known. Therefore, in the monograph, the author considered it expedient to mention the theory of a relativistic magnetron of planar geometry developed by V.E. Nechaev, M.I. Petelin, M.I. Fuchs, and present the original theory of the relativistic magnetron of cylindrical geometry. These theories make it possible, on the basis of simple physical considerations, to estimate the possibilities and to indicate the optimal parameters of a relativistic magnetron without the limitation of the relativistic factor.

High cathode-anode voltage and high current in a relativistic magnetron lead to the appearance of specific effects associated with the destruction of anode blocks under the action of thermal shock and significant current losses caused by the influence of the azimuthal magnetic field. These issues are also given attention in this chapter.

To date, a large amount of experience has been accumulated and many papers have been devoted to the experimental studies of the RM [8, 10–72]. The data on the input and output characteristics of devices and some design features are presented in Table 1.1 and in the Appendix to chapter 1, on the basis of which one can determine the characteristic parameters of relativistic magnetrons and compare them with classical devices.

1.1. Design of relativistic magnetrons

In order to highlight the general structural features of the RM, as well as differences from classical devices, below is a brief review of experimental work, including a description of individual subsystems of relativistic magnetrons and installations in general.

Typical designs of relativistic magnetron generators of direct geometry are shown in Fig. 1.1 *a*, *b*, and the RMs of the inverted geometry are presented in Fig. 1.1 *c*, *d*. Cathodes *2* are installed coaxially to the anode block *1*. The region between the cathode and the anode forms the interaction space of electrons with microwave waves of the resonator system. The anode block consists of identical resonators separated by lamellae (segments) and connected with

Table 1.1.

Parameter	Classical magnetron	Relativistic magnetron
Cathode type	Thermoemission	Explosion emission
Voltage	Less than 50 kV	100–1500 kV
Current	Less than 0.1 kA	3–100 kA
Duration of voltage pulse	1 to 20 μs or continuous mode of work	30 ns–1.5 μs
Steepness of the voltage pulse front	up to 100 kV/μs	up to 100 kV/ns
Duration of microwave radiation pulse	Complies with duration of voltage pulse	20 ns–1.2 μs
Power	Less than 10 MW	100–10 000 MW
Radiation wavelength	0.8–60 cm	3–30 cm
Efficiency	50–82%	Less than 30%

the interaction space by the coupling slots. Typically, the cathode is connected to a high-voltage electrode *5* of a negative polarity supply source, the anode is grounded. For inverted magnetrons, a multi-cavity anode is located inside the cathode and is connected to a positive electrode *9* of the power source. The RM elements are placed in the vacuum chamber *4*. To limit the loss (end) current from the interaction space, a vacuum chamber (drift tubes) of large diameter [71] is applied. Such devices allow approximately 20% increase in output power. For this purpose, the classical devices use end screens on the cathode, but for the relativistic magnetrons they are not used to avoid 'parasitic' electron emission and breakdown [19, 20].

Electrons in a relativistic magnetron are emitted from the cathode surface by creating an electric field strength between the cathode and the anode that exceeds the critical value for excitation of explosive electron emission accompanied by evaporation of the material and the formation of a cathode plasma. The axial magnetic field formed by the magnetic system *3* causes the electrons to move along the cycloidal trajectories and prevents their rectilinear movement on the surface of the anode. Microwave energy from the resonators of the anode block is emitted into the free space by means of a waveguide *6* or diffraction *8* output of power and antenna *7*. The inverted magnetron systems use special microwave output devices

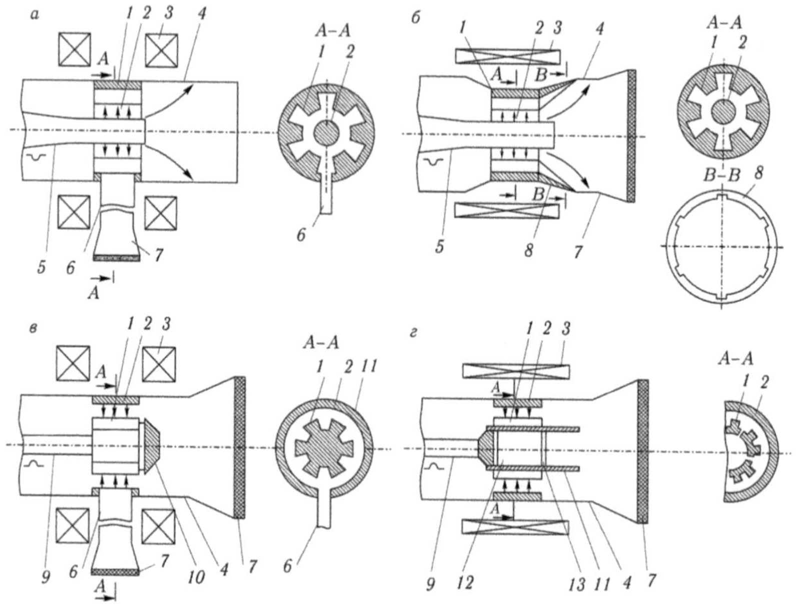

Fig. 1.1. Typical RM designs are: *a*) RM of direct geometry with waveguide power output; *b*) RM of direct geometry with diffraction power output; *c*) inverted relativistic magnetron; *d*) inverted coaxial relativistic magnetron; *1* – anode block; *2* – cathode; *3* – magnetic system; *4* – vacuum chamber; *5* – cathode holder; *6* – waveguide output of power; *7* – antenna; *8* – diffraction power output; *9* – anode holder; *10* – device for output of microwave radiation; *11* – filter types of waves; *12* – stabilizing resonator; *13* – the diaphragm.

(coupling elements) *10* or filters for the types of waves are applied for inverted coaxial generators occupied with the internal stabilizing resonator *12* through the diaphragm *13*.

1.2. Voltage diagram, modes of oscillations of a relativistic magnetron

In the static case (all physical quantities are constant in time), for the magnetron diode in the absence of microwave fields one can write the energy conservation laws and the generalized electron momentum conservation law in the following form [73]:

$$\gamma = 1 + e\left(U + \Phi_c\right) / mc^2; \tag{1.2.1}$$

$$\gamma r^2 \dot{\varphi}(r) = eH(r^2 - r_c^2) / 2mc, \tag{1.2.2}$$

where e, m are the charge and mass of an electron; c is the speed of light; U is the voltage applied between the cathode and the anode; Φ_c is the potential of the electron cloud; r is the radius; r_c is the radius of the cathode; $\dot{\varphi}(r)$ is the angular velocity of azimuthal rotation; H is the magnetic field. Since

$$\gamma = (1 - V^2 / c^2)^{-1/2}, \quad V = \sqrt{\dot{r}^2 + (r\dot{\varphi})^2}, \tag{1.2.3}$$

and in the static case $\dot{r} = 0$, $\dot{r}$ is the radial velocity, equation (1.2.2) is transformed to the form

$$\frac{eH}{mc^2}\frac{r^2 - r_c^2}{r} = \sqrt{\gamma^2 - 1}. \tag{1.2.4}$$

From the last expression we can find the value of the critical voltage (the mode when the vertex of the cycloid of the electron trajectory touches the surface of the anode and $\Phi_c = 0$ at $r = R$, R is the inner radius of the anode block resonators) for the non-relativistic and relativistic energy ranges:

$$U_{\text{cr.non}} = \frac{mc^2}{2e}\left(\frac{R^2 - r_c^2}{2R}\right)^2\left[\frac{eH}{mc}\right]^2; \tag{1.2.5}$$

$$U_{\text{cr.rel.}} = \frac{mc^2}{e}\left[\sqrt{1 + \left[\frac{eH}{mc}\right]^2\left[\frac{R^2 - r_c^2}{2R}\right]^2} - 1\right]. \tag{1.2.6}$$

When the cathode–anode voltage exceeds the level of 100 kV for typical dimensions of the cathode–anode gap of the relativistic magnetrons of $d \sim 1$–2 cm, the manifestation of the relativistic effect and the difference in the parabolas of the critical regime become noticeable, as illustrated in Fig. 1.2.

The multiresonator relativistic magnetron refers to the generators of the Cherenkov type. Cherenkov synchronism eliminates the cyclotron mechanism, and for unperturbed cyclotron rotation the interaction of electrons with a slow electromagnetic wave is realized under the following condition:

$$V_\varphi = \frac{cE_r}{H} \approx V_p = \frac{r\omega}{n}, \tag{1.2.7}$$

where E_r is the static electric field strength between the cathode and the anode; V_φ is the azimuthal rotation speed of electrons in the interelectrode gap; V_p is the phase velocity of the electromagnetic wave; ω is the frequency of the oscillations of the azimuthal mode with the number n.

This resonance is achieved when the Buneman–Hartree condition [74] for the cylindrical geometry of electrodes and taking into account relativistic effects is fulfilled:

$$U_{\text{th.rel}} = H\beta_p \frac{R^2 - r_c^2}{2R} - \frac{mc^2}{e}\left(1 - \sqrt{1-\beta_p^2}\right), \qquad (1.2.8)$$

where $\beta_p = V_p/c$. This condition means that when the voltage between the cathode and the anode is below the threshold value $U_{\text{th.rel}}$ synchronous motion of electrons with an electromagnetic wave is not possible at any radius.

The voltage diagram (Fig. 1.2) shows the dependence (1.2.8), as well as the corresponding dependence without regard for relativistic effects. Only in the region of voltages and magnetic fields between

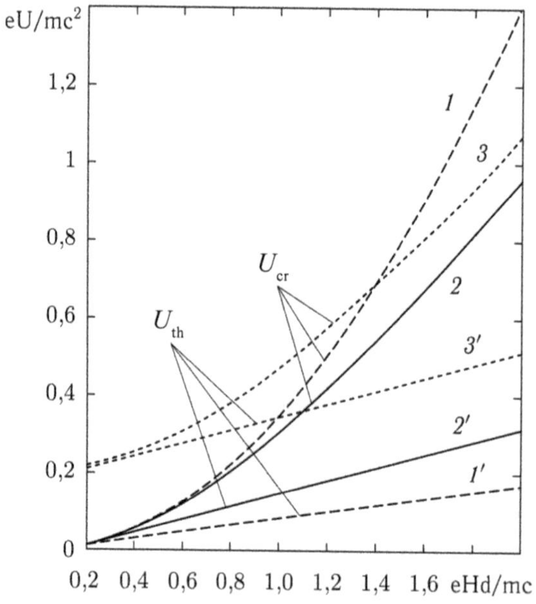

Fig. 1.2. The voltage diagram of the RM without taking into account (curves *1*, *1′*) and taking into account (curves *2*, *2′*) the relativistic effects and the influence of the azimuthal magnetic field flowing through the cathode (curves *3*, *3′*).

the critical mode parabola and the Buneman–Hartree curve is it possible to generate microwave oscillations in a certain mode.

The change in the operating mode range of the RM can also be caused by axial currents. The high-current effect in the RM is manifested in the appearance of an axial drift of electrons under the action of the azimuthal magnetic field ($H_\varphi = \mu_0 I_z/2\pi d_{\text{cat}}$, where d_{cat} – distance from the cathode), created by the current flowing along the cathode I_z. In this case, the ratio for the value of the critical voltage acquires the following form:

$$U_{\text{cr}} = \sqrt{\left(\frac{mc^2}{e}\right)^2 + H^2\left[\frac{R^2 - r_c^2}{2R}\right] + H_\varphi^2} - \frac{mc^2}{e}. \tag{1.2.9}$$

The lower limit is determined by the Buneman–Hartree threshold voltage:

$$U_{\text{th}} = H\beta_p\left[\frac{R^2 - r_c^2}{2R}\right] - \frac{mc^2}{e}\left(1 - \sqrt{1 + H_\varphi^2}\sqrt{1-\beta_p^2}\right). \tag{1.2.10}$$

In Fig. 1.2, the dependences of these voltages on the axial magnetic field calculated for I_z = 10 kA are given as an example. Since the azimuthal magnetic field interferes with the radial movement of electrons to the anode, this leads to an increase in the operating voltages of the relativistic magnetron.

The oscillatory system of the relativistic magnetron, like the classical magnetron, consists of N coupled resonators interacting with the electron flux by means of gaps (slots) located along the surface of the anode block. The resonance condition in such a system is the integer number of wavelengths that fit along the length of the block. That is, when traversing along the entire circumference of the inner surface of the anode block, the total phase shift should be a multiple of 2π: $N\varphi = 2\pi n$, where n = 0, 1, 2,... It follows that the phase difference of the oscillations in the resonators can take only the discrete values determined by the relation: $\varphi = \dfrac{2\pi n}{N}$. Thus, the types of magnetron oscillations can be characterized by the number n and the magnitude of the phase shift φ. For a six-cavity magnetron (N = 6) we obtain (Table 1.2):

Table 1.2

n	0	1	2	3	4	5	6
	0	$\pi/3$	$2\pi/3$	π	$4\pi/3$ $(-2\pi/3)$	$5\pi/3$ $(-\pi/3)$	2π (0)

In a general case, the magnetron anode block has N types of oscillations. Beginning with the value $n = N$, a further increase in n does not lead to types of oscillations that are physically different from those that correspond to the values of n from 0 to $(N - 1)$. At $n = 0$, the oscillations in all resonators occur in phase. When $n = N/2$, the neighbouring resonators oscillate in antiphase, that is, with a phase shift of π.

The modes of oscillation of the anode block generally have different frequencies. The oscillations with the numbers $n = 1$ and $n = (N-1)$ differ only in the sign of the phase difference. The configuration of the high-frequency electromagnetic field for these modes is the same, their resonant frequencies coincide. These modes are degenerate. A similar situation occurs with the oscillation modes $n = 2$ and $n = (N-2)$, and so on up to $(N/2-1)$ and $(N/2+1)$. With complete symmetry of the anode block with N types of oscillations only the modes $n = 0$ (2π-type) and $n = N/2$ (π-type) remain non-degenerate. Thus, the N-resonator anode block has $(N/2+1)$ different resonant frequencies.

The natural frequencies can be estimated using the equivalent circuit method (see Fig. 1.3). For a vane-type magnetron, the eigenfrequency for the n-th mode is determined by the expression [6]

$$\omega_n = \frac{\omega_0}{\sqrt{1 + \dfrac{C_1}{2C_0\left(1-\cos\varphi\right)}}},$$

where $\omega_0 = 1/\sqrt{L_0 C_0}$, L_0, C_0 are the parameters of a separate resonator (inductance and capacitance); C_1 is the capacitance between the segment of the anode block and the cathode.

The presence of azimuthal asymmetry of the fields (due, for example, to the elements of power output) removes the degeneracy, which leads to the formation of doublets with closely arranged frequencies. It is known from practice that it is difficult to ensure stable operation of the generator at one of the frequencies of the doublet. For this reason, the π-type, which does not degenerate, is usually chosen as the basic (working) form of oscillations of the magnetron.

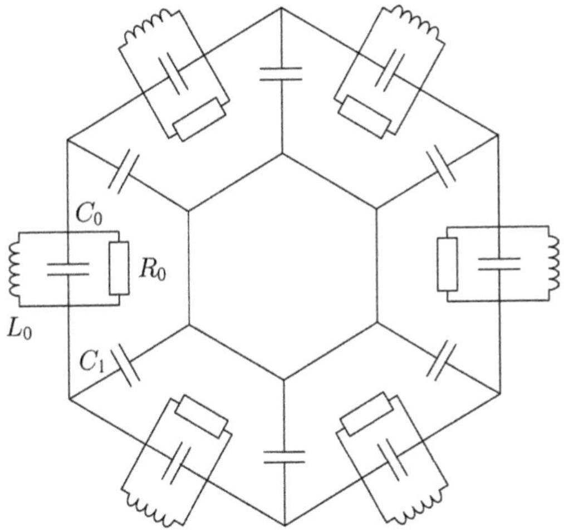

Fig. 1.3. Equivalent circuit of a six-resonator magnetron.

Various methods are used, based on the organization of additional connections, to separate the types of oscillations by frequencies and stabilize the main working mode.

The classical and most common method is the use of straps [4]. Straps are called conductive elements that are directly introduced into the construction of the anode block and connect those segments whose oscillations for the main modes must be in phase. Non-phase oscillations of the resonators create a potential difference at the ends of the straps. Significant currents flowing through these straps, suppress the resulting parasitic oscillation, and the reactive nature of the conductivity of the straps leads to a frequency shift.

Another fairly effective method of stabilization is based on the use of an external high-Q resonator and the phenomenon of pulling in the generation frequency [75]. In such a magnetron, the oscillatory system consists of two coupled resonant systems: a system whose field interacts with the electron beam, and a stabilizing resonant system.

Direct use for the relativistic magnetron of the above methods of stabilizing the generation process is associated with a number of problems. First, a high level of generated power makes it impossible to use the straps. Secondly, for the relativistic magnetron, a significant change in the feeding fields during the pulse is characteristic, which leads to the fulfillment of excitation conditions for different modes of oscillation [24]. Thirdly, the short duration of the microwave pulse of a relativistic magnetron limits the possibilities for using an external high-Q resonator.

1.3. Elements of relativistic magnetrons

Magnetic systems of the RM. To create a longitudinal magnetic field in the operation conditions of single voltage pulses or with a low repetition rate it is necessary to use electric pulse systems. The synchronization device ensures the operation of the (RM) pulsed power supply at the maximum of the current pulse of the magnet.

Superconducting magnetic systems [29, 56, 58], as well as direct current magnets [61, 63, 66, 69] are used for the operation of devices in the pulse-periodic mode. In experiments [61, 63], a magnet was used, consisting of four separate coils. In each coil there were three layers of turns from a hollow copper trough for water circulation, separating four layers of continuous copper coils. Such a magnet formed a magnetic field by induction to 0.8 T and allowed the relativistic magnetron to operate in a packet mode for 10 seconds with a 30-minute pause to cool the coils.

The distribution of the magnetic field along the axis of the magnet must be as homogeneous as possible. Otherwise, electrons in different sections of the anode block will rotate with different azimuthal velocities and the synchronism condition with the microwave wave will be satisfied only for part of the flow. For the RM with the radial output of microwave power to obtain a uniform magnetic field, a system of two or four Helmholtz coils is used (see Fig. 1.1, *a*, *c*). When power is axial output along the axis of the system, solid solenoids are used (see Fig. 1.1, *b*, *d*).

Vacuum systems of the RM. Various vacuum systems are used to create a vacuum in the interaction space. Traditionally, oil–vapour diffusion pumps (ODP) or turbomolecular pumps are used, complete with pre-pumping plate-rotor pumps. It is also possible to use freezing nitrogen traps, which can significantly reduce the contamination of the pumped volume by oil vapors. Advantages of the ODP are the ease of operation and the small weight and size.

The experiments were carried out [62, 66] using the cryogenic vacuum pumps which required replacement of the accelerator's insulators from organic materials to ceramic ones. The authors of [63] note an increase in the output power of the magnetron by a factor of 1.3 with vacuum improvement from $2 \cdot 10^{-5}$ to $8 \cdot 10^{-6}$ Torr.

Anode blocks of the RM. In various studies, multi-cavity (from 6 to 16 resonators) anode blocks of length from $\lambda/3$ to 2λ were used. For most relativistic magnetrons the slowing-down systems of length $L \sim \lambda$ are used. Long systems can reduce the density of the anode current, which is hundreds of amperes per square centimeter, and short ones improve the separation of frequencies between oscillation modes. Usually, vane-type resonators are used, more rarely slit and 'slit-hole' type. The material for the anode blocks was stainless steel, aluminum, copper, and stainless steel, coated with copper. With almost identical output parameters, the anode blocks made of stainless steel showed greater durability due to higher erosion resistance to electron beam bombardment. The authors of [54] used fully copper anode blocks and copper with molybdenum coating to obtain high-energy pulses. In the latter case, microwave pulses of longer duration were observed, which the authors attribute to the high melting point of molybdenum, but the pulsed power was reduced by 10–20% in comparison with copper anode blocks.

In practice, anode blocks are used closed, half-open and open geometry. The difference is in the use of rings (end caps) at the ends of the anode block. In this case, the value of the loaded Q-factor, the generated frequency, and, more importantly, the axial distribution of the high-frequency field, change. For a closed system, the field has a maximum in the centre of the block, for a half-open system on the edge where the ring is missing. As a rule, closed systems provide higher output parameters [14–16], however, when using high-current power supplies half-open anode blocks are preferable [24]. As noted earlier, a high axial current along the cathode produces an azimuthal magnetic field that carries electron spokes to the edge of the anode block to the maximum of the microwave field for a half-open anode block.

In experiments with the vane-type cavity resonator systems, the number of resonators did not exceed eight. Different resonator systems were used to increase the efficiency and output power of the device, due to the increase in the volume of the interaction space and the increase in the number of resonators up to 12 [24] and 16 [47]. Such systems are designed to solve the problem of mode competition in the magnetron. However, in the experiments of Ref. 47, simultaneous excitation of two $\pi/2$ and $\pi/3$ modes was observed, as well as of jumping from $3\pi/8$ to π- or to $7\pi/8$-type during a pulse. In experiments [24], due to technical difficulties, it was not possible to excite the π-type, while at the same time for the half-open system, high values of power and efficiency were obtained at the −1-harmonic of the $\pi/6$-type oscillation.

It should be noted that the determination of the mode of oscillation was carried out by comparing the experimental values obtained with the 'cold' measurements and the calculated values of the oscillation wavelengths, and by calculating the magnitude of the electromagnetic wave slowing-down wave. Direct measurement of the phase shift between neighbouring resonators of the RM was carried out in Ref. [48], where the value of the phase difference is experimentally recorded equal to π.

The Q_n of the resonant system is determined by the type and the geometric dimensions of the resonators. This parameter is very important, since the duration of the RM voltage pulse is tens of nanoseconds. Thus, a sufficiently small loaded quality factor of the electrodynamic system is needed so that the oscillation rise time in it is comparable with the duration of the current pulse. In [13], considering perturbations of the near-cathode flux of electrons as a result of their interaction with a synchronous wave and using the theory of resonator excitation, the following expression is obtained for the increment of rise of oscillations:

$$\operatorname{Im}\omega = -\sqrt{\left(\frac{\omega}{4Q_n}\right)^2 + \frac{E_r S_c \chi_{\mathrm{HF}}^2}{2nV}} - \frac{\omega}{4Q_n}, \tag{1.3.1}$$

where ω is the oscillation frequency; χ_{HF} is the coupling coefficient between the azimuthal component of the microwave electric field in the region of the cathode layer and the mean microwave electric field in the resonators $\bar{E}$; S_c is the area of the cathode; V is the volume of the resonator; n is the number of the mode of oscillation. Thus, in the mode of nanosecond pulses, when the rise time of oscillations should be small, it is necessary to take structures in which the energy of the high-frequency field is stored directly in the interaction space. For voltage pulse lengths of the order of one microsecond, magnetrons with stabilizing resonators can be used, for which $\chi_{\mathrm{HF}} \ll 1$.

The rise time of the oscillations can be estimated, according to [13], as

$$\tau_{\mathrm{rise}} = \frac{(N/2+7)}{\operatorname{Im}\omega}. \tag{1.3.2}$$

Let us estimate $\operatorname{Im}\omega$. For the parameters of the 6-resonator magnetron

(λ = 10 cm, d = 0.84 cm, R = 2.15 cm, r_c = 1.1 cm, L = 7.2 cm), we can calculate the volume of the resonator system consisting of the volume of the interaction space: $V_1 = \pi L(R^2 - r_c^2)$ and the volume of the resonators. The latter in the case of vane-type resonators is $V_2 = \pi LlR$, where l is the depth of the resonators; in the case of resonators of the 'slit-hole' type

$$V_2 = \pi L\frac{N}{2}\left(R\frac{d_1 + d_2}{N} + \rho_1^2 + \rho_2^2\right),$$

where d_1 and d_2 are the depth of the slits, ρ_1 and ρ_2 are the radii of cavity holes. The area of the cathode is $S_c = 2\pi r_c L$. The azimuthal component of the high-frequency electric field in the cathode region is:

$$E_y = \bar{E}\,\mathrm{sh}\,pR' / \mathrm{sh}\,pd, \tag{1.3.3}$$

where

$$R' = E_N / (H_N^2 - E_N^2) \tag{1.3.4}$$

– the Larmor radius of rotation of electrons;

$$E_N = eE_r / mc^2; \quad P_N = eH / mc; \quad p = \pi\sqrt{\frac{1}{D^2} - \frac{4}{\lambda^2}}$$

– transverse wave number for the working π-type of oscillations; $D = 2\pi R/N$ is the period of the anode block.

Thus, the formulas for the rise time of oscillations for 6- and 12-cavity relativistic magnetrons will be written as follows:

$$\tau_{\text{rise}} \approx 0.79 / (\sqrt{U}\,\mathrm{sh}\,pR'); \tag{1.3.5}$$

$$\tau_{\text{rise}} \approx 1.73 / (\sqrt{U}\,\mathrm{sh}\,pR'). \tag{1.3.6}$$

Figure 1.4 shows the dependence of the rise time of the oscillations on the magnitude of the voltage U for different values of the magnetic field.

Thus, for relativistic magnetrons, according to the estimate (1.3.5)–(1.3.6), the transient time is a few nanoseconds.

In [82, 83], the influence of the loaded Q-value on the output power of the devices was considered from the results of numerical calculations. Calculations show that a Q-factor reduction of 3 times leads to a proportional increase in the efficiency of the instrument.

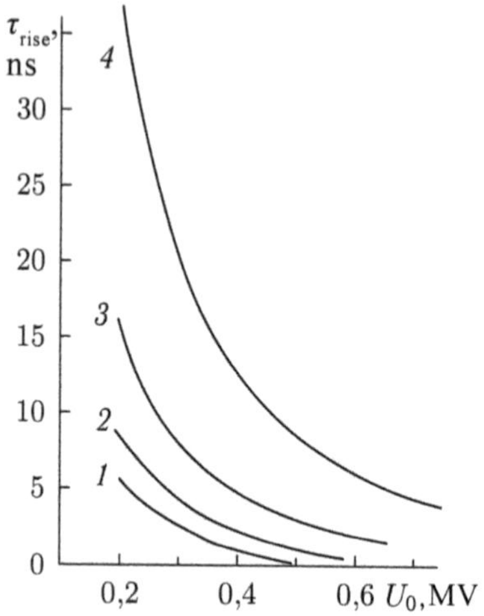

Fig. 1.4. Dependence of the rise time of the oscillations on the magnitude of the voltage for different values of the magnetic field strength *H*, kOe: curve *1* – 2.5, curve *2* – 3.0, curve *3* – 4.0, curve *4* to 6.0.

A small value of the loaded *Q* is provided by using high-frequency energy output devices from generators with a coupling factor. Structurally, such devices are performed in the form of resonant windows (communication slots) with a large size (see Fig. 1.1 *a*), or from several resonators, or in the form of a diffraction output (see Fig. 1.1 *b*).

Output of microwave power from the RM. Traditionally, the output power is a smooth waveguide transition from the coupling gap in the resonator to the size of a standard waveguide [11–26]. In the output waveguide, the lower mode of the rectangular waveguide H_{10} is excited, which allows efficiently transporting, radiating, converting microwave energy into other waveguide modes.

It is known [4] that when waveguide leads through one resonator, the symmetry of the high-frequency field is violated, degeneracy in the structure of synchronous waves is removed, doublets are formed in the spectrum, which can lead to hopping of modes of oscillation. In order to preserve the symmetry of the distribution of microwave fields in the interaction space, in experiments [16], energy was simultaneously extracted from all the resonators of the anode block into equal loads. However, the expected increase in power did not occur and the recorded level of ~4.6 GW was achieved by increasing, for example, in comparison with [8] the voltage level. Further development of this method of emission [28] has shown the influence of the number of output waveguides on the characteristics of RM. A 6-cavity relativistic magnetron with a number of waveguide outputs from 1 to 6 was used. The total power, depending on the number of outputs, varied as follows: 1.4, 2.8, 3.3, 2.0, 2.5 and 3 GW. As can be seen from the above dependence, the use of 3 outputs is most effective, which is connected with the formation of a symmetric distribution of high-frequency fields compared with case 1, 2, 4 and 5 outputs and better optimization of the loaded *Q*-factor of the slowing-down system in comparison with the case of 6 outputs. These results were influenced by the type of power source – an accelerator with a low-impedance forming line, which allows a significant change in the load impedance with an increase in the RM anode current without a noticeable decrease in voltage.

When the microwave is radiated from a 3 cm relativistic magnetron [10, 23], the output unit is made as a conical horn formed by smoothly expanding continuations of the resonators and creating a diffraction coupling of the magnetron with a cylindrical waveguide. Such systems provide, in addition to an increased dielectric strength of the output, also the selection of modes of oscillations by the axial index due to the decrease in the quality factor of the mode with increasing its longitudinal number. Diffraction power output also makes it possible to reduce the loss current of a relativistic magnetron due to the natural expansion of the electron drift region. Uniform loading of all resonators creates conditions for more stable operation at the chosen azimuthal mode of oscillations. However, when the power output is applied, a wave of the H_{n1} type (where $n \geq 3$) is excited in the output waveguide. Such waves have zero field strength along the axis, which is not always convenient for practice and requires the use of wave-type converters. In addition, when using a diffraction output, deflecting magnets are required to discharge the electrons of the end current to the wall of the output horn to prevent electrons from entering the output dielectric window. At the same time, with the deposition of electrons on the wall, a collector plasma arises, which, propagating at a speed of $2 \cdot 10^6$ cm/s, overlaps the output window and screens the microwave radiation. Therefore, in devices operating with microsecond-duration power pulses and having electrodynamic systems of small dimensions, the duration of the microwave pulse can be shortened. To avoid the effect of shielding the microwave radiation it is advisable to use electrodynamic structures that allow spatially to separate the beam and power output areas. In a relativistic magnetron this is naturally done in the construction of the anode block with a radial waveguide output of power from the resonators through the resonance window (see Fig. 1.1 *a*).

Cathodes of the RM. The characteristics of the cathode material have a significant effect on the work of the relativistic magnetron. In the experiments of various scientific groups, the effect on the output characteristics of the following parameters of the cathode was investigated: the material, the length and the diameter, the position in the interaction space of the device.

By how often one or another cathode material is used, materials can be arranged in the following order: stainless steel, graphite, copper, brass, molybdenum, metal-dielectric. Stainless steel cathodes are used with powerful high-current electron accelerators. The operation of such power sources is characterized by a large entrainment of the cathode material during the pulse, and the use of stainless steel makes it possible to minimize this process. Graphite cathodes are used in relatively low-power installations, when the cathode is required to ensure the uniform formation of explosion-emission centres. Graphite has the advantage of decreasing the delay time of the appearance of current, and provides stable emission characteristics [78].

In [55, 59] experiments with RMs are described using the so-called velvet cathode (graphite cloth is applied to the surface of the cathode) and it is noted that the onset of current emission occurs at lower electric field strengths than for graphite, due to which the duration of the microwave pulse is 200 ns, i.e., 2 times higher than for a metal cathode.

In most experimental studies, the cathode is executed in the form of a cylinder located coaxially to the anode block along its entire length. In experiments [25, 28], a washer cathode in the form of a stainless steel disc with a diameter exceeding the diameter of the cathode holder was studied in detail. It is noted that such a cathode produces more power than a cylindrical cathode. These data contradict the results of most other experiments in which the cathode of cylindrical geometry provides a higher output parameter. Apparently, the construction of a long cylindrical cathode is still more promising.

The diameter of the RM cathode determines the impedance of the diode and its matching with the internal resistance of the power source. Usually the dependence of the generated microwave power of a relativistic magnetron on the cathode diameter has a resonant character. In addition, the ratio of the diameter of the cathode to the

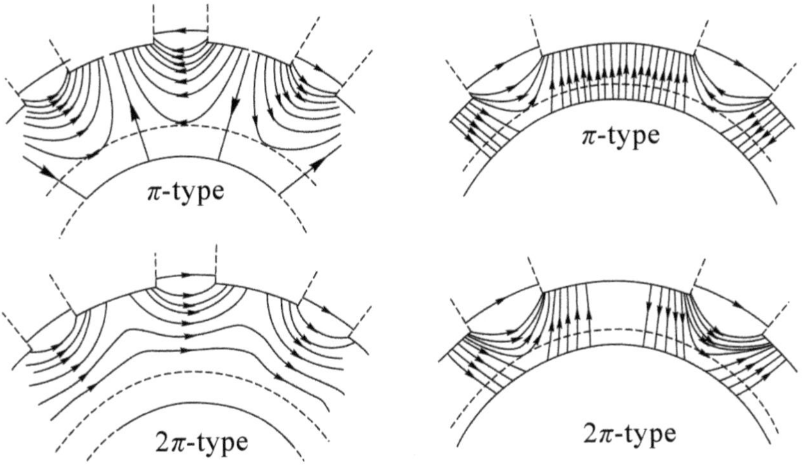

Fig. 1.5. Distribution of high-frequency electric fields of the π- and 2π-type oscillations for a wide and a narrow interelectrode gap.

diameter of the anode determines the efficiency of the interaction of the high-frequency field with the space charge, specifying the 'sagging' of the microwave field in the interelectrode gap. So, in the case of a 10-cm relativistic magnetron with a wide gap (7–16 mm), the optimum for interacting with electrons is the π-type of oscillations due to the approximately equal distribution of the azimuthal and radial components of the microwave field (Fig. 1.5). For a narrow gap (4–6 mm) [14, 16], the 2π-type becomes optimal, which is far inferior in frequency from other types. However, with gaps of the order of 5 mm, it is necessary to ensure good centring of the cathode, since radial displacement of the cathode by approximately 1 mm, as experiments have shown, causes a 3-fold drop in the generated power. In addition, it is dangerous in terms of destruction of the anode block.

It is noted in [84] that the expansion of the cathode plasma into the interaction space of the instrument is the main reason for shortening the duration and reducing the power of the generated microwave pulse. To reduce this effect, a relativistic magnetron was tested for a 30 cm wavelength band. This six-cavity magnetron with two power outputs from the opposite resonators was tested with different types of cathodes at a voltage of about 500 kV with an output power close to 1 GW. The length of the emission surface of the cathodes was 110 mm, the diameter of the cathodes was 74 mm. The following cathodes, shown in Fig. 1.6, were tested:

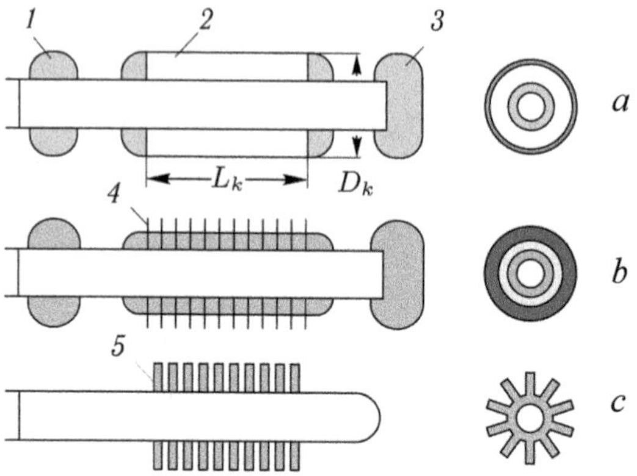

Fig. 1.6. Constructions of cathodes for relativistic magnetron: *a*) 'smooth' cathodes: from velvet, graphite, perforated stainless steel, *b*) disc cathode, *c*) multi-pin cathode. *1*, *3* – end screens, *2* – emitting surface, *4* – graphite discs, *5* – points made of stainless steel.

1) velvet cathode with an external perforated tube (47% geometrical transparency of the pipe) made of stainless steel with a hole diameter of 2 mm;
2) a graphite cathode of the same dimensions;
3) cathode with thin graphite discs;
4) multi-point stainless steel cathode (pin-cathode).

The first two cathodes had a smooth emission surface, the disc cathode was an intermediate variant between cathodes with a smooth surface and a multi-point cathode. The cathodes were tested both with end screens (as in classical magnetrons) and without them. For classical magnetrons, a method for increasing their efficiency is known, consisting in the use of screens installed at the edges of cathodes outside the interaction space [4]. The end screens have an outer diameter that exceeds the diameter of the cathode, and sometimes exceeds the diameter of the anode, which prevents electrons from escaping from the interaction space under the action of space charge forces. For relativistic magnetrons, the end screens were not used earlier because of the fear of the formation of explosive electron emission (EEE) on surfaces protruding relative to the surface of the cathode. The formation of a plasma and its drift in a longitudinal magnetic field with a characteristic rate of explosive electron emission of the order of 10^7 cm/s can lead to a plasma closure of the gap between the end screen and the end of the anode block and the shorting of the power source. It should be added that the RM cathodes are connected to the power supply only on one side, unlike the classical magnetrons, which, as a rule, have a symmetrical cathode supply. Therefore, in RM, in addition to losses due to space-charge forces, there is an additional factor of current loss from the interaction space under the action of the azimuthal magnetic field, which arises from the current flowing through the cathode. In the crossed radial electric field between the cathode and the anode and the azimuthal magnetic field of the flowing current, a power acts on the electrons, pushing them out of the interaction space.

As shown by experiments [84], in the case of using end screens on both sides of the cathode, a much higher microwave power was achieved than using any of the cathodes without screens. It is characteristic that at low magnetic fields the behaviour of the power dependences on the magnetic field for cathodes with and without end screens coincided, the differences were observed with increasing magnetic field. For cathodes with end screens, the values of the synchronous magnetic field are higher, while the generated power is one and a half to two times higher, reaching 850 MW (for example, for a velvet cathode). Higher magnetic fields were required because of the increased cathode–anode voltage. This effect is caused by a decrease in the current loss from the interaction space and, as a consequence, an increase in the impedance of the relativistic magnetron. The results of tests of a relativistic magnetron with different types of cathodes are shown in Fig. 1.7.

In addition, the authors drew attention to the dependence of the duration of the microwave pulse on the generated power (Figure 1.8), recorded for different values of the magnetic fields. It was noted that the behavior of these dependencies does not depend on the presence or absence of end screens. Only for a disk cathode, which is significantly different from other cathodes, was found a completely different behavior of the curve $P(\tau)$. This made it possible to conclude that the cathode emitting surface had a significant effect on the duration of the generated microwave pulse.

Among the cathodes studied, the velvet cathode demonstrated the best output characteristics and durability. After 1000 'shots', the authors did not notice any significant change in the characteristics of the magnetron, which is probably due to the fact that the surface of the graphite is closed partly with metal from above. However, considerable deterioration of the graphite surface of the cathode was observed. The lifetime of such a cathode of the order of 1000 pulses is extremely small for practical application in pulsed-periodic

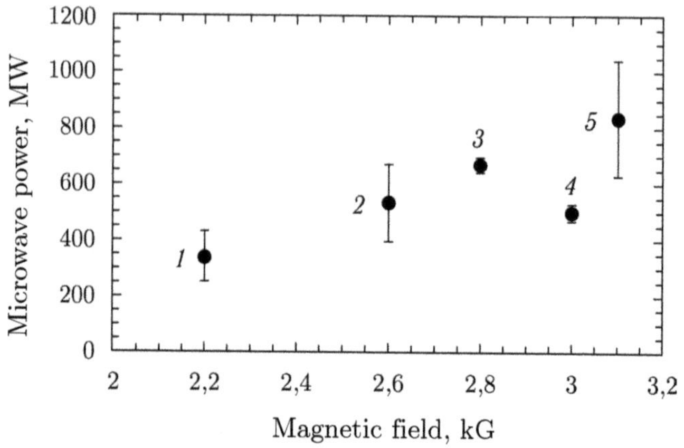

Fig. 1.7. Pulse power of the RM when using: a multi-point cathode without (*1*) and with (*4*) end screens, 'smooth' cathodes without (*2*) and with (*5*) end screens, a disc cathode without (*3*) end screens.

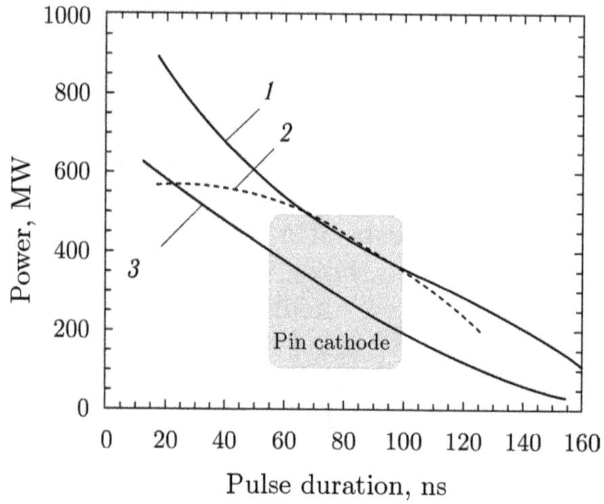

Fig. 1.8. Dependence of the duration of the microwave pulse at half-power level on the output power for velvet cathodes with (*1*) and without (*3*) end screens, disk cathode (*2*) with end screens.

RMs. Therefore, the authors tested the cathode without velvet-fabric, only with a perforated metal tube. The RM with this cathode showed slightly smaller values of the power and duration of the microwave pulse at a significantly higher lifetime. Thus, the use of end screens in relativistic magnetrons by analogy with classical devices leads to a noticeable increase in the generated power without decreasing the pulse width. The experimental results show that the use of cathodes

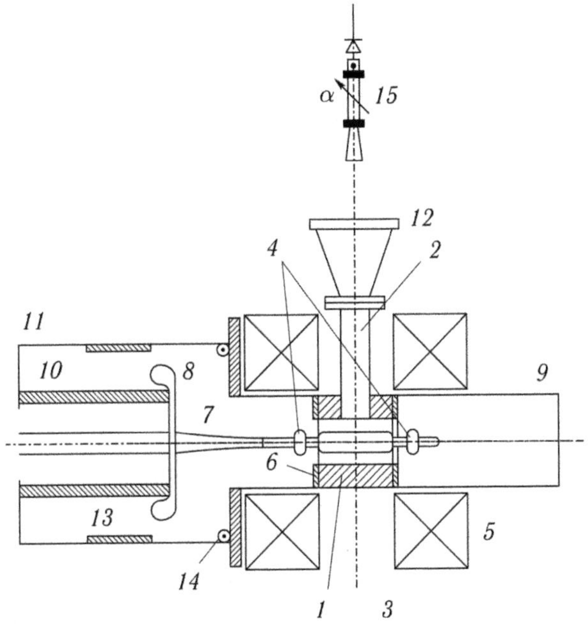

Fig. 1.9. Scheme of the experiment: *1* – anode block of the RM; *2* – waveguide output of microwave power; *3* – cathode; *4* – end screens; *5* – magnetic system; *6* – end caps of the anode block; *7* – cathode holder; *8* – high-voltage flange; *9* – drift tube; *10* – insulator of the linear induction accelerator (LIA); *11* – LIA vacuum chamber; *12* – horn antenna with vacuum window; *13* – capacitive voltage divider; *14* – Rogowski coil; *15* – microwave detector with receiving antenna and attenuator.

providing a uniform formation of a plasma surface makes it possible to achieve not only greater power, but also longer duration.

The effect of the diameter of the end screens and their location relative to the edges of the cathode on the output parameters of the generator was investigated in [85]. The experiments were performed on a linear induction accelerator LIA 04/6. The scheme of the experiment is shown in Fig. 1.9. The RM of the 10-cm wavelength range had a 6-cavity anode block *1* with an inner diameter of 43 mm and a length of 72 mm with a waveguide output of power *2* and a graphite cathode *3* of a diameter of 20 mm with a length of 72 mm and an end screen coaxially to the anode. A direct current created a magnetic field by induction to 0.59 T. End caps *6* with a thickness of 5 mm are installed at the ends of the anode block. The cathode with the help of the cathode holder *7* is connected to the high-voltage flange *8* of the LIA, to which a negative voltage pulse was applied. The diameter of the high-voltage flange and the distance to the anode block were initially chosen from the results of calculations of the magnetic field lines, so that the lines emerging from the RM interaction space were 'closed' on the surface of the high-voltage flange. This reduces the loss of electrons from the interaction space. To limit the loss current from the opposite side of the anode block, a drift tube *9* with an inner diameter of 184 mm was used. The larger the diameter of the drift tube, the lower the limiting transport current in it. However, the outer diameter of the tube is limited by the internal diameter of the magnetic system. An excessive increase in diameter leads to an unjustified increase in the magnetization volume and, correspondingly, to an increase in the energy consumption for the creation of a magnetic field.

The end screens were made of duralumin in the form of washers with an outer diameter of 22, 24, 28 and 32 mm 10 mm thick and an outer radius of cylindrical surface rounding of 5 mm. The screens could move relative to the edges of the cathode, for this purpose a thread was made on the surface of the cathode holder, and on the side of the drift tube the cathode holder protruded beyond the anode block by 70 mm.

The characteristics of a high-voltage pulse of the LIA were recorded with a capacitive voltage divider *13*, the Rogowski coil *14* for measuring the total current of the accelerator, and a lamp detector *15* with a calibrated waveguide attenuator placed in the 'far' zone for recording microwave pulses were used.

To determine the effect of cathode end screens on the microwave pulse parameters of the RM, experiments with a cathode without end screens, with one screen from the side of the high-voltage flange, with one screen on the side of the drift tube and with two screens, were carried out at the initial stage of the study. The diameter of the end screens was 24 mm, and the distance between the screens and cathode was 10 mm on each side (up to the end caps of the anode block – 5 mm each).

Figure 1.10 shows the dependence of the power and duration of the microwave pulses on the value of the magnetic field induction, taken with an equal charging voltage of the primary power source of the LIA. As can be seen from Fig. 1.10 *a*, the use of two metal screens located at both ends of the cathode leads to the greatest increase in the power of microwave pulses in comparison with the case without screens. However, the positive contribution of screens to this trend is different. The screen mounted on the side of the drift tube leads to a larger increase in power. This allows one to conclude that the main current losses from the interaction space of the RM occur in the drift tube. The measures taken to reduce losses to the surface of the high-voltage flange have already been described. The above dependences also show that in the case of the use of end screens, the region of synchronous magnetic fields is expanded, at which the maximum microwave power level is recorded. Analysis of the oscillograms showed that when using screens, an increase in the output voltage of the LIA is observed in comparison with the cathode without screens due to a decrease in current losses. In turn, this requires an increase in the synchronous magnetic field and is

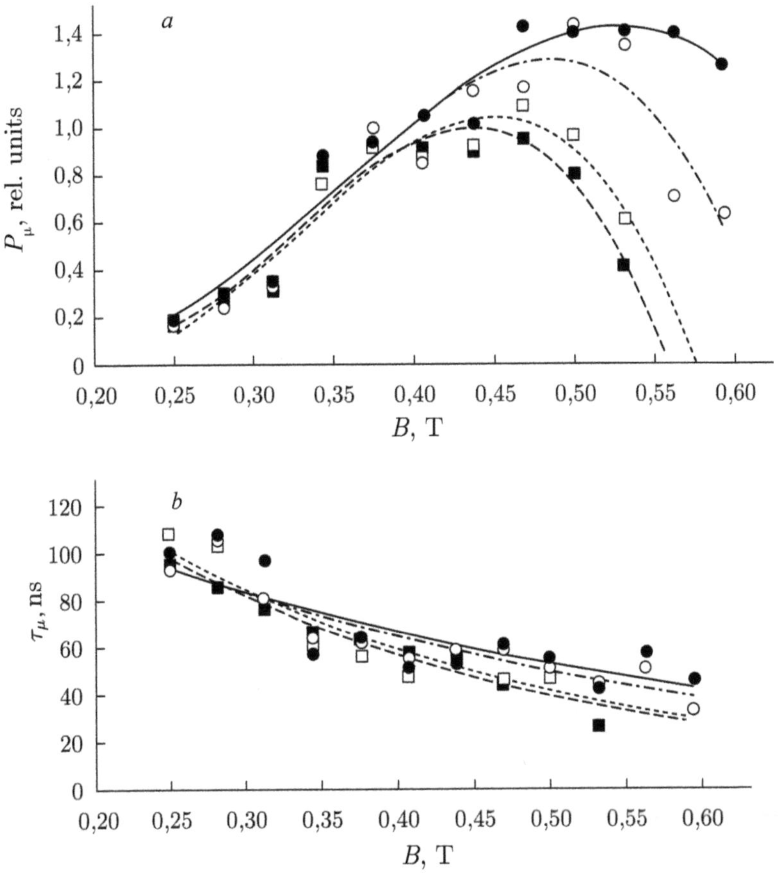

Fig. 1.10. Dependence of the power (*a*) and the duration (*b*) of the microwave pulse of the RM on the value of the magnetic field induction at a single charge voltage of the primary power source of the LIA equal to 800 V, and different versions of the cathode: ■ – cathode without end screens; □ – cathode with screen diameter 24 mm on the side of the high-voltage flange; ○ – cathode with a screen 24 mm in diameter on the side of the drift tube; ● – a cathode with two screens 24 mm in diameter.

accompanied by a noticeable increase in power. If a synchronous magnetic field with an induction value of $B = 0.42–0.44$ T is optimal for a cathode without end screens, then this value is already $B = 0.50–0.57$ T for a cathode with end screens. Since such a noticeable increase in the RM power with end screens is recorded, it can be concluded that the loss current from the interaction space in RM without screens is a significant amount of the total current of the source power supply.

Figure 1.10 *b* shows the dependence of the duration of the microwave pulses on the magnitude of the induction of the magnetic field. An approximate equality of the duration of the pulses is observed in the presence and absence of screens at the cathode, despite a significant difference in the power levels of the output pulses. Usually, the increase in the power of microwave radiation is accompanied by a decrease in the pulse duration, which can be observed from the above dependences, where the radiation pulse is longer for small magnetic fields and, correspondingly, for the lower power of the microwave radiation.

The next stage of the experiments consisted in studying the effect of the diameters of the end screens on the output power of the RM. Figure 1.11 shows the dependence of the microwave pulse power of the magnetron on the value of the magnetic field induction at the screen diameters of 22, 24, 28 and 32 mm and an equal distance from the cathode of 10 mm. As a result of the experiments it was established that the effect of increasing the power from the use of large-diameter end screens (28 mm, 32 mm) is not as significant as for small-diameter screens. Apparently, this is due to an increase in the strength of the electric field and the formation of an additional current of losses from their surface.

The results also show that the main current losses from the RM interaction space originate from a thin cathode layer, on the surface of which the microwave field strength is small. In this case, the electrons captured in the spokes of the space charge and located at a great distance from the cathode are in the powerful microwave fields of the anode block, exceeding the defocusing forces.

Therefore, further experiments were carried out with end screens 22 mm in diameter at different distances from the cathode. In this experiment, the screens symmetrically approached both cathode edges for a distance of 15, 10, 5, and 0 mm. The results are shown in Fig. 1.12, from which it follows that the optimal distance providing both the maximum output characteristics of the generated microwave pulses and the best pulse repeatability is 10 mm. In this case, the screens are located far enough from the anode block so that the electric field strength on their surface is low and there is no parasitic explosive electron emission. At the same time, small losses of electrons from

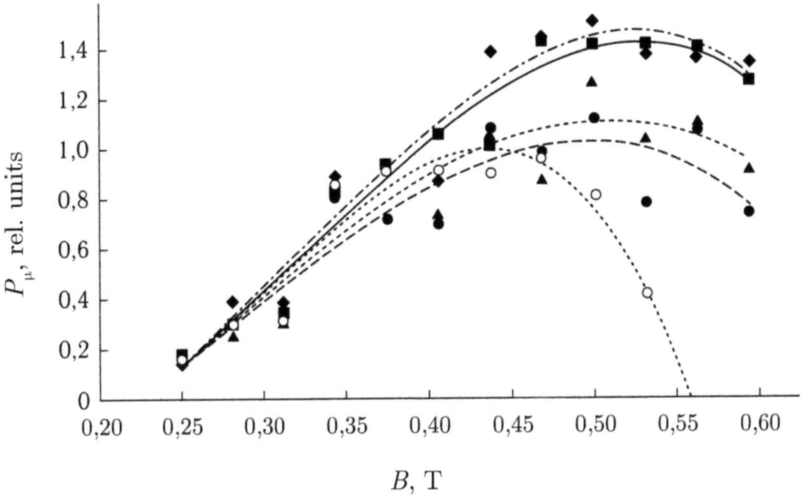

Fig. 1.11. Dependence of the microwave pulse power of the RM on the value of the magnetic field induction for different diameters of the end screens of the cathode, but the same distance from the edges of the cathode, equal to 10 mm: □ – diameter of screens 22 mm; ■ – 24 mm; ▲ – 28 mm; ● – 32 mm. Cathode without screens – ○.

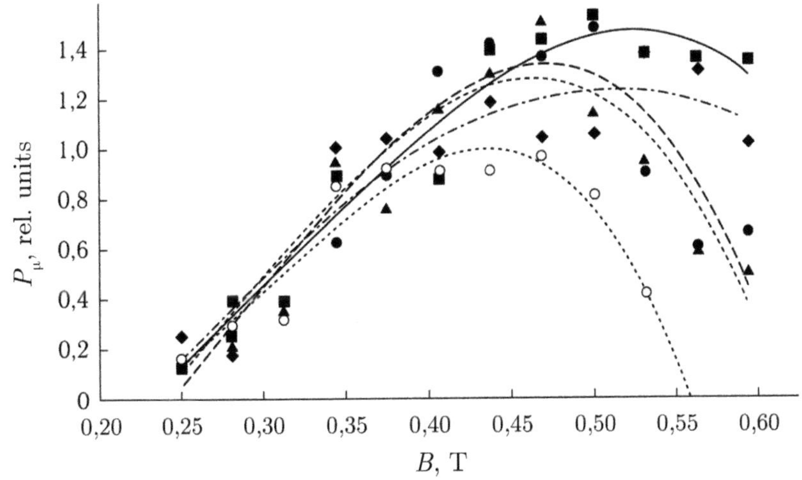

Fig. 1.12. Dependence of the microwave pulse power of the RM on the value of magnetic field induction at cathode end screens with a diameter of 22 mm, located at different distances from the edges of the cathode: □ – 15 mm; ■ – 10 mm; ▲ – 5 mm; ● – 0 mm. Cathode without screens – ○.

the interaction space are achieved. Large distances from the screens to the cathode reduced the effect of their application, which is due to the weakening of their screening effect. However, in any of these cases, regardless of the distance, the generated microwave pulse power significantly exceeds this parameter for a RM with a cathode without end screens.

The final stage of the research was to determine the influence of the power parameters on the output characteristics of the RM. The voltage and current in the cathode–anode gap of the RM were varied by regulating the charging voltage of the primary energy storage of the LIA. Accordingly, the output power of the microwave pulse varied, and consequently, the value of the space charge of electrons in the spokes also changes. The results are shown in Fig. 1.13, from which it follows that the positive effect of the use of end screens is manifested in a wide range of voltages and currents.

As a result of investigations of a relativistic magnetron with cathode end screens, it was established that:

1) the use of screens increases the radiation power of the RM by 40–50%, reaching 400 MW, while maintaining the duration of the microwave pulse;
2) a greater influence on the amount of generated power is exerted by a screen located on the side of the drift tube;

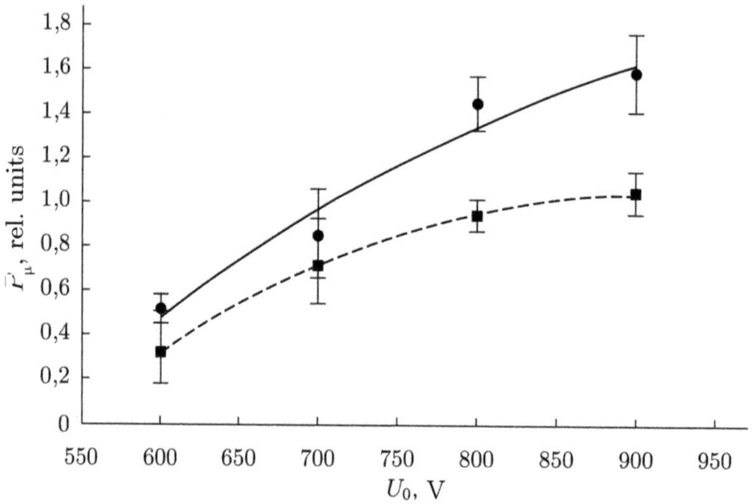

Fig. 1.13. Dependence of the microwave power of the RM on the value of the charging voltage of the primary energy storage of the LIA: ■ – cathode without end screens; ● – cathode with end screens 22 mm in diameter and a distance of 10 mm from the edges of the cathode.

3) the most effective use of screens whose diameter slightly exceeds the diameter of the cathode;
4) there is an optimal position of the screens relative to the cathode at which the maximum increase in the power of the microwave pulses is observed, while maintaining high stability of their amplitude and duration;
5) the increase in the parameters of the RM power pulse leads to a proportional increase in the power of the microwave pulses and maintains a ratio of ~40–50% compared to the RM power without end screens on the cathode.

Ferroelectric plasma cathode. The results of experimental studies of a 10-cm relativistic magnetron with a ferroelectric plasma source on the cathode surface are presented in [86]. This type of cathode was called 'active' by the authors. The cathode plasma was obtained with an additional pulse (~3 kV, 200 ns) fed to the electrodes of a ferroelectric plasma cathode (FPC) before the accelerating pulse of a linear induction accelerator. The authors compared the power and microwave characteristics of a magnetron using a ferroelectric plasma cathode and a cathode with explosive electron emission (the 'passive' cathode in the terminology of the authors of the article).

The explosive-emission cathode is usually a cylinder with a smooth or corrugated surface and is made of stainless steel or graphite. These cathodes can be considered as passive cathodes, since the electron source is the plasma of explosive electron emission. The parameters of the plasma (the delay in the formation of the plasma, its density and temperature, homogeneity and the rate of expansion) depend on the parameters of the accelerating pulse, namely, on the amplitude and time of its increase. For example, the existence of an electric field threshold ($E < 10^6$ V/cm) to generate explosive electron emission causes a delay in the emission of electrons. The latter leads to a delay in the appearance of the electron flux in the region of the RM interaction and the delay in the generation of the microwave pulse relative to the onset of the voltage pulse.

To avoid the drawbacks of the passive cathode, the article suggests an active cathode with controlled excitation of plasma before the accelerating pulse. The authors believed that the use of such a plasma source in a relativistic magnetron can significantly reduce the delay in the onset of microwave generation, and also improve the parameters of the microwave pulse.

The experimental setup on which the studies were conducted was made at the Tomsk Polytechnic University and transferred under contract to Technion (Haifa, Israel). The installation includes: a linear induction accelerator with an output voltage of 400 kV and a current amplitude of ~4 kA with an accelerating pulse duration at half-height of ~150 ns, a relativistic magnetron of 10 cm of the wavelength range and a magnetic system. Vacuum in the experimental chamber and magnetron was maintained at a level of ~$5 \cdot 10^{-5}$ Torr using two oil–vapour diffusion pumps.

In the experiments, plasma and explosive-emission cathodes were investigated. The latter was made of a stainless steel tube with an external diameter of 20 mm, which has azimuth symmetrically distributed grooves (Fig. 1.14 *d*). The FPC was made of a ferroelectric tube (dielectric constant $\varepsilon_r \sim 80$) with an outside diameter of either 20 mm or 16 mm, the thickness and length of the tube were 2 mm and 60 mm, respectively. The inner electrode was made of brass, and the contact between this electrode and the internal surface of the ferroelectric tube was produced by means of a conducting silver paint. The outer electrode of the FPC was ~1 mm wide. The copper strips were glued to the surface of the ferroelectric tube at a distance of ~1.5 mm from each other:

1) the strips of the upper electrode of the FPC were attached to the cathode holder;
2) FPC was inserted inside a stainless steel tube, either with grooves or with rods (Fig. 1.14 *c*).

In the latter case (see Figs. 1.14 *b*, *c*), the FPC was made of a ferroelectric tube with an external diameter of 16 mm, and the upper strips had electrical contact with a stainless steel tube with an internal diameter of 18 mm. The driving pulse with an amplitude of ~3 kV with negative or positive polarity and a half-height duration of 150 ns was obtained using a generator based on a cable forming line (impedance of the line 12.5 Ω) and a thyratron. This impulse was applied to the FPC electrode with the use of a high-voltage cable passing inside the high-voltage electrode of the LIA. For this purpose, the LIA feature was used – the possibility of grounding one of the ends of a high-voltage electrode. The study showed that the FPC design without screening rods (or strips) is preferable because of its simplicity and better parameters generated by the relativistic magnetron of microwave radiation; therefore, the results of the magnetron work with this design of the plasma cathode will be presented below.

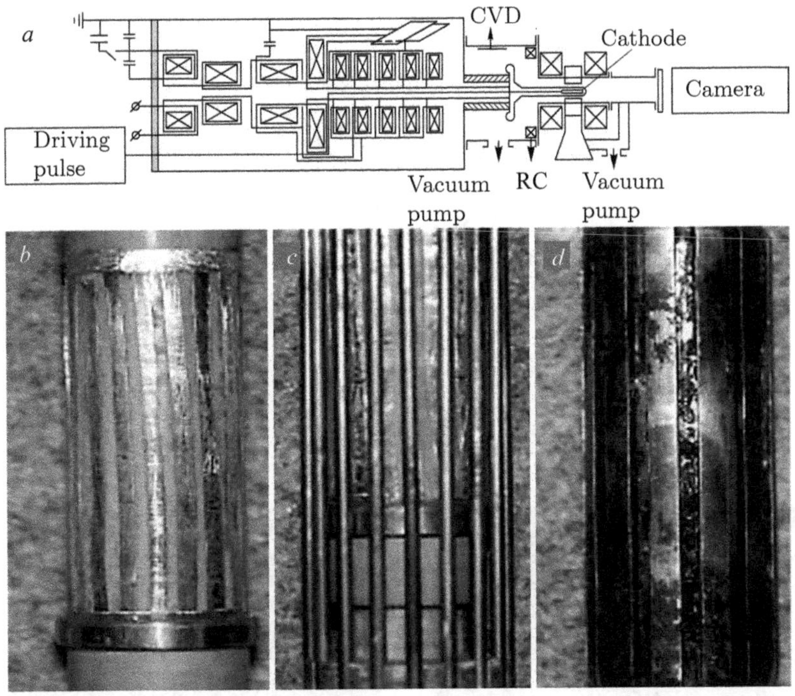

Fig. 1.14. *a*) Experimental installation; *b*)–*d*) the appearance of the cathodes studied: *b*) only FPC; *c*) FPC inside the cathode rod; *d*) a passive cathode (RC – Rogowski coil; CVD – capacitive voltage divider).

The main features of the FPC operation are described in [87] which shows that the use of the driving pulse leads to the formation of unfinished surface discharges formed at the 'triple' points (metal–insulator–vacuum). These discharges are narrow plasma channels located along and outside the ferroelectric surface. The formation of a plasma is accompanied by the emission of electron and ion fluxes. Electrons for the anode current of a relativistic magnetron are selected from the preformed plasma.

Typical pulses of the voltage and current of the LIA in the single mode are shown in Fig. 1.15. As can be seen from the figure, the use of FPC allows to avoid the time delay of electron emission and to achieve a better match between the impedance of the magnetron and the LIA. Without a driving impulse, the voltage and current pulses with FPC were similar to the corresponding pulses with a passive cathode for which the voltage amplitude is higher and a time delay between voltage and current is observed. The latter indicates the explosive electron nature of the plasma formed on the surface of the FPC, when the electric field becomes $>10^6$ V/cm.

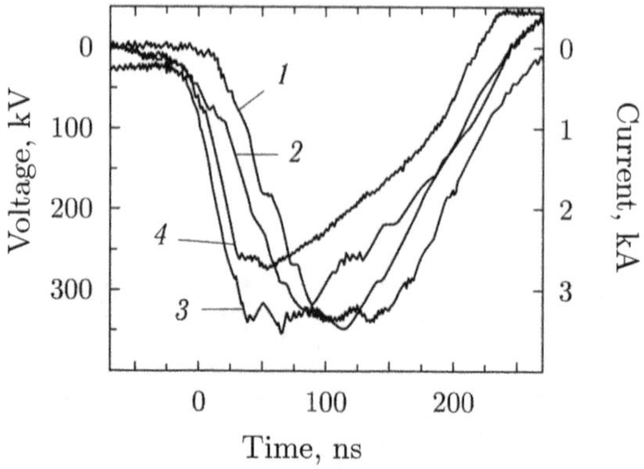

Fig. 1.15. Typical impulses of voltage and current with cathode with EEE (*1* – current, *3* – voltage) and with FPC (*2* – current, *4* – voltage).

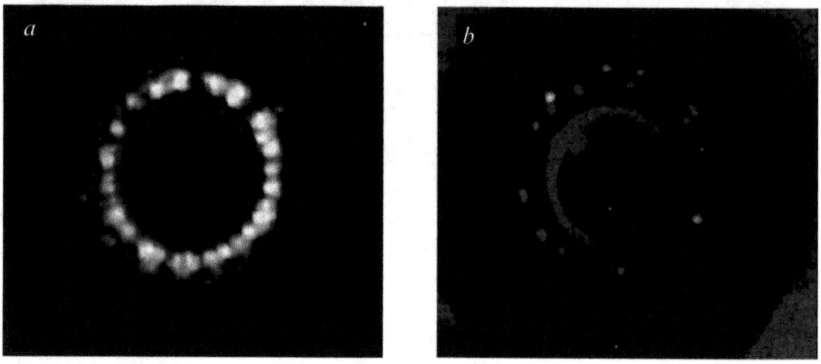

Fig. 1.16. Images of the FPC of the cathode *a*) c (on the left) and *b*) without (on the right) the driving pulse. The frame duration is 10 ns. The time delay of the image relative to the beginning of the accelerating pulse is 10 ns.

In addition, a photograph was taken from the side of the FPC and the passive cathode, which shows intense light radiation from the plasma formed between the rods near the ferroelectric surface already at the beginning of the accelerating pulse (Fig. 1.16). Without using the driving pulse, weak light emission was obtained only at the outer edges of the rods with a time delay of ~40 ns.

The use of the FPC leads to a significant change in the work of the relativistic magnetron. Figure 1.17 shows typical microwave oscillations of a relativistic magnetron obtained with an FPC and a passive cathode under the same conditions (accelerator charging voltage, magnetic field induction, vacuum level). It can be seen that the microwave pulse begins with a delay of ~20–30 ns relative to

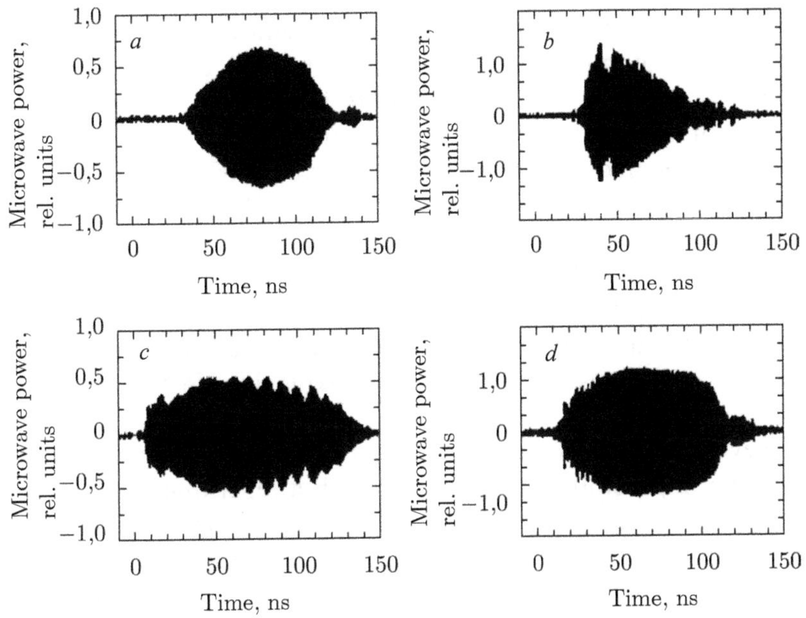

Fig. 1.17. Typical microwave pulses of a relativistic magnetron at two values of the charge voltage of the LIA and the induction of the magnetic field *a*) and *b*) pulses obtained with the passive cathode, *c*) and *d*) pulses obtained from the FPC.

the beginning of the accelerating pulse in the case of the use of the FPC and ~30–40 ns for the passive cathode, respectively. In addition, in the case of the FPC application, the amplitude of the voltage and current at the beginning of the generation of microwave radiation is ~100 kV and ~1 kA, whereas in the case of the passive cathode the microwave radiation begins at ~340 kV and ~1 kA.

The most significant difference in the parameters of the microwave pulses was obtained at low voltages supplied from the LIA (regulated by the value of the charge voltage of the primary energy storage U_{ch}). Namely, at a voltage of 180–250 kV and a current of 2.5–3 kA, the duration of the microwave pulse was ~45% greater than in the case of using a passive cathode with a slightly larger (several percent) amplitude of microwave oscillations. With a voltage of 260–330 kV at the cathode, current 3.5 kA, the duration of the microwave pulse was again higher (~30%) than with the passive cathode, and the amplitude of the microwave oscillations was higher by 12%. The electronic efficiency of a relativistic magnetron, estimated as the ratio between the microwave energy and the electron beam energy, is much higher when using a FPC. It reaches 30% and 20% at low and high output voltage of LIA, respectively. The energy of the microwave pulse with a FPC was 1.6 J and 4 J at low and high voltage, while the energy of the microwave pulse with a passive cathode was only ~1 J and ~2 J. Obviously, these results indicate the advantages of the active plasma cathode for a relativistic magnetron.

It can be concluded that the use of a ferroelectric plasma cathode allows to avoid delays in the appearance of electron emission, to achieve a better correspondence between the impedances of the magnetron and the LIA, and also to significantly increase (~30%) the duration of the microwave pulse with an energy increase of ~ 100%. At the same time, the efficiency of a relativistic magnetron reaches 30%, while using a cathode with explosive electron emission, the efficiency does not exceed 20%.

However, a significant drawback was found when using the FPC, namely, the impossibility of operating with a repetition rate of more than 1 Hz due to the deterioration of the vacuum in the interaction space of the relativistic magnetron. Passive cathodes with explosive electron emission are free from this drawback. In addition, the resource of the work of the FPC is not clear. Another important conclusion is that with the increase in the voltage applied to the cathodes of the relativistic magnetron from the power source, the effect of the FPC is reduced. Since the explosive electron plasma begins to prevail in this case in comparison with the plasma formed by the driving pulse on the surface of the ferroelectric.

A 'transparent' cathode. Recently, much attention has been paid to the development of various methods for increasing the rate of rise of oscillations and increasing the efficiency of relativistic magnetrons. The results of numerical simulation of the output characteristics of relativistic magnetrons are presented in [88] using the so-called 'transparent' cathode. A 'transparent' cathode is made of individual metal strips assembled on a cylindrical surface and serving as separate electron emitters. Effective preliminary grouping of electrons to excite the operating mode of the magnetron is ensured by the correct choice of the number and azimuthal position of the cathode strips. A strong azimuth high-frequency field in the cathode region captures the pre-grouped electrons into a rotating electron spokes, forming an anode current. This process provides a faster start of microwave oscillations than the use of smooth cylindrical cathodes. Moreover, a strong high-frequency field in the flow of electrons above the cathode, for any thickness of the stream, gives the best efficiency.

Computer simulation of a magnetron with cylindrical and 'transparent' cathodes, shows the advantages of the latter.

For any magnetron, the time to the onset of microwave oscillations is determined by two factors, each equally important, namely: the initial conditions that give a primary impulse to the development of dynamic processes, and the rate of increase. For a magnetron with uniform electron emission from the cathode surface, the initial noise determines the initial conditions for the development of instabilities in the electron flux. A typical noise level is approximately 10^{-10} of the electron energy [89]. The time of the development of instability in the electron beam from the initial noise leading to the disintegration of the symmetric electron beam and the formation of its azimuthal modulation is equal to at least ten cyclotron periods. Such a slow start the formation of an alternating current and a small variable tangential component of the electric field in the electron

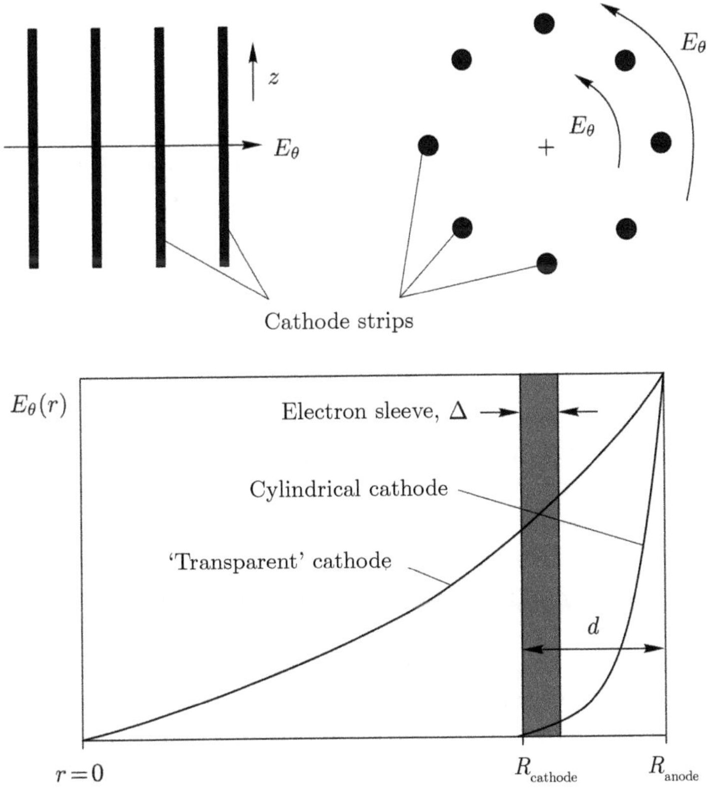

Fig. 1.18. The location of the 'transparent' cathode with respect to the planes $z\vartheta$ and $r\vartheta$ (above). Dependence of the azimuthal electric field of the synchronous wave on the radius for the 'transparent' and cylindrical cathode (bottom).

flux, which is responsible for the intensity of the increase, leads to the slow appearance and growth of microwave oscillations. The paper [88] proposes a simple method of simultaneous provision of a rapid start of oscillations and improvement of the electronic efficiency by increasing the synchronous electric field in the centre of the interelectrode space. This can be done by removing the longitudinal strips from a thin-walled tubular cathode to make the cathode 'transparent' for a field with an azimuthal orientation in which the axial component of the electric field is absent, $E_z = 0$, that is, for transverse electric waves that are used as working modes in magnetrons. This method is similar to a transparent grid consisting of wires perpendicular to the polarization of the direct wave, as shown in Fig. 1.18 (top).

In magnetrons with a 'transparent' cathode, the azimuthal high-frequency field is distributed according to a modified Bessel function of order n (where n is the azimuthal index of the operating wave),

$$E_\vartheta = E_\vartheta(r_a)\frac{I_n(gr)}{I_n(gr_a)}. \tag{1.3.7}$$

In the case of a cylindrical cathode, the radial distribution of the azimuthal electric field of the operating mode in the magnetron is represented as:

$$E_\vartheta(r) = h_\Theta E_{\mathrm{HF}}\frac{\mathrm{sh}\left[p(r-r_c)\right]}{p\,\mathrm{sh}(pd)}. \tag{1.3.8}$$

(for more details, see Chapter 2, Section 2.1.1). In this case h_Θ is the transverse wave number, E_{HF} is the effective amplitude of the synchronous harmonic on the surface of the anode block and the field (1.3.7) on the metallic surface of the cathode tends to zero. However, for a 'transparent' cathode, the field (1.3.8) passes through the cathode strips and tends to zero only at the centre of the cathode, as shown in Fig. 1.18 (bottom). Therefore, in accordance with this distribution, the electromagnetic energy in magnetrons with a cylindrical cathode accumulates mainly in resonators of the resonant system, so the average amplitude of the E_{HF} field on the anode surface depends little on the changes in the cathode configuration.

The field distribution at the 'transparent' cathode provides a higher field amplitude in the electron flow region than with the cylindrical cathode, thus giving a faster oscillation start, according to equation (1.3.2). Due to the redistribution of electromagnetic fields, the 'transparent' cathode also ensures the rapid formation of high-frequency current, due to the presence of 'cathodic and magnetic priming.'

What is the 'cathodic' and 'magnetic priming.' The cathode strips act as emitting centres, placed periodically on the perimeter of the cathode. The number N and the position of the cathode strips can be selected to provide a 'cathodic priming' on the desired operating mode. When the voltage is applied, an azimuthally modulated electron beam is immediately formed. The currents along the longitudinal cathode strips I_z form an azimuthal magnetic field locally around the strips:

$$B_{\varphi N} = \mu_0 I_{zN} / 2\pi d_N . \tag{1.3.9}$$

Here d_N is the distance from the centre of the strip, and I_{zN} is the longitudinal current of the strip N, which gradually decreases along the strip to the axial loss current from the interaction space. This results in an azimuthally modulated periodic magnetic field, which provides a 'magnetic priming'. This implementation of the 'magnetic priming' differs from that described in section 1.8 below, where it is implemented by means of an azimuthal perturbation from externally applied axial magnetic fields B_{0z}.

The configuration of the longitudinal cathode strips can be chosen arbitrarily (for example, they can be in the form of cylindrical rods or tapes formed by cutting a thin-walled tubular cathode, etc.). A cathode consisting of longitudinal radiating strips can be considered 'transparent' even when the space inside the cathode is completely or partially filled with an insulator. A 'transparent' cathode may also comprise a metal rod with a diameter smaller than the internal diameter of the cathode, which is constructively necessary for some cathodes.

It should be noted that a cathode with individual rods located on a cylindrical surface forming a slowing-down system along with an identical multi-cavity coaxial anode centrally located in the resonator was applied in the so-called nigotron.

The field of synchronous waves in the region of the electron flux depends little on its thickness when using a 'transparent' cathode, as shown in Fig. 1.18 (bottom). Thus, it is possible to improve the output characteristics of the magnetron by increasing the applied voltage and the magnetic field. The 'transparent' cathode also partially reduces the effect of undesirable additional electron drift in strong fields of the operating wave, which reduces the effectiveness of magnetrons. In magnetrons with such cathodes, the amplitude of the operating wave varies much less in a narrow cathode–anode gap than in a magnetron with a cylindrical cathode, as shown in Fig. 1.18 (bottom). Thus, the spread of the drift velocities of the electrons is smaller, which facilitates the correction of any discrepancy in the synchronism conditions by changing the external magnetic field.

Plasma appearing on a 'transparent' cathode with explosive electron emission is not an obstacle to the rapid onset of oscillations, because the plasma velocity v_p through the magnetic field is approximately $4 \cdot 10^5$ cm/s, i.e., the time of formation of a continuous plasma ring around the cathode is approximately

$$t_p = \pi r_c / N_{cath} v_p \sim 0.5 \cdot 10^{-6} \text{s}.$$

At the same time, in contrast to a magnetron with a cylindrical cathode in which the plasma can propagate exclusively along the radius outwards to the anode, the plasma that appears on the cathode strips of the 'transparent' cathode propagates in all directions, thereby reducing the rate of closure and its density, both in multi-point and disc cathodes.

Nevertheless, for relativistic magnetrons designed to work with pulses of long duration, $\sim 10^{-6}$ s, there is a concern that the solid plasma ring formed around the cathode may be opaque to the operating wave, which excludes the possibility of increasing the efficiency of the magnetron by simultaneously increasing applied voltage and magnetic field. To avoid the problems associated with the cathode plasma, the use of cathode materials with a low level of plasma formation, such as in graphite (velvet) or oxide cathodes, will be optimal.

In order to demonstrate the advantage of 'transparent' cathodes over cylindrical cathodes in relativistic magnetrons, a three-dimensional fully relativistic MAGIC code was used for computer simulation of a magnetron A6 with different cathodes.

The choice of the magnetron configuration A6 is determined by previous detailed studies. The experimental and calculated characteristics of the magnetron A6 with a cylindrical cathode have been published in many articles, see, for example, [15, 89, 90]. The resonator system of this magnetron consists of 6 identical sector spaces with an axial length of 7.2 cm, a minimum radius (anode radius) equal to r_a = 2.11 cm and a maximum resonator radius of

r_{cav} = 4.11 cm, an angular width of resonators of 20° and a radius of the solid cathode r_c = 1.58 cm. The best output characteristics for the magnetron A6 – radiation power $\tilde{P}$ = 450 MW and electronic efficiency η = 7% – were demonstrated when the space of the resonant system was closed by metal end caps at both ends, and the magnetron operated at an applied voltage U = 350 kV in a 2π-type mode [15].

In computer simulation, the characteristics of the magnetron A6 with a cylindrical cathode and a cylindrical cathode with a 'cathodic priming' were compared with the characteristics of a magnetron with a 'transparent' cathode. To find the best output characteristics, the work of a relativistic magnetron with different number, arrangement and width of cathode strips, as well as with different voltage and time of voltage increase was studied.

When modelling a cylindrical cathode with a 'cathodic priming', the cathode had 6 emitting zones, which facilitated the formation of 6 electron spokes and, thus, the excitation of the 2π-mode in a magnetron. The radiating zones had an angular width of 30°. The design of the 'transparent' cathode was chosen in such a way that it consisted of 6 separate cathode strips with an angular width of 5°, 10° and 15° each, whose azimuthal position (as well as the position of the emitting zones for the 'cathodic priming') is optimal for achieving maximum radiation power. With this comparison, the same applied voltage U = 350 kV and the voltage rise time were used, as in [90]. In the simulation, an external axial magnetic field B_{0z} was used, which provides a high output power for all three cathode designs.

Comparison of the output power (Fig. 1.19, left) and the anode currents (Fig. 1.19, right) for three different types of cathodes shows that the time for achieving the maximum output power of the magnetron with a 'transparent' cathode is noticeably smaller than for the other two designs.

Figure 1.20 shows the motion of electrons in the $r\vartheta$ plane for three types of cathodes at different times. It is clear from the figure that the formation of six electron spokes, which is necessary for operation on the 2π-mode, occurs most rapidly with the use of the 'transparent' cathode. The lower pictures in Fig. 1.20 correspond to the time at which the electron spokes for the three types of cathodes are considered fully formed.

As shown in Fig. 1.20, the appearance of a continuous electron sleeve above the cylindrical cathode occurs very quickly – for the

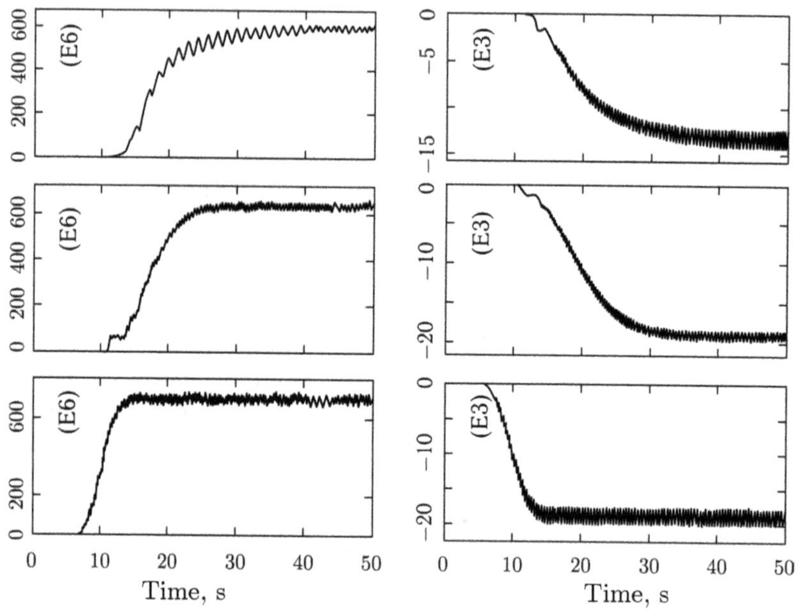

Fig. 1.19. Radiation power (left) and anode flow (right) in the magnetron A6 with *a*) a cylindrical cathode, *b*) a cylindrical cathode with 'cathodic priming' and *c*) a 'transparent cathode' at $U = 350$ kV, $t_U = 10$ ns and $B = 0.52$ T.

first 4 ns. However, the visible preliminary bunching associated with the high-frequency flux appears only after 12 ns (see the beginning of the process in Fig. 1.19 *a*). This time is due to the development of the instability of space-charge waves from the original noise. The modulation of the electron flux corresponding to the formation of six electron spokes for the excitation of the 2π-mode is fully formed within 19.6 ns (Fig. 1.20, left), and the maximum radiation power is attained only after 40 ns (Fig. 1.19 *a*).

When a cylindrical cathode is used with separate electron emission bands periodically located on the cathode surface (Fig. 1.20, centre), the high-frequency current begins almost simultaneously with the appearance of electron emission. At the same time, the development of the generation process is slow (Fig. 1.19 *b*) because of the slow growth rate. The modulation of the electron beam completely ends up at 15.5 ns, that is, faster than in a magnetron with a cylindrical cathode. In addition, Figs. 1.19 *a*, *b* show that the application of the 'cathodic priming' provides a faster achievement of the maximum radiation power compared to a cathode with uniform electron emission.

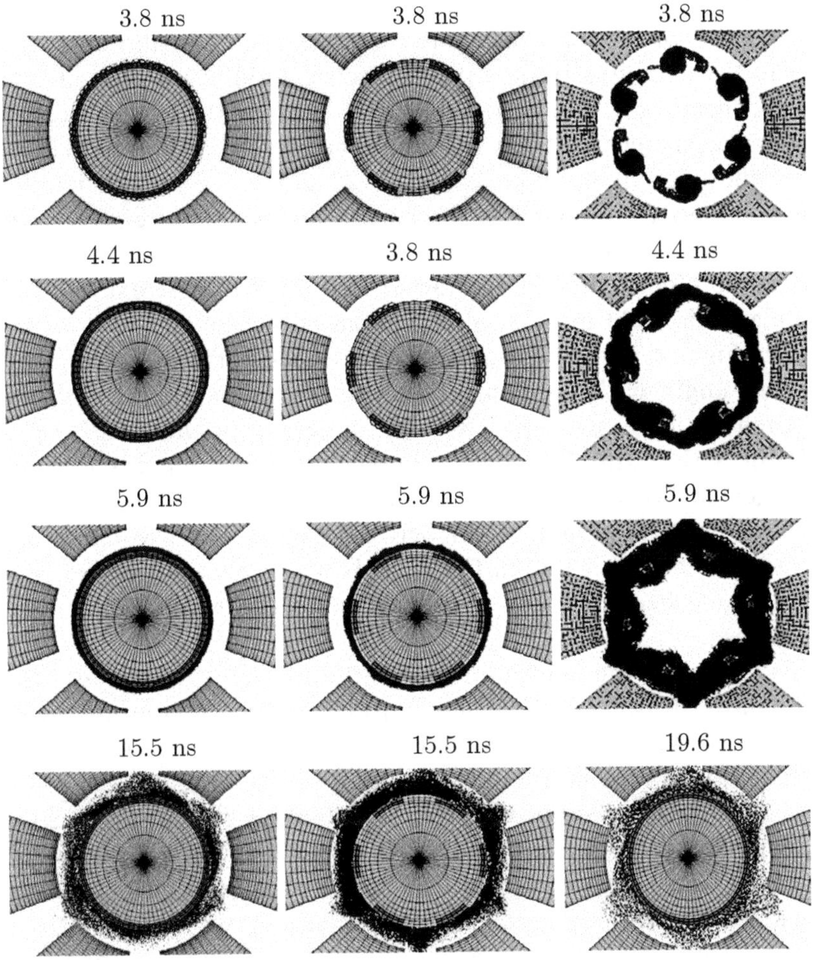

Fig. 1.20. The formation of electron spokes in an A6 magnetron with a cylindrical cathode (left), 'cathodic priming' with 6 emission zones (centre) and a 'transparent' cathode with 6 strips (right) with an applied voltage U = 350 kV and a rise time t_U = 10 ns.

A 'transparent' cathode provides a strong priming effect, which leads to a faster start and a higher level of coherent oscillations (see Fig. 1.20, right and Fig. 1.19 *c*). The modulation of the electron beam is completed approximately 3 times faster than when a cylindrical cathode is used. The maximum radiated power is reached afer 14 ns, while in a magnetron with a cylindrical cathode this occurs only after 40 ns.

In the construction of a 'transparent' cathode, where the distance between the electrodes is small compared to their radii, $d \ll r_a$, r_c, the arrangement of the cathode strips relative to the anode block can significantly affect the output characteristics of the magnetron due to the azimuthally modulated electric field at the cathode. The dependence of the output power on the location of six cathode strips of the 'transparent' cathode was investigated. The azimuthal width of the cathode strips was varied at 5°, 10° and 15°. For each width of the cathode strip, as well as the position of the strip relative to the anode block resonator (the strips rotated relative to the structure of the anode block), modelling was carried out for different values of the magnetic field to achieve maximum output power.

The highest power was achieved when the cathode strips formed a small angle with the central axis of the lamella, and the width of each cathode strip was 5°. The increase in the width of the cathode strips did not have a significant effect on this optimum arrangement. However, as the width of the cathode strips increases, the difference in output power between the best and worst cathode locations becomes less significant. This behaviour is expected, since the cathode becomes less 'transparent' as the width of the strips increases.

As soon as the width of the cathode strips increases, they eventually form a cylindrical cathode. The efficiency of the magnetron A6 was calculated for a 'transparent' cathode and a cylindrical cathode in a wide range of applied voltage $U = 0.35–1.5$ MW with optimal axial magnetic field values for each voltage. A 'transparent' cathode consisted of six cathode strips with an angular width of each strip equal to 10°. In these simulations, a voltage rise time of 1 ns was used. The results are shown in Fig. 1.21. As follows from the figure, the use of a 'transparent' cathode in an advantageous position leads to a greater efficiency with a tendency

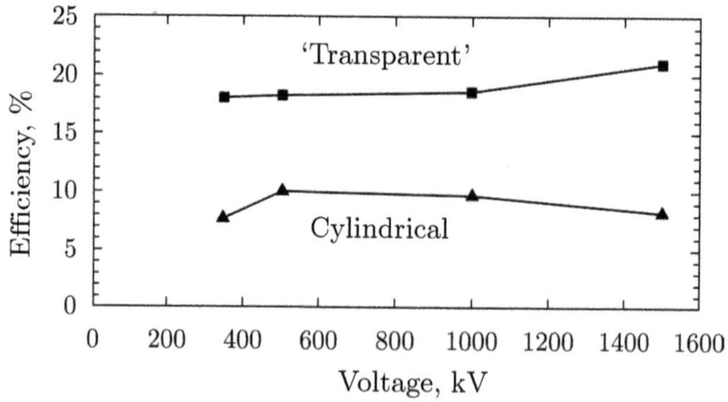

Fig. 1.21. The electronic efficiency of RM A6 when operating on a 2π-type oscillation with a cylindrical and 6-band 'transparent' cathode.

to increase with the growth of the applied voltage than the use of a cylindrical cathode.

Thus, for three-dimensional modelling by the particle-in-cell method, the output characteristics of a relativistic magnetron can be increased if the 'transparent' cathode is used. Such a cathode provides improved conditions for rapid conversion of the potential energy of electrons into electromagnetic energy. Simultaneously, the 'transparent' cathode provides a 'cathodic priming' for the desired operating mode by a specific choice of the number of cathode strips. Optimal azimuthal positions of the strips relative to the structure of the anode block are observed to achieve the highest efficiency. It is shown that in a magnetron with the 'transparent' cathode, the efficiency can be improved, and at the same time a rapid start of the oscillations is ensured in comparison with a solid cathode magnetron under equal conditions.

Diagnostics of pulses of microwave radiation. The generated microwave power is estimated using lamp or semiconductor detectors with calibrated volt–watt characteristics, calorimeters, directional couplers built into the output paths of the generators. When using a waveguide output, the RM power is estimated from the detector's recorded readings, taking into account the attenuation of the microwave radiation between the receiving and radiating antennas. The amount of attenuation between the antennas is calibrated during 'cold' measurements in a wide frequency range. This is possible, since the lowest type of oscillations H_{10} is excited in the output waveguide (antenna). In separate experiments, measurements are made of the radiation pattern in the far field with its subsequent integration if the mode of oscillation is unknown or there are several of them. Frequency measurements are taken using sets of transverse waveguides, bandpass (rejection) filters, heterodyne devices with automated processing, frequency analyzers of spectra, dispersion lines. The measurement technique is described in more detail in Ch. 4 (section 4.7.1).

1.4. Design, choice of parameters of the resonator system

Table 1.1 presents the most characteristic differences of relativistic magnetrons from their classical counterparts. In this section, the structural features of the devices due to high voltages and currents are analyzed.

High voltages lead to an increase in the radius of rotation of the electrons. In this case, we should expect a decrease in the electronic efficiency of the relativistic magnetron in accordance with the estimation formula [79]:

$$\eta_e = 1 - \frac{2r_c}{d}\gamma_p, \tag{1.4.1}$$

where r_c is the cyclotron rotation radius of the electrons;

$$\gamma_p = (1 - V_p^2 / c^2)^{-1/2}.$$

To obtain high values of the electronic efficiency, it is advisable to increase the interelectrode gap to the maximum possible. However, an increase in the interelectrode gap hinders the interaction of the microwave fields of the anode block with the rotating electron layer, especially at the beginning of the supply pulse when the fields are still small. When designing relativistic magnetrons, it should be taken into account that it differs from the classical one by the presence of a cathode plasma and by the motion of its emission boundary, by the shortening during the pulse of the magnitude of the interelectrode gap and by closing by the plasma of the gap when long enough pulses are used. It is clear that a decrease in the influence of this effect is achieved by increasing the interelectrode gap. Finally, the RM is characterized by the destruction of anode blocks under the action of anode current electrons. When using a microsecond high-current electron accelerator (HCEA), the closure by the plasma of the cathode–anode gap and the flow of short-circuit current cause additional destruction of the anode block.

Another difference between magnetrons is the appearance of a loss current caused by the edge electric field of the anode block and the azimuthal magnetic field, which causes a decrease in the level of generated power and requires the use of drift tubes of large diameter in comparison with the diameter of the anode block [71]. The electrons emitted from the surface of the cathode can partially return back to the interaction space in the event that the cathode-injected current exceeds the limiting transport current in the drift tube. Falling into the region of the correct phases, the returned electrons can take part in energy exchange with the microwave field, increasing the efficiency of the device.

The study [80] describes a special configuration of the cathode, which makes it possible to reduce the end current of the magnetron. First, the length of the cathode is recommended to be greater than the length of the anode block, i.e., the end of the cathode should be located in the region of the drift tube, and secondly, the cathode end profile is rounded to form an uniform electron beam, rather than a coaxial one. In this case, the current injected into the drift tube is minimal and is determined according to [81] by the expression

$$I_{\text{end}} = \frac{mc^3}{e} \frac{(\gamma - \gamma_b)\sqrt{1 - 1/\gamma_b^2}}{1 + 2\ln\left(R_{\text{tube}} / r_c\right)}, \tag{1.4.2}$$

where $\gamma_b = \sqrt{2\gamma + 0.25} - 0.5$ is the kinetic energy of the electrons; R_{tube} is the radius of the drift tube. As follows from (1.4.2), it is desirable to increase diameter of the drift tube to the maximum possible. However, the outer diameter of the tube is limited, since it determines the internal diameter of the magnetic field coils, and hence the power consumption of the magnetic system.

High voltage and high current in a relativistic magnetron impose restrictions on the wavelength of the generated oscillations. It was shown in Ref. [82] that the microwave power in the wavelength range from 3 to 10 cm falls approximately as $\lambda^{2.5}$, which is associated with a decrease in the size of the anode blocks in the transition to shorter wavelengths. On the contrary, for 30-cm relativistic magnetrons, the dimensions of the instruments become so significant that it is necessary to use cumbersome magnetic systems. These considerations follow from the similarity relations that can be obtained using the Buneman–Hartree relationship (1.1.8) for the n-th mode. It follows from the expression that for constant values of U and r_c/R

$$H_{\text{th}} R^2 \frac{1}{\lambda} \sim \text{const.} \tag{1.4.3}$$

Taking into account that the wavelength of the mode of oscillation is proportional to the circumference of the anode block, it is possible to obtain: $H_{\text{th}} \sim 1/\lambda$. Thus, if the wavelength is reduced (for example, 3 times) at a constant voltage and a constant ratio of the radius of the cathode to the inner radius of the anode block, it is necessary to reduce the geometric dimensions of the device by a factor of 3 and to increase the magnetic field by a factor of three. However, as will be shown below, using relativistic magnetrons with stabilizing resonators, effective modes of operation of devices are also possible in the 3-cm wavelength range.

Thus, for traditional RM designs, the 10-cm wavelength range is preferable. Reflex triodes also operate in this range [83]. Having certain advantages associated with the absence of a magnetic system, they are considerably inferior to the relativistic magnetron in terms of the efficiency of converting the electron energy to electromagnetic radiation, the stability of the amplitude and frequency parameters of the pulses, and the durability of the elements [92].

The choice of the length of the magnetron L, as well as the number of the resonators N, is determined by the method of selecting the modes for the purpose of isolating one working mode. The length of the anode block is limited by the value $L_{max} \sim \lambda$. At such an anode block length, the 'parasitic' mode of oscillations closest to the π-type in the longitudinal index has a frequency:

$$f_{\text{par}} = f_{\pi}\sqrt{1+(\lambda/2L)^2} \tag{1.4.4}$$

and will be detuned by at least 10% of the phase velocity of the wave. It is advisable to use an output device matched with a standard waveguide of 7.2 × 3.4 cm^2 (for a 10-cm wavelength band), which predetermines the length of the anode block of 7.2 cm. The use of a wider waveguide increases the distance between the coils of the Helmholtz pair, which means it increases the energy consumption of its supply system.

Since the output power is proportional to the number of resonators, it is natural to increase them to the maximum possible. However, the requirement of a single-mode generation regime imposes limitations. With a large number of resonators, the modes of oscillations adjacent to the operating mode may enter the synchronization band. There are also restrictions on N and below to ensure $R/r_c \leq 2$, otherwise the large curvature of the interaction space will complicate the process of bunching of the electrons and lead to the breakdown of the anode current. Thus, in relativistic magnetrons of the 10 cm wavelength, N can vary from 6 to 12.

The internal diameter of the anode block of the magnetron (or the outer diameter of the anode block of the magnetron of the inverted geometry) is estimated from the empirical formula [75]:

$$R = N\lambda\sqrt{(1-\eta_e)\cdot U \cdot 10^{-7}}. \tag{1.4.5}$$

Given the wavelength of the generated oscillations $\lambda \sim 10$ cm, the number of resonators $N = 8$–12, the limiting value of the electronic efficiency $\eta_e \sim 0.8$, and the expected value of the cathode–anode voltage $U \sim 500$ kV, this formula gives $R \sim 2$–3 cm.

The ratio of the width of the lamellae to the width of the cavity gap is generally chosen depending on the requirements for the dispersion characteristic of the resonator system (the dependence of the phase velocity of the wave on the mode of oscillation) and the value of the characteristic resistance (reactance of capacitance and the inductance of the circuit at the resonant frequency). However, as the experience of constructing classical magnetrons shows, this ratio can vary within wide limits without significantly affecting the quality of their work. Therefore, for a more uniform distribution of the anode current electrons along the lateral and cylindrical parts of the lamellae, relativistic magnetrons with a slit width equal to the width of the lamellae were usually made.

Above are the most general considerations regarding the choice of the parameters of the anode block. Regardless of the specific resonator structure of the anode, its main function is to create microwave fields of a given frequency and configuration that are able to effectively interact with the space charge of the electron beam. If it turns out that the excitation voltage of the nearest mode is adjacent to the operating voltage zone or falls on its boundaries, then for stable operation of the magnetron $\Delta\omega$ these oscillations must be spread in frequency by an amount greater than the bandwidth of the slowing-down structure in the loaded state.

1.5. Calculation of the resonator system of a relativistic magnetron by the field theory method

Calculation of the geometry of the anode block and the interaction space is carried out by the methods of field theory [4]. The method is based on finding the input conductivities in the interaction space and in the space of the resonators, equating them at the interface and on the basis of these data the composition of the dispersion equation.

From the solution of Maxwell's equation in the interaction space and in the space of resonators for the H-wave ($E_z = 0$), taking into account the boundary conditions: $E_\varphi = 0$ on the surfaces of the cathode and anode segments (lamellae), the input conductivities of these regions are determined as follows:

$$Y = \int_{-\vartheta}^{\vartheta} E_y^* H_z r d\varphi \Big/ \left[\int_{-\vartheta}^{\vartheta} E_\varphi^r d\varphi \right]^2, \qquad (1.5.1)$$

where $r\vartheta$ is the angular size of the gap; E_φ = const at the interface between the interaction space and the resonators.

For the vane-type resonators, the dispersion equation has the form [4]

$$\sum_{p=-\infty}^{\infty}\left(\frac{\sin\chi\vartheta}{\chi\vartheta}\right)^2\frac{Z_\chi(Kr_m)}{Z'_\chi(Kr_m)}=-\frac{\pi}{N\vartheta}\frac{J_0(Kr_m)N_1(Kr_m)-J_1(KR)N_0(Kr_m)}{J_1(Kr_m)N_1(KR)-J_1(KR)N_1(Kr_m)}. \quad (1.5.2)$$

For slit–hole type resonators

$$\sum_{p=-\infty}^{\infty}\left(\frac{\sin\chi\vartheta}{\chi\vartheta}\right)^2\frac{Z_\chi(Kr_m)}{Z'_\chi(Kr_m)}=$$

$$=-\frac{\pi}{N\vartheta}\left[\frac{J_0(Kr_m)}{J'_0(Kr_m)}+2\sum_{s=1}^{\gamma}\left(\frac{\sin(\chi\vartheta)}{\chi\vartheta}\right)^2\frac{J_p(Kr_m)}{J'_p(Kr_m)}\right], \quad (1.5.3)$$

where $\chi = n + pN$, p is the spatial harmonic number; $K^2 = k^2 - (m\pi/L)^2$ is the transverse wave number; $k = 2\pi/\lambda$, m is the number of variations of the electric field in the axial direction; r_m is the outer radius of the anode block resonators,

$$Z_\chi(Kr)=J_\chi(Kr)-\frac{J'_\chi(Kr_c)}{N'_\chi(Kr_c)}N_\chi(Kr),$$

$$Z'_\chi(Kr)=J'_\chi(Kr)-\frac{J'_\chi(Kr_c)}{N'_\chi(Kr_c)}N'_\chi(Kr),$$

$J'_\chi(Kr)$, $N'_\chi(Kr)$ are derivatives with respect to the argument of the Bessel and Neumann functions. In the absence of axial variations of the RF field along the magnetron axis (open or half-open anode block) K is replaced by k ($m = 0$). The appearance of axial variations of the RF electric field is associated with the boundary conditions at the ends of the anode block. To change these conditions in practice, short-circuiting rings are used at the ends of the anode block, which makes it possible to obtain an open, closed and half-open configuration of the resonator system.

Solving equations (1.5.2) or (1.5.3) allows us to find their roots and maximize the ratio: $\beta_\pi/\beta_{N/2-1}$, where β_π, $\beta_{N/2-1}$ is the wave slowing-down magnitude of the π-type and the nearest to it ($N/2-1$)-th type of the oscillations. In calculations of the anode blocks used at the

Tomsk Polytechnic University, the number of spatial harmonics taken into account in the interaction space is limited to $p = \pm 7$, while the accuracy of the function of the sought roots is $<10^{-3}$.

By choosing the geometry of the resonator structure, it can be achieved that the separation of the modes and their harmonics in terms of the magnitude of the phase velocity will be acceptable. Figures 1.22 and 1.23 show the dependences of the separation of the modes of oscillations of a 6-cavity magnetron for the π-type and the neighbouring ($N/2–1$)-type along the wavelength

$$\Delta\lambda = |\lambda_{N/2-1} - \lambda_\pi|$$

and the phase velocity

$$\Delta\beta = |\beta_{N/2-1} - \beta_\pi|$$

(the cathode–anode gap is chosen to be 0.9 cm). As can be seen from the graphs, with increasing D_r/R ($D_r = r_m - R$ is the depth of the

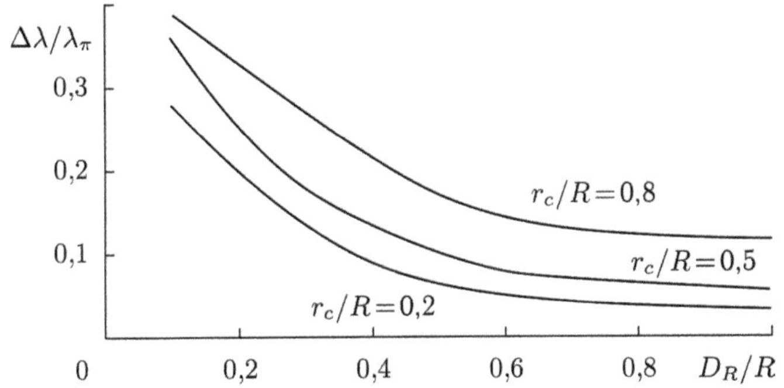

Fig. 1.22. Dependence of the separation of oscillation modes over the wavelength $\Delta\lambda/\lambda_\pi$ on the relative depth of the resonators D_R/R (R = 2.15 cm, N = 6).

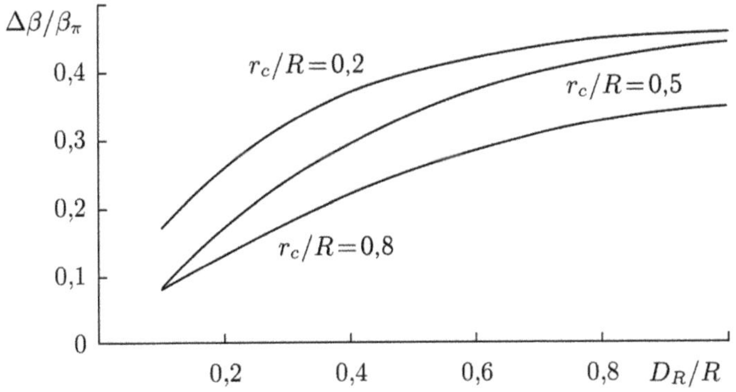

Fig. 1.23. Dependence of the separation of the oscillation modes over the wavelength $\Delta\beta/\beta_\pi$ on the relative depth of resonaors D_R/R (R = 2.15 cm, N = 6).

resonator), wavelength separation $\Delta\lambda/\lambda_\pi$ becomes smaller and the phase velocities separation $\Delta\beta/\beta_\pi$ increases.

Consequently, when the magnetron is operating in an area where the wavelength separation $\Delta\lambda/\lambda_\pi$ is small, the generation stability can be ensured by a good phase-velocity separation and vice versa. Thus, in the operation of a magnetron with a pulsed power supply from the point of view of suppressing 'parasitic' modes, it is preferable to use a region with good phase-velocity separation, i.e., when $D_r/R \approx 1$ to take the maximum possible r_c/R (for better frequency separation) – Figs. 1.17 and 1.18.

In [11, 12, 51, 69, 70] anode blocks with 6 and 8 vane-type resonators were used for various purposes. This type of slowing-down system has a high characteristic impedance, that is, it has a higher intrinsic quality factor and a circuit efficiency. Figure 1.24 shows the dispersion curves calculated for the cathode–anode gap of 0.9 cm, as well as the results of 'cold' measurements of the natural frequencies of the resonator system. It can be seen from the figure that the device has a fairly good separation according to the frequency of the modes.

As shown by the results of experiments, further increase in the power of the relativistic magnetron, as in classical instruments, is impossible without increasing the working volume and complicating their resonator systems. Calculation of resonant systems that ensure reliable operation of devices, especially such as different resonator and coaxial ones, is aimed at finding an acceptable combination based on a directional search of a large number of parameters. The

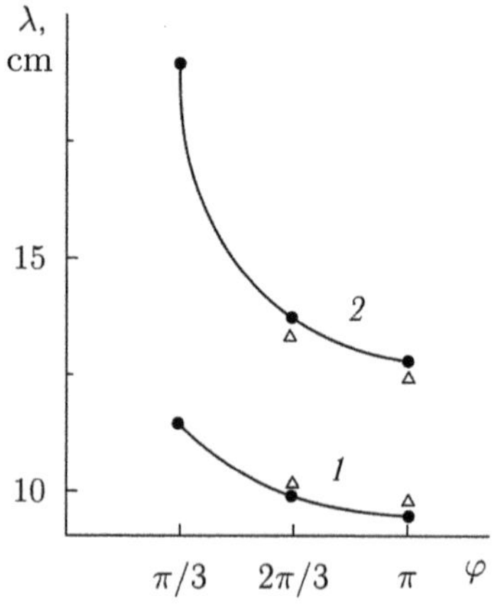

Fig. 1.24. Dispersion characteristics of a relativistic magnetron: curve *1* – with one axial variation, curve *2* – without axial variation, Δ – results of 'cold' measurements.

problem of choosing a compromise version of the resonator system of the relativistic magnetron acquires in this sense an extreme character and is very labour-consuming and, therefore, must be solved with the use of computer technology. A whole group of researchers from the Tomsk Polytechnic University worked on this task. These results in more detail will now be described in greater detail.

1.6. Optimization of the parameters of the resonator system of the relativistic magnetron

In [92], the choice of the parameters of the geometry of the slowing-down system of the relativistic magnetron is posed as the problem of optimal (greatest) separation of the roots of the dispersion equation and is solved by numerical methods of search optimization.

The authors considered a resonant system of a magnetron with different resonator types of the 'slit–hole' type (Fig. 1.25). Such a system of resonators was chosen for the following reasons:

1) ease of manufacture;
2) increased heat transfer;
3) in the process of calculations, it is possible to change the characteristic impedance of the slowing-down system by varying more geometric parameters than for other resonator systems.

The 12-cavity anode block proposed on the basis of calculations has an increased volume of interaction space, a smaller magnitude of the slowing-down of the electromagnetic wave. Both factors are aimed at increasing the output parameters of the magnetron due to

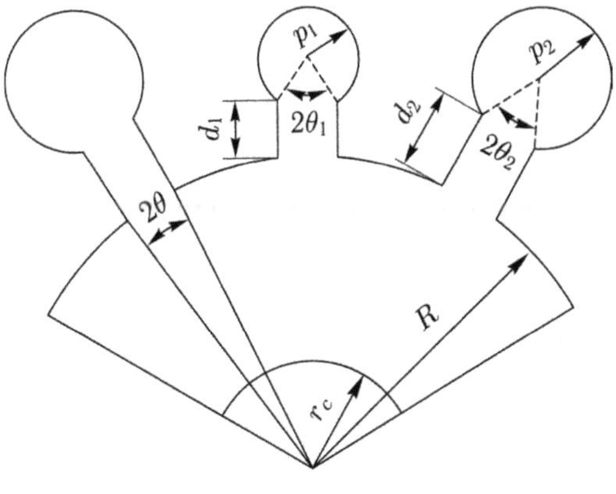

Fig. 1.25. Resonant magnetron system with 'slit–hole' type resonators.

the growth of electronic efficiency. At the same time, a significant number of resonators intensify the competition of oscillation modes. As is well known in classical electronics, it is possible to increase the number of resonators of a magnetron while maintaining a satisfactory frequency separation of the types of oscillations when using different resonator systems or straps. The use of straps is impractical since the magnetron has a long anode. This is due to the fact that high-frequency fields rapidly weaken with an increase in the length of the anode block. In addition, a high power level of the high-frequency field can lead to breakdowns in the straps. Therefore, a different resonator system was chosen.

Investigation of the resonator system of the magnetron is reduced to the solution of the dispersion equation obtained by the field theory method [4]. The construction of a magnetron with the given characteristics of the operating mode leads to the formulation of a problem characteristic of non-linear mathematical programming (MP) [94]. In this formulation, the problems of maximum separation of the operating type of oscillations from the neighbouring one, minimization of the phase velocity, stabilization of a given wavelength of the working type of oscillations, minimization of the magnetron dimensions were formulated and solved. The chosen desired parameters of the resonator system are the radii and depths of the resonators, the radii of the anode and the cathode, the ratio of the width of the cavity gap to the period of the system. The MP problem was solved by one of the methods that do not require the calculation of the derivatives, namely by the method of random search with the change in the metric of the test space [88]. Numerical experiments showed its higher efficiency in similar problems in comparison with the known methods of Box and Rosenbrock and the simplex method [94].

An example of a specific calculation of the resonator systems of relativistic magnetrons with the maximum separation of the operating type of oscillations from the neighbouring one is given in [95]. The calculation was carried out for the wavelength of the π-type of the closed block $\lambda_\pi = 9.75$ cm for the number of resonators $N = 12$, 14, 16 and 18. The spatial harmonics up to ±7 numbers and the number of azimuthal variations of the field in the resonators ≤7 were taken into account. The minimum separation of the operating mode with respect to the frequency is determined by the width of the resonant band of the system

$$|\lambda_\pi - \lambda_\gamma| / \lambda_\gamma > Q_{\text{load}}^{-1},$$

where λ_γ is the wavelength of the natural oscillation of the system. The loaded quality factor of relativistic magnetrons can be in the order of magnitude assumed to be $Q_{\text{load}} \approx 100$. The value of the phase velocity is chosen, on the one hand, from the need to work with not too great induction of the magnetic field (0.5–1.5 T) and, on the other hand, from the desire to obtain acceptable values of efficiency. Therefore, for the phase velocity of the π-type, a region of $0.1 < \beta_\pi < 0.4$, and the phase velocity separation with the neighbouring modes

$$\left(|\beta_\pi - \beta_\gamma|\right) / \beta_\pi \geqslant 0.2.$$

A condition was imposed on the magnitude of the zero harmonic of the resonator magnetron ('contamination') $E_{\varphi 0}/E_{\varphi\pi} < 0.2$ and a restriction on the geometry resulting from the technological requirements was introduced.

The following important feature of the problem was revealed in the process of research. The range Ω of parameters turned out to be non-convex. It contains at least two subregions ($\Omega = \Omega_1 \cup \Omega_2$), each of which has a local maximum. In one of these sub-regions, Ω_1, the solution is very stable to the variation of the geometry parameters, but the wavelength separation is 3–4% worse than in region Ω_2. In region Ω_2, a ~7% separation is achieved, but the insignificant (0.1%) variation in the geometry of the magnetron leads to a sharp violation of the limitations on the magnitude of the separation at phase velocities. With an increase in the number of resonators, the ratio of the volumes of these regions decreases, and there is no noticeable deterioration in the separation along the wavelengths. It is interesting to note that, in practice, the geometry parameters in subdomain Ω_1 are usually given as the starting point. In subdomain Ω_2, it is possible to obtain the initial point only formally, with the aid of a random number generator (in some cases much more computer time is spent on searching for such a point than on the actual solution of the problem).

As a result of the calculation, the following geometrical parameters of the 12-cavity system were chosen in the subregion Ω_1: $R = 3$ cm, $r_c = 2$ cm, $d_1 = 0.6$ cm, $d_2 = 1.3$ cm, $p_1 = 0.55$ cm, $p_2 = 0.85$ cm, $N\vartheta = \pi$. The main calculated dispersion characteristics of such a system for open and closed anode blocks are given in

Table 1.3 (types of oscillations with $\beta < 1$ are indicated). The form ($n = 1, -1$ harmonic) closest to the operating π-type of oscillations is quite satisfactory, is divided by wavelengths $\Delta\lambda_\gamma/\lambda_\pi \approx 5\%$ and by phase velocities $\beta_\gamma/\beta_\pi \approx 26\%$ for the open block and $\Delta\lambda_\gamma/\lambda_\pi \approx 2.7\%$, $\beta_\gamma/\beta_\pi \approx 10\%$ for the closed block.

Table 1.3 also shows the results of the 'cold' measurements λ_{cold} of the resonator system. The radiation wavelengths λ_{rad} of the relativistic magnetron were also measured in those modes that could be carried out experimentally [24].

From the comparison of the data given one can draw a conclusion about the good correspondence between the mathematical model and the manufactured sample of the resonator system, which was used in a powerful high-efficiency magnetron with an output power of $P \approx 8$ GW, with an electronic efficiency of ≈47% (see 1.9.5).

In conclusion, we should note some features of calculating the resonator system of a relativistic magnetron. It is known that in high-current accelerators, the deviations in the value of the voltage during one pulse and from pulse to pulse are not less than 10%, the same percentage characterizes the deviations of the pulsed magnetic field. This leads to significant fluctuations in the magnitude of the azimuthal electron drift velocity. In addition, it follows from the calculations that resonant systems with a high phase velocity of the wave ($\beta_\pi \sim 0.3$) have a much worse separation of the types of oscillations in frequency and phase velocity, and also a greater

Table 1.3

	Closed system				Open system			
	Range of systems with large resonators				Range of systems with small resonators			
n	1	2	3	6 (π)	6 (π)	1	2	3
λ_{calc}	12.87	11.53	11.26	9.75	13.17	12.55	10.9	10.17
β	1,464	0,817	0.558	0.322	0.238	1.501	0.865	0.617
γ_{-1}	–5	–4	–3	–6	–6	–5	–4	–3
β_{-1}	0.293	0.409	0.558	0.322	0.238	0.300	0.432	0.617
λ_{cold}	12.6	11.6	11.3	9.7	13.1	12.6	11.1	10.4
λ_{rad}	–	11.5	–	10.5	–	12.7	9.6	–

relative 'pollution' in comparison with the non-relativistic case ($\beta_\pi \ll 1$). Thus, in order to successfully select a multi-cavity system, a careful analysis of numerous combinations of parameters is required in order to maximize the working area of the feeding fields.

The dimensions of the anode blocks of the 10-cm wavelength range manufactured at the TPU in different years for various applications are presented in Table 1.4.

A relativistic magnetron of the 30-cm wavelength range was investigated for a number of practical applications in the late 80's [96]. An anode block 9.8 cm in diameter with six different vane-type resonators was made of stainless steel. Microwave power was output through the communication slot in one of the resonators and a smooth waveguide junction and was radiated by a horn antenna. The magnetron was installed on the 'LUCH' accelerator complex and was tested at voltages of 100–500 kV, with a magnetic field of 0.1–0.7 T. The anode block of the closed and half-open configuration with various types of cathodes was investigated: cylindrical, washer,

Table 1.4

Magnetron geometry	Direct			Inverted			
Number of resonators	6	8	12	8	10	12	12
Anode diameter, mm	43	43	60	60	48	90	90
Cathode diameter, mm	16–22	16–22	22–50	80	68	110	110
Depth of resonators, mm	21.5	21.5	–	18	14	31	31
Width of slit, mm	7.5	5.62	7.83	11.8	7.5	15	12
Opening angle of resonators	30–40°	22.5°	–	22.5°	18°	20°	15°
Radius of holes in large and small resonators, mm	–	–	8.5 5.5				
Height of slits in large and small resonators, mm	–	–	13 6				

with symmetrical feed. The maximum power of 212 MW with an efficiency of ~6% was obtained with a closed anode block using a cylindrical cathode with a diameter of 58 mm. The pulse duration of the microwave radiation was 130 ns and was approximately equal to the duration of the voltage pulse (150 ns). The magnetron operated on the π-type of oscillations with a frequency of 1200 MHz. The operating characteristics of the magnetron are shown in Fig. 1.26.

As a result of the experiments it was obtained that the level of power generated by the relativistic magnetron with a cylindrical cathode turned out to be 1.5–2 times higher than for a washer cathode. A magnetron with a cathode with a symmetrical supply was also investigated (that is, the voltage applied to the cathode from the two ends of the cylinder, and not from one, as usual). It was assumed that such a power supply should provide a more uniform emission and elimination of the undesirable azimuthal magnetic field produced by the current flowing through the cathode. However, the symmetrical power supply did not lead to an increase in power, The power of $P \sim 168$ MW was recorded at an efficiency of 5.2%. In these experiments, the level of side oscillations of the magnetron, depending on the type of cathode, was also investigated. For different cathodes of the same diameter in the closed anode block, the ratio of

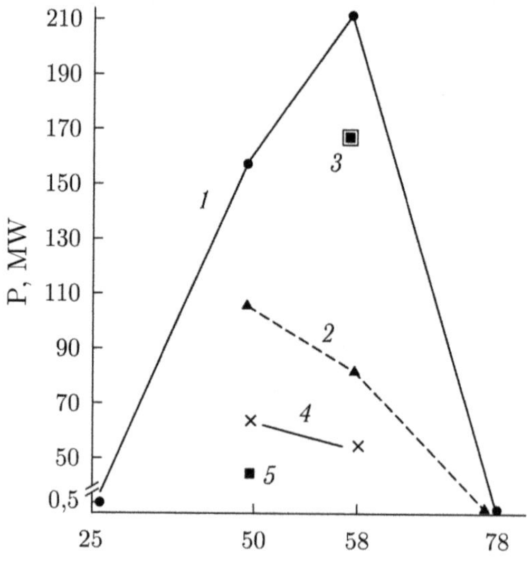

Fig. 1.26. Dependence of output power on the type of cathode and the type of resonator system. Closed block: curve *1* – cylindrical cathode, curve *2* – washer cathode, curve *3* – cathode with symmetrical power supply. Half-open block: curve *4* – cylindrical cathode, curve *5* – washer cathode.

power at one frequency to the level of side-oscillation power (third harmonic and above) was −35.3 dB for a cathode with a symmetrical supply, −32.5 dB for a washer, −33 dB for a cylindrical cathode. In the transition from the closed anode block to the half-open one (for cathodes with a diameter of 50 mm), the ratio of the powers at the fundamental frequency and the side oscillations for the cylindrical cathode decreased by 6 dB, and for the washer cathode increased by 2 dB.

1.7. Synchronization of relativistic magnetrons

By the mid-80s of the last century, the registered power levels of the RMs reached the limit levels for the electrical strength of resonators and radiation output devices. An increase in the volume of the resonators V leads to mode competition, since the number of modes is proportional to $(V/\lambda)^3$. Therefore, in order to further increase the power (up to 10 GW and above), research has begun in various scientific centres on the creation of sources based on phased (synchronized) relativistic generators. It was assumed that the most natural way to ensure the coherent operation of many powerful generating devices (autogenerators) is their synchronization due to mutual coupling. With the spatial formation of radiation, this should lead to an increase in the power flux density proportional to the square of the number of sources. By that time, a significant theoretical and experimental experience was accumulated in the construction of power summation systems on low-power devices.

The beginning of studies of synchronization processes in magnetron systems can be attributed to the 1940s of the 20th century [97]. A number of general requirements were advanced when creating multi-generator microwave systems, regardless of the specific implementation of the device and the range of operating frequencies. These included the high efficiency of summation of power, high quality of the output signal spectrum, and also the most important condition – the stability of the operating mode. The stability of the operating mode is understood as the ability of the system to retain the specified, primarily phase characteristics of the coherent mode with admissible deviations in the amplitude and frequency characteristics of partial self-oscillators, with changes in loads, the quality of the elements of communication circuits (channels), and the parameters of power supplies,

The first reports on the experimental study of powerful self-oscillatory systems began to appear in the late 80's and early 90's. [52–55]. Researchers in the United States managed to synchronize two relativistic magnetrons of a 10-cm wavelength band. The magnetrons were fed from a common accelerator through a branched magnetically isolated transmission line and connected to each other by a waveguide with the length multiple of the working wavelength. The RMs operated at 2.5 GHz for π-type and at 3.8 GHz for 2π-type and had a frequency detuning of less than 50 MHz. Their power was ~1.5 GW. In the experiments, a doubling of the power flux density in the far zone was recorded when each magnetron was operating on its radiating antenna. Phase synchronization was observed only 5 ns after the start of generation. Later, the same authors attempted to synchronize modules, including four to seven short-pulse relativistic magnetrons, and also considered various ways to combine generators in modules [52]. To construct such sources, it was proposed to use magnetrons with power outputs from several resonators. In [54], a study was reported of the degree of coherence of magnetron oscillations by means of phase measurements. In the experiments, an acceptable stability of the synchronous regime was not achieved: the synchronization band did not exceed 50 MHz, the coherent regime was observed only in half of all pulses. This, apparently, was the result of suboptimal interactions of generating devices connected to each other by different waveguide communication channels, which probably led to the curtailment of interesting studies. In addition to the synchronization of magnetrons, the same authors studied the external synchronization of one and two vircators by a relativistic magnetron [64]. The devices, fed from a single source, had a common magnetic field system. Summation of the power of individual generators occurred in space in the far zone. During the pulses (~40 ns), stable phase synchronization was obtained and the effect of power amplification in the vircators was recorded upon pumping by the microwave field of the RM.

More optimal interaction of the generators was realized in experiments conducted at the TPU [98–101], where the mutual synchronization of two relativistic magnetrons was carried out. Unlike the results of the American studies, the stability of the synchronous regime was ensured by the inclusion of a dissipative element in the waveguide channel of communication – the common load. The approach is based on a general idea of the influence of the dissipative elements of the interconnection chains on the existence and stability of synchronous oscillations of self-excited generators. The introduction into the communication channels of generating devices of loads or special dissipative elements provides the so-called resistive coupling [102–104]. Such a connection is optimal from the point of view of the stability of synchronous in-phase, antiphase or close to them modes. A distinctive feature of the communication channels with the dissipative elements is a wide area of adjustment of its phase parameters. This leads to low criticality of coherent systems in tuning, and with strong links – to a wide synchronization band, which is very important for practice. Chapter 6 will discuss a modified model of a relativistic magnetron with external coupling of resonators, including a distributed power output from the communication channel.

The proposed variants were based on symmetric and antisymmetric communication circuits [104]. Figure 1.27 shows the simplest version of a four-terminal connection, which realizes a strong interaction between oscillators. With the help of this circuit the autogenerators are directly connected to each other and to the load R_{load} by segments of the transmission line with wave conductivity g_0 and with different, in general, electric lengths Θ_1 and Θ_2.

The circuit in Fig. 1.27 allows us to interpret quite clearly the amplitude–phase relations in the system of two coherent self-oscillators. Indeed, in the setting area

$$\Theta_\Sigma = \Theta_2 + \Theta_1 \approx 2n_1\pi, \ n_1 = 1, 2, ...,$$

in a system of identical self-oscillators a synchronous regime with a phase difference of oscillations is stable $\Delta\varphi = \Theta_2 - \Theta_1 \equiv \Theta \neq 0$; In this case, the oscillations enter the phase, that is, the summation of the capacities is realized. In a symmetric system ($\Theta = 0$) in the setting area $\Theta_\Sigma \approx (2n_1 + 1)\pi$ the mode of antiphase oscillations is stable

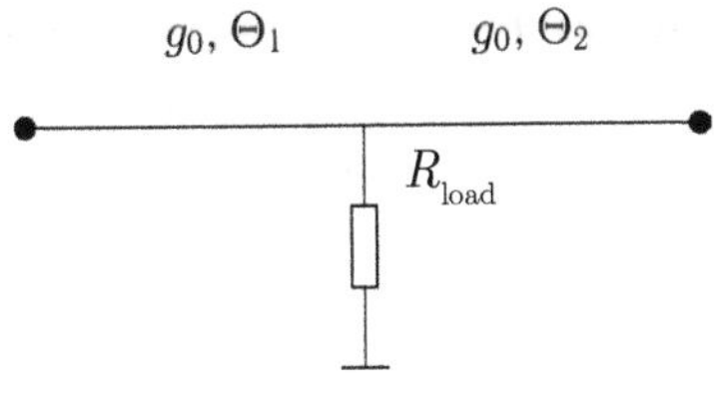

Fig. 1.27. The scheme of the simplest circuit of mutual coupling of self-excited generators.

$\Delta\varphi = \pi$. When the amplitudes are equal, there is a complete subtraction of powers in the common load, and the generators operate in the idle mode. Since the tuning area for both options (especially for the second one) is large, then the criticality of the parameter selection Θ_Σ is very low. In practical schemes on semiconductor self-oscillators of centimeter and millimeter ranges, the synchronization bandwidth reaches 10–12% [107]. The extreme simplicity of practical implementation makes it possible to use these circuits as the basis for constructing more complex multi-generator coherent systems, in particular, with the distribution of oscillations along the load system for the formation of directional radiation [98–100].

The experiments were carried out at the installation (Fig. 1.28), which allowed to investigate various variants of the RM connection, including the mode of summation or subtraction of powers in the common load [98–100]. Two RMs of a 10-cm wavelength range with identical vane-type resonators were used. Two LIAs were used as power supplies of magnetrons. (The characteristics and advantages of the LIAs are described in Chapter 3). The accelerators were powered by a common power supply system; the LIA was synchronized by a control discharger; the spread in the operation of the accelerators did not exceed 5 ns.

The RMs had two or one output of power from the resonators of the anode block and operated at a frequency of 3 GHz. The magnetrons were connected to one another and to a common load (antenna A_2) by waveguide sections, that is, according to the scheme in Fig. 1.28. The total length of the communication line was selected in the course of various experiments.

In the study of the magnetrons with two outputs, the second outputs were loaded with matching antennas A_1 and A_3. Approximately half the power of each magnetron was output through these antennas. The control of microwave radiation from the side antennas made it possible to easily diagnose the operation of the magnetrons both in joint operation and in an autonomous mode.

When the total electrical length of the communication channel is approximately $2n_1\pi$ (the length of the waveguide communication lines was 7–8 wavelengths) the summation of the powers in the common load A_2 occurs in a symmetrical system. The total power in this load was 1.9 times higher than the power of a single magnetron and reached 190 MW (Fig. 1.29). With the total electrical length of the communication channel $(2n_1 + 1)\pi$ in the system the antiphase mode with subtraction of oscillations in the common load A_2 is

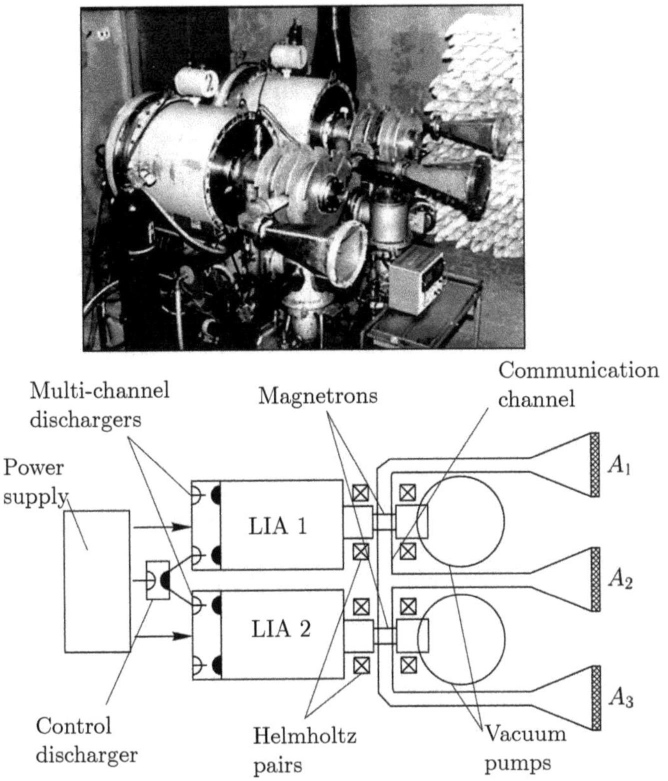

A_1, A_2, A3 – output antennas

Fig. 1.28. Appearance and installation scheme for experiments with the phase synchronization of relativistic magnetrons.

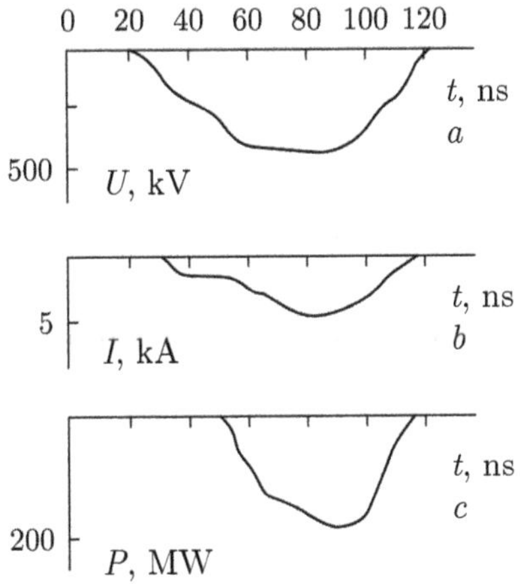

Fig. 1.29. Pulse forms: *a*) voltage, *b*) current, *c*) microwave power.

resistant; the radiation level through this antenna was 25% of the power of an individual magnetron. The high enough radiation level was due to the unevenness of the partial powers of the magnetrons and the detuning of their natural frequencies [99, 100]. In the case of magnetrons with two outputs, the radiation in an antiphase mode was output through the side antennas and, when the phase shift was compensated, was summed in space. It should be noted that it was not possible to determine the limiting permissible frequency detuning of the magnetrons (synchronization band), but in some experiments, stable synchronization took place at detunings of the order of ~60 MHz.

1.8. Ripped-field magnetron

It should be noted that to date the magnetron is not the only representative of relativistic devices with crossed fields. MILOs (magnetically insulated line oscillator) are widely studied theoretically and experimentally [108]. In such devices, the azimuthal magnetic field necessary for synchronism is created by a high-current electron beam propagating in the gap between the external ground electrode and the internal potential electrode. Naturally, such systems do not require the use of magnetic systems, but electron beams with a current of 15–25 kA are necessary. For short-wave wavelengths, magnetron systems such as Gyromagnetron (an electron beam rotating in its own electric and external magnetic fields, interact with a slow-wave system at cyclotron frequency harmonics) are investigated [108]. Also known is a free-electron laser with crossed fields, otherwise known as a rippled-field magnetron [110].

In this device, the electron beam interacts with the resonator system in crossed radial electric and axial magnetic fields, and also with an azimuthally periodic magnetic field added to them. The azimuthally periodic magnetic field is formed by permanent magnets located at the anode and the cathode. The device is a hybrid of a magnetron (electrons rotate in crossed fields) and a free-electron laser (electrons are subjected to the action of an azimuthally periodic magnetic field, acquiring wave-like motion).

The device is shown schematically in Fig. 1.30 and consists of a smooth cylindrical anode of radius r_a and a smooth coaxial cylindrical cathode of radius r_c. The electrons emitted by a thermionic or explosive-emission cathode from a metal surface are simultaneously acted upon by two constant or quasi-permanent fields

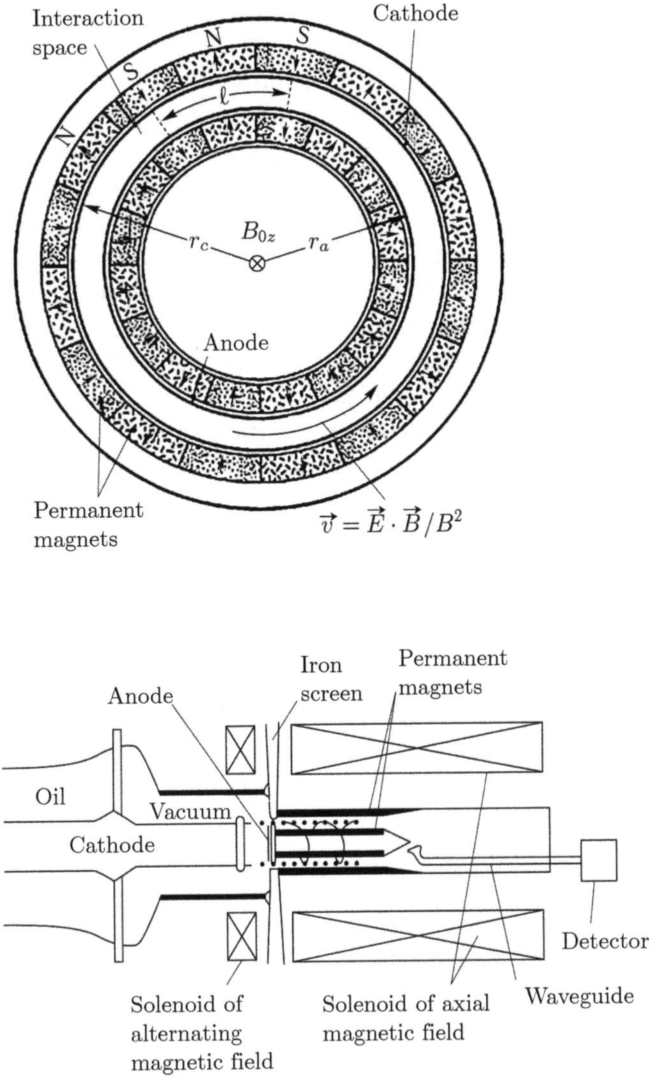

Fig. 1.30. Scheme of magnetron with alternating magnetic field: cross-section (top), flat approximation (bottom).

directed perpendicular to each other: a uniform axial magnetic field B_{0z} produced by a solenoid and a radial electric field E_r obtained by applying a voltage U between the electrodes. Overlapped on the fields E_c and B_{0z} is an azimuthal periodic field of the form $B_\omega \cos(N\Theta)$, where B_ω is the field amplitude, $N = 2\pi r_c/l$ is the number of spatial periods. In Fig. 1.30 the periodic field is directed mainly in the radial direction and can be created, for example, by Sm–Co magnets inserted into the anode block and cathode (from the electron beam, the magnets are separated by thin non-magnetic metal electrodes). This is the 'magnetic priming', which is mentioned in the section devoted to the simulation of a relativistic magnetron with a 'transparent' cathode. The resulting periodic force acting on the electrons is directed along the z axis, which is also the direction of polarization of the generated radiation. The radiation is propagated in the azimuthal direction and is related to TE by the coaxial waveguide mode determined by the cathode–anode gap. When the cathode–anode gap is small in comparison with the anode radius, it is possible to imagine a cylindrical device in the plane approximation, which is shown on the bottom part of Fig. 1.30.

The characteristics of a magnetron with a periodic field, calculated for radiation with λ = 1.3 mm, are as follows: a voltage of 1022 kV, an axial magnetic field B_{0z} = 1.03 T and a periodic magnetic field B_{ω} = 0.228 T. This value is real for a cathode–anode gap of 0.5 cm when using Sm–Co magnets. The period of this field is 1 cm and the total number of periods 26. The time increment corresponds to a spatial power increase of 2.3 dB/cm. Consequently, the wave will amplify by 70 dB as it moves around the interaction region. Since the interaction space is an integral part of the coaxial waveguide, the electromagnetic radiation generated therein is output axially, for example, by means of an appropriate coaxial mode transformer into a waveguide mode.

The study [110] reported the results of measurements of millimeter microwave radiation for a magnetron with a rippled magnetic field. Narrow-band (less than 2 GHz) radiation was observed in the wavelength range 7–9 mm with a power of about 300 kW (individual pulses up to 1 MW).

1.9. Control of the output parameters of microwave radiation

In this section, the effect of the dimensions of individual elements of relativistic magnetrons and power supply parameters on the amplitude and frequency characteristics of generated microwave pulses is analyzed and methods of their control are considered. At the same time, it becomes possible to evaluate the validity of the following assumptions used in calculating the resonator systems of magnetrons: the neglect of the effect of space charge, the two-dimensionality of the problem, the limited number of harmonics taken into account, and the accuracy of calculating the roots of the dispersion equation.

1.9.1. Effect of voltage on the output characteristics of the microwave radiation of the RM

In [111] experiments with 6-resonator relativistic magnetrons are described. The slowing-down system closed by the end caps was used, which ensures the maximum output parameters of the device.

The frequency of the π-type oscillations was calculated by the field theory method (section 1.5), and its results are given in Table 1.5, where the results of 'cold' measurements of a relativistic magnetron without (f', Q') and assembled with a power output (f'', Q'') are also presented. The 'cold' measurements of the device were carried out according to the scheme of the four-terminal network.

Table 1.5

f_{calc}, MHz	f', MHz	Q'	f'', MHz	Q''	f_{meas}, MHz	Q_{meas}
					at voltage 1 MV	
3145	3257	345	3160	121	3075	51

As can be seen from Table 1.5, the calculated and measured values of the frequencies are close, which indicates the possibility of using the above assumptions. The use of an asymmetric waveguide power output somewhat changes the frequency of the π-type of oscillations. The reactive conductivity introduced by the electron beam also leads to a decrease in frequency in comparison with the 'cold' system.

The radiation characteristics of a relativistic magnetron were investigated at the Tonus-1 HCEA at voltages of 0.6–1 MV, a total current of 6–20 kA, and a voltage pulse duration of 60 ns. A set of equipment for measuring the parameters of microwave pulses included: calibrated lamp microwave detectors of the 10-cm wavelength band, cryogenic semiconductor detectors of the 3-cm and 8-millimeter wavelength band.

Figure 1.31 shows the dependences of the carrier frequency, the power of the microwave pulse, and also the electronic efficiency of the relativistic magnetron on the value of the voltage. The increase in the power of the electron beam with practically unchanged electronic efficiency of the device leads to a noticeable increase in the output

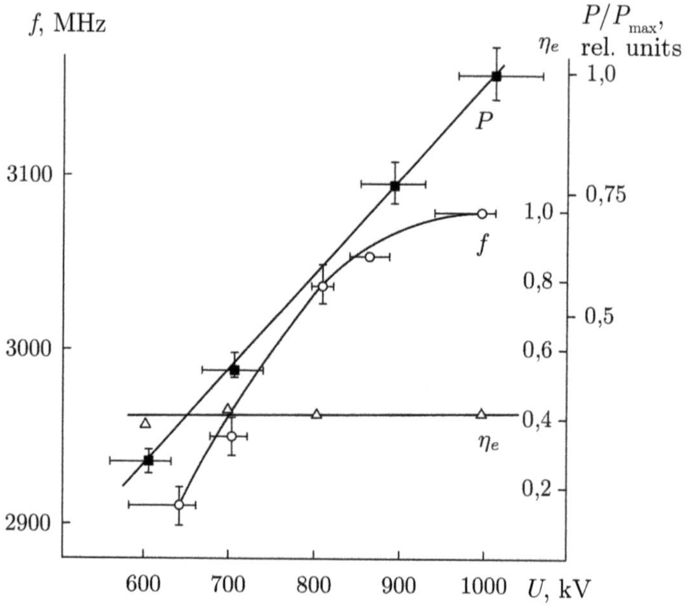

Fig. 1.31. Dependences of the carrier frequency (f), the power of the microwave pulses (P) and the electronic efficiency (η_e) of the RM on the value of the cathode-anode voltage.

power and to the approximation of the carrier frequency of the generated microwave pulse to the frequency of the 'cold' system.

Let us consider the physical aspect of the applied voltage influence on the frequency properties of the relativistic magnetron. The increase in voltage is accompanied by an increase in the radius of the cyclotron rotation of the electrons, the growth of the anode current generated by the microwave power, and the synchronous magnetic field. An increase in the diameter of the cathode electron layer causes a change in the generated frequency in accordance with the relation

$$f_n = f_0 \Big/ \sqrt{1 + \frac{C_1}{2C_0(1 - \cos(2\pi n / N))}}, \tag{1.9.1}$$

where C_0 is the equivalent capacitance of a single cavity; C_1 is the concentrated capacitance between one segment and the cathode; f_0 is the resonance frequency of a single resonator [112]. On the other hand, the change in the electric field intensity is reflected in the average angular velocity of rotation of the electrons and, consequently, on the angle of the mismatch Θ_1 ($\Theta_1 < 0$) between the amplitudes of the 1-st harmonic of the induced current and the high-frequency voltage on the resonators.

The increase in voltage leads to an increase in the generated frequency, approaching it to the natural frequency of the resonator system. The opposite effect of the two factors explains the behaviour of the dependence $f(U)$, depicted in Fig. 1.31. At low voltages, the value of the electronic frequency shift (EFS) is determined to a greater extent by a change in the phase mismatch, as evidenced by the steepness of the EFS curve. This effect is due to a rapid change in the angle of mismatch Θ_1 at a large value of the latter. As the voltage increases and the angle Θ_1 decreases, the effect of the charge of the electron cloud on the generated frequency increases, which reduces the steepness of the EFS curve (saturation is observed on the curve $f(U)$).

The generation band of the relativistic magnetron did not exceed 60 MHz at the –3 dB level for the entire voltage range. Within the sensitivity of the recording equipment used in the experiments (the detection threshold of the second harmonic was –28 dB, the third harmonic –35 dB), the presence of harmonic components was not recorded. It should be noted that this fact allows us to optimistically consider the possibility of using relativistic magnetrons for nonlinear radar.

1.9.2. Effect of the diameter of the cathode on the radiation frequency of the RM

The effect of the diameter of the cathode on the carrier frequency of the radiation of the π-type of oscillations was found in experiments with the RM powered by a linear induction accelerator. The results of the studies are shown in Fig. 1.32. The decrease in the generation frequency with increasing cathode diameter is explained by the change in the cathode–anode capacitance in accordance with expression (1.9.1). Measurements of the radiation spectrum of the magnetron made it possible to notice a smooth decrease (drift) of the frequency (~100 MHz) during the pulse. This effect is especially noticeable for long microwave pulse durations (~120 ns). The signals from the reference microwave detector were synchronized with the detector signals with a filter installed in front of it, which was tuned to different transmission frequencies. With a decrease in the transmission frequency of the filter, the maximum of the signal

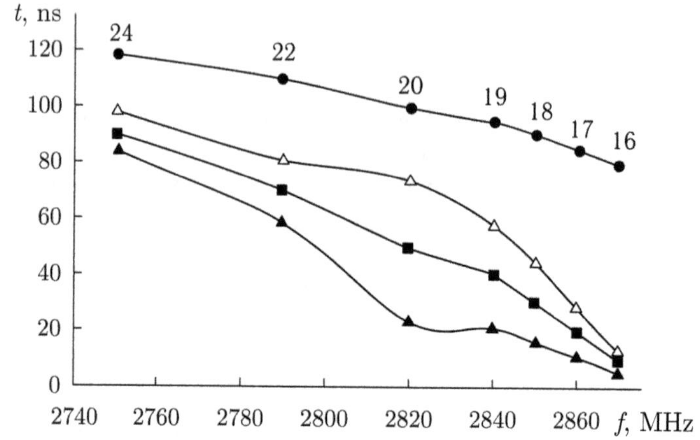

Fig. 1.32. Dependence of the central radiation frequency of the RM (●) on the diameter of the cathode from 16–24 mm. The change in the central frequency of radiation of a relativistic magnetron with a cathode of 19 mm in diameter (■) and a generation band at −3 dB (▲– Δ) during a pulse.

travelled to the trailing edge of the microwave pulse. This frequency behaviour is explained by the expansion of the outer boundary of the cathode plasma and the increase in the interelectrode capacitance of the slowing-down system. A comparison of the experimental curves in Fig. 1.32 allowed us to estimate the speed of radial motion of the cathode plasma in a magnetron diode, which amounted to 4–5 cm/μs. The results obtained are in good agreement with the data of [72], in which another estimation technique was used.

1.9.3. Mechanical frequency tuning of the microwave radiation of the RM

The results of studies of the possibility of mechanical tuning of the radiation frequency of the RM are described in Refs. [68, 70]. In the experiments in [68], the range of frequency tuning for 10- and 30-cm relativistic magnetrons exceeded 30%, which was achieved by changing the length of the anode block when the profiled end cap was moved.

Another method involves changing the inner diameters of the end caps [70, 113, 114], based on the following. The resonance frequencies of the modes of oscillations of the closed (f_{3n}) and open (f_{0n}) anode blocks are in the ratio

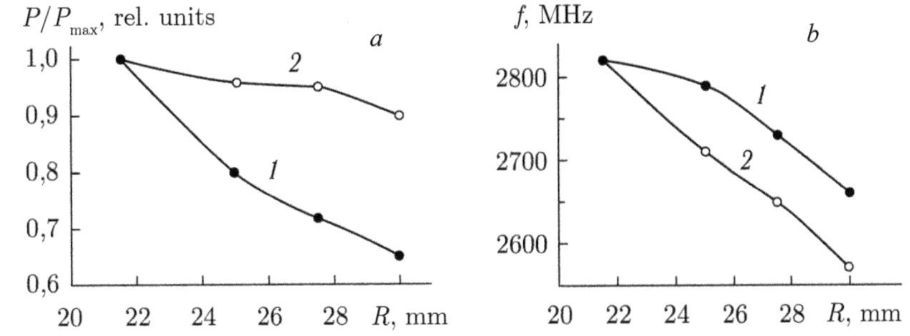

Fig. 1.33. Dependences of microwave radiation power (*a*) and radiation frequency of a relativistic magnetron (*b*) when the inner radii of two end caps (curve *2*) or one cap change (curve *1*).

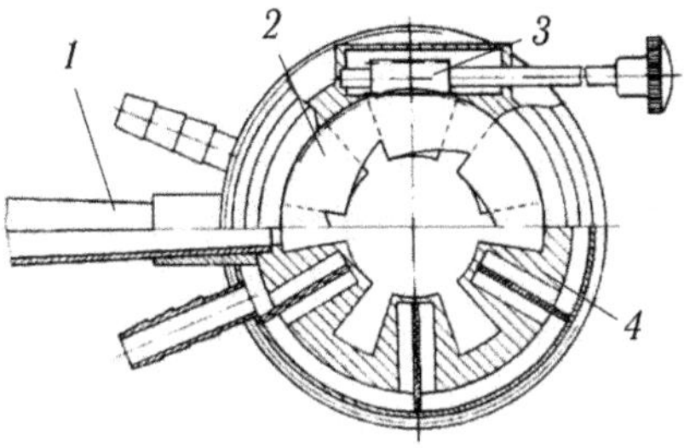

Fig. 1.34. The device for mechanical frequency tuning of RM radiation: *1* – waveguide output of power; *2* – the end cover; *3* – mechanism for turning the end cover; *4* – anode block.

$$f_{3n} = f_{0n}\sqrt{1+(c/f_{0n}2L)^2}. \qquad (1.9.2)$$

In specially designed studies for the tuning of the radiation frequency of a relativistic magnetron, end caps with different internal radii were used. Figure 1.34 shows the dependence of the power and the central frequency of the radiation when the inner radii of two end caps or one cap change (the inner radius of the second cap is equal to the inner radius of the anode block).

It can be seen from the presented experimental dependences that it is expedient to change the internal radii for both end caps to adjust the frequency of magnetron radiation. In this case, the reduction in the power level does not exceed 10%. If the internal radius of only one end cap changes, the frequency range is less, and the power reduction is greater. The latter is explained by the displacement of the antinode of the microwave electric field from the geometric centre of the anode block, the deterioration of the energy exchange of electrons with the high-frequency field and, possibly, the violation of the excitation conditions for the H_{10} wave in the waveguide power output. Figure 1.34 shows an embodiment of a device for mechanically tuning the radiation frequency of a relativistic magnetron. The main elements of the device are two end caps of the anode block with special cutouts and two devices for turning the covers around the axis of the device. The tests of this device demonstrated the possibility, without dis vacuumization of the RM, of tuning of the generated frequency in the above-mentioned limits.

1.9.4. Regenerative amplifier

At the first stages of the development of relativistic high-frequency electronics, experiments were devoted to the study of autogenerator devices (travelling wave tube, backward-wave tube, cyclotron-resonance maser, magnetrons) operating at fixed frequencies having a relatively narrow bandwidth and characterized by unstable operation. By analogy with low-current microwave electronics, interest has arisen in the study of amplifiers based on devices with crossed fields, which have high efficiency and gain.

The investigated device [21, 22] was made on the basis of a 6-cavity magnetron of a 10-cm wavelength band. The amplified signal was fed to its input by means of a smooth waveguide transition (Fig. 1.35). The same transition served also for the output of the amplified

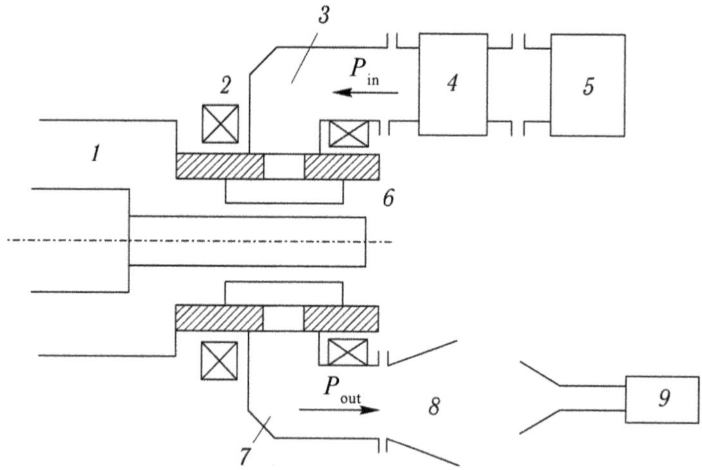

Fig. 1.35. The scheme of the experimental setup: *1* – accelerator 'Tonus-1'; *2* – magnetic system; *3*, *7* – waveguide transitions; *4* – ferrite valve; *5* – excitation generator; *6* – magnetron; *8* – transmitting antenna; *9* – system for measuring the parameters of the amplified signal.

signal. The dimensions of the windows connecting the waveguide transitions with the anode block were selected experimentally. The reflection coefficient of the input path in the operating frequency range, measured at a low power level, was 0.42–0.5. The magnetron was fed from a high-current electron accelerator Tonus-1, and excitation by the 10 cm magnetron generator with tunable frequency. The duration of the microwave pulse of the exciting generator was much higher than the duration of the anode voltage pulse and was ~2 μs. The signal power was $P_{\text{in}} \approx 40$ kW and was constant when the frequency was tuned.

As a result of the experiments, a gain factor of ~30 dB was recorded at an electron beam power of $P_0 \approx 5$ GW. The low efficiency of the device (~1%) can be explained by the closure of the resonator system and the need to increase the dissynchronism between the electron velocity and the phase velocity of the wave in order to eliminate self-excitation. In this mode of operation, in the absence of an input signal, the output signal had a random noise character.

A change in the oscillation frequency of the master oscillator in the operating band (2700–3000 MHz) led to a similar change in the frequency of the output signal of the amplifier, and both the amplification factor and the efficiency of the instrument did not change significantly.

Thus, the conducted studies indicate the possibility of creating microwave amplifiers with crossed fields using high-current relativistic electron beams. The amplifier studied has a relatively high gain and reproduces the frequency of the input signal, whose power is much less than the power of the high-current beam ($P_{in}/P_0 \approx 10^{-5}$).

1.9.5. Effect of the internal diameter of the anode block on the output characteristics of the RM

The influence of the internal diameter of the RM interaction space on the generated power can be estimated by comparing the experimental results obtained on the Tonus-1 HCEA using 6- and 12-cavity anode blocks with a cathode in the interaction space and with external injection of the electron beam (RM with external injection are discussed in detail in chapter 5). As previously noted, the advisability of developing the transverse dimensions of devices is associated with the possibility of increasing the electronic efficiency, reducing thermal loads on the surface of the anode block, increasing the pulse length of the radiation when feeding of microsecond voltage pulses,

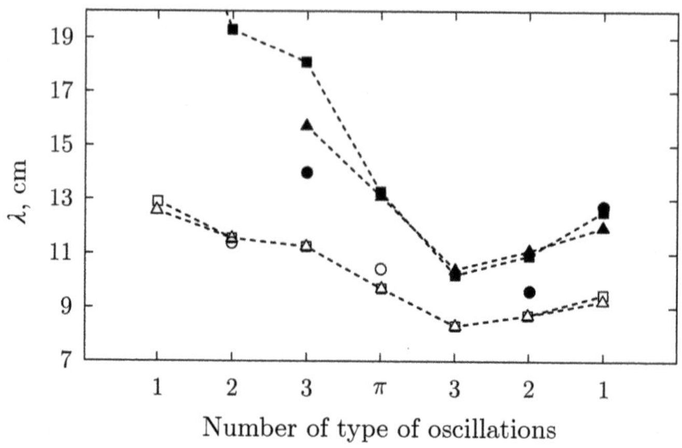

Fig. 1.36. Dispersion characteristics of a 12-resonator relativistic magnetron: ■,□ – calculated values of the wavelengths of the modes for open and closed anode blocks; ▲, Δ – results of 'cold' measurements of half-open and closed anode blocks; ●, ○ – measured values of wavelengths in experiments on HCEA.

etc. For this, it is necessary to increase the number of resonators of the anode block. However, the use of a slowing-down system in a magnetron with a large number of resonators sharply increases the competition of modes of oscillation. A satisfactory separation of the types of oscillations in frequency and in the threshold excitation voltage is realized when using a different resonator-based slowing-down system. The slowing-down system, calculated using the technique described in section 1.5, had 12 slit–hole resonators. The geometric dimensions of the anode block are shown in Table 1.4. The results of calculating the dispersion characteristics with and without allowance for the axial variations of the high-frequency field are shown in Fig. 1.36, where the points taken at 'cold' measurements and in the operating mode are also marked. It can be seen that the system under investigation has a completely satisfactory separation of the types of oscillations in frequency. With regard to the separation of the modes by the slowing-down of the electromagnetic wave, the smallest separation value for the closed anode block between the π-type and (−1) harmonic of the first type of the short-wave part of the spectrum (the group of small resonators) was 9%, and for the half-open anode block between the π-type and (−1) harmonic of the first type of the short-wave part of the spectrum is 21%. In experiments on the Tonus-1 HCEA, the RMs with closed and half-open resonator systems were investigated. Figure 1.37 shows the

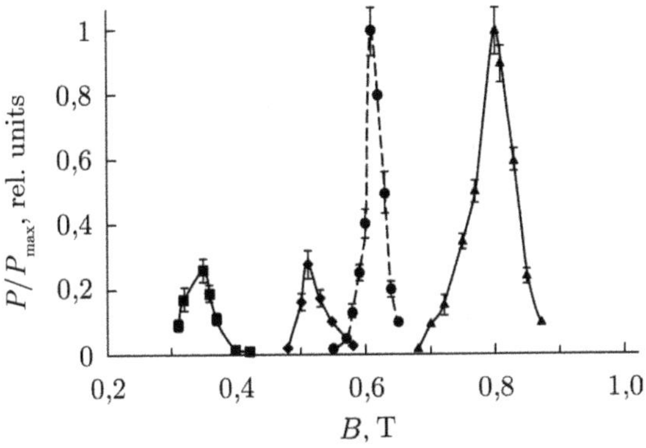

Fig. 1.37. Dependence of the microwave radiation power of the relativistic magnetron with 12 different resonators on the induction of the magnetic field for half-open (solid curve) and closed (dashed curve) resonator systems.

dependence of the microwave radiation power on the magnitude of the magnetic field induction for these two types of systems.

When using a half-open system in the range of magnetic fields that the electropulse magnet used in experiments could create, three maximum microwave oscillations were detected (Fig. 1.37). The following radiation was registered: 1) on the mode of oscillations $n = 3$ with a wavelength of 14 cm and an efficiency of ~14%; 2) on the (−1)-th harmonic of the mode of oscillation $n = 2$ of the short-wave part of the spectrum with a wavelength of 10 cm and an efficiency of ~1%; 3) excitation of the (−1)-th harmonic of the mode of oscillation $n = 1$ of the short-wave part of the spectrum with a wavelength of 12.7 cm. The maximum power in the latter case was about 8 GW with an electron efficiency of ~47% (the magnetron efficiency estimates were performed without taking into account the losses of the end current). In the range of magnetic fields from 3.5 up to 5.2 kG the generation of two types of oscillations ($n = 3$ and 2) was observed simultaneously, which due to the dispersion of the waveguide transmission line of 40 m in length was clearly visible on the oscilloscope screen. Two of these types of oscillations have close values of the electromagnetic wave slowing-down $\beta_p = 0.348$ and 0.432 respectively, which, apparently, leads to their simultaneous excitation.

For a closed system, other types of oscillations were not observed except for the main π-type with a wavelength of 10.5 cm with an efficiency of ~4% (P_{max} ~ 660 MW), and the maximum microwave power is located in a narrower range of magnetic fields, indicating a higher quality of the closed system, which is confirmed by 'cold' measurements.

It follows from the experiments that for a half-open system the level of generated power is higher, and this is due to a change in the distribution of the high-frequency field in the interaction space of a relativistic magnetron and the longitudinal drift of electron spokes into the region of the maximum high-frequency field.

An analysis of the results of experiments with 12 different resonator relativistic magnetrons makes it possible to draw the following conclusions:

1) when constructing the RM it is importnt to achieve the greatest possible separation of the modes from the slowing-down of the electromagnetic wave (at least 9%);
2) the use of half-open resonator systems leads to a decrease in this parameter, which makes it difficult to excite the working mode of oscillations;
3) the higher efficiency of using half-open systems in comparison with closed systems, which is typical for devices with high current, should be noted.

1.10. Resonance compression of microwave pulses at the output of a relativistic magnetron

Some positive properties of relativistic magnetrons, preserved from classical analogs, can be successfully applied to solve a number of scientific technical problems, for example, in radar, including nonlinear. Below are the results of experimental studies on increasing the pulsed power of RM using resonant compressors.

The principle of resonant compression of microwave pulses is based on the accumulation of high-frequency energy in the resonator from a pulsed microwave source and its rapid outputting in the form of shorter and more powerful microwave pulses than those entering the resonator [115]. Typically, industrial magnetrons are used as sources, generating microsecond microwave pulses up to several megawatts. In this experiment, the compression device is powered by a relativistic magnetron.

Interest in the study of the compression of microwave pulses of the RM is due to the following. First, the peak radiation power can be increased by a factor of W in accordance with the relation

$$W = \eta_c t_1 / t_2,$$

where η_c is the efficiency of the compression device, t_1 and t_2 are the duration of the pulses at the input and output of the compressor, respectively. In comparison with classical magnetrons, the RMs have a relatively short pulse duration ($t_1 \sim 10^{-7}$ s), which limits the obtaining of a high power gain at the compressor output. On the other hand, when the storage resonator is excited by such pulses in the coupling mode as shown in Ref. [116], it is possible to obtain a higher efficiency (reaching 80%) than for microsecond excitation when $\eta_c \approx 40\%$. Secondly, the connection to a magnetron of a high-quality storage resonator of a compressor can lead to the stabilization of its frequency due to pulling [116]. At the same time, the possibility of the compressor working in conjunction with the relativistic magnetron was not obvious, since the microwave pulses have a higher level of amplitude modulation and frequency instability (in comparison with the classical devices), so that their spectrum can go beyond the operating band of the resonator of the compressor. Therefore, the task of demonstrating the principal possibility of the relativistic magnetron working together with the microwave pulse compressor was very relevant.

The scheme of the experimental setup is shown in Fig. 1.38 [118, 119]. The relativistic magnetron operated at a frequency of 2840 MHz and had a radiation power of up to 200 MW with a pulse duration of ~120 ns and a repetition frequency of 10 Hz. The compressor was made of waveguides with a cross section of 7.2 × 3.4 cm^2 and was a double waveguide tee with symmetrical short-circuited lateral arms. The excitation of the compressor resonant system was carried out through a communication hole in the wide wall of the waveguide. The energy was extracted through the H-arm of the tee after operation in the self-breakdown mode of a gas-discharge microwave switch placed from a short-circuited wall at a quarter wavelength distance in the waveguide. The minimum duration of the generated pulses was determined by the time of the double travel of the wave with the group velocity along one of the arms of the tee.

Before installing the compressor in the relativistic magnetron circuit, the communication hole was enlarged so that the loaded quality factor of the storage resonator decreased from 7600 to 750 (the compressor with a loaded Q-factor of 7600 is used in experiments with a classical magnetron). To provide electrical strength, the internal cavity of the compressor and the waveguide channel were filled with nitrogen at a pressure of $7 \cdot 10^5$ Pa. The separation of the vacuum and gas-filled parts of the installation was

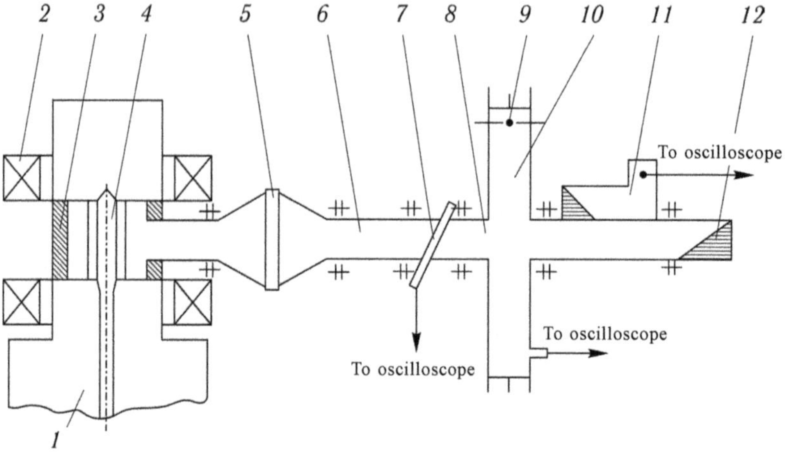

Fig. 1.38. Experimental scheme of RM studies with a microwave compressor: *1* – vacuum chamber; *2* – magnetic system; *3* – anode block; *4* – cathode; *5* – dielectric window; *6* – ferrite valve; *7* – directional coupler; *8* – communication hole; *9* – the discharger of the compressor; *10* – the resonator of the compressor; *11* – directional coupler; *12* – matched load.

carried out by a dielectric window made of plexiglass measuring 17 × 18 cm. The microwave pulses in the storage resonator of the compressor and at its input and output were monitored by microwave probes, directional couplers, attenuators and an oscilloscope.

After the compression, the output microwave pulses entered the matched load. During the research it was found that the duration of the process of energy storage in the resonator depends on the length of the input waveguide tract between the relativistic magnetron and the compressor. With a tract length equal to 3 m, an increase in the field strength in the resonator lasted no more than 30 ns. Approximately in the same time in the input microwave pulses recorded in the tract from the RM, the amplitude decay and strong amplitude–frequency modulation began. After the switch was triggered, pulses with a peak power of up to 480 MW and a duration of ~5 ns at half-power level were registered at the compressor output. In this case, the pulsed power of the magnetron did not exceed 120 MW, which is due to the presence of a reflected microwave wave from the compressor.

After increasing the length of the waveguide tract up to 10 m the duration of the pulses from the magnetron in the tract and process of resonator excitations reached 120 ns. But the maximum amplitude of the signal of the resonator, which characterizes process of accumulation of energy at it, corresponded only to ~50 ns. It is possible that this is associated with the drift of the frequencies of the magnetron due to the movement of the external boundary of the cathode plasma. When the discharger of the compressor was switched on, microwave pulses with the duration of ~5 ns with the peak power up to 1100 MW were produced from the resonator, with the power of the output pulses of the RM being 180 MW. At such power levels the strength of the electric field in the resonator and the tract reached ~300 and ~170 kV/cm respectively. A decrease if the duration of excitation of the resonator of the compressor leads to enhancing of the electrical strength of the system. However, the problem of the electrical strength of the storage resonator and the input waveguide tract remains actual. It can be solved not only by selecting isolating environment, but also by the application of constructions of the entrance tract and the compressor using oversized waveguides [120].

Figure 1.39 shows typical envelopes of microwave pulses recorded at resonator compressor and on output compressor when length tract was 10 m.

Analysis of the experimental results showed that when powerring the resonator by the relativistic magnetron, as from a conventional classical magnetron, it is required organize between them a certain communication and reduce the wave reflected from the compressor at the elementary period of its excitation for effective pulling of the frequencies by the high-quality resonator of the compressor. This can be done by using decoupling devices or, for example, when feeding two resonators from a relativistic magnetron through a bridge circuit, as was done in Ref. [121].

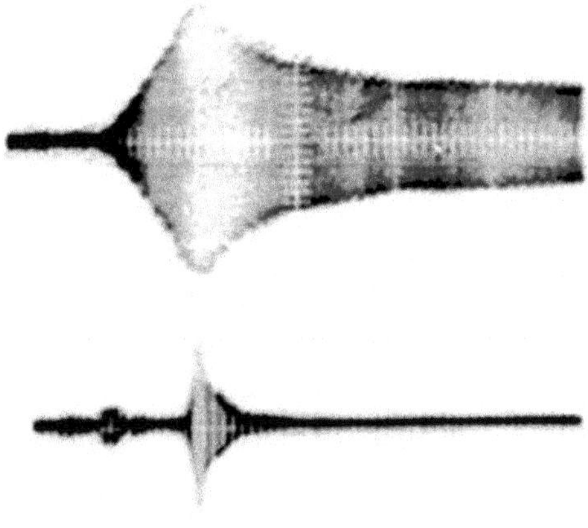

Fig. 1.39. Typical microwave pulses in the compressor resonator and at the compressor output.

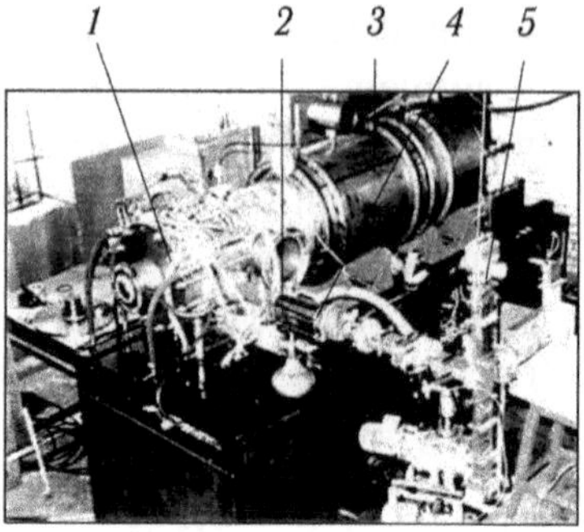

Fig. 1.40. Appearance of the experimental setup consisting of RM and microwave compressor: *1* – magnetic system; *2* – waveguide output of RM power; *3* – LIA; *4* – ferrite valve; *5* – microwave compressor.

In subsequent experiments, a ferrite valve was installed to the tract between the magnetron and the compressor, which made it possible to eliminate the reflected wave (Fig. 1.38). The appearance of the experimental setup is shown in Fig. 1.40.

Since the other elements of the installation remained the same, the microwave output pulses of the compressor had similar parameters (1100 MW, 5 ns).

In this experiment, the possibility of compression of microwave pulses at the output of a relativistic magnetron with an increase in power by a factor of 6 at a repetition rate of 10 Hz is shown for the first time. The result obtained demonstrates the high stability of the RM frequency parameters, sufficient for operation on the resonant load.

This chapter briefly describes the principle of the RM operation and the differences between similar devices and the classical analogs. The analysis of RM constructions is carried out, the results of experimental studies performed in various organizations are described. The engineering method for selecting the dimensions of RM elements is presented. An exact calculation of the geometry of the anode block and the interaction space is carried out by the methods of field theory. The method is based on finding the input conductivities in the interaction space and in the space of the resonators, equating them at the interface and on the basis of these data the composition of the dispersion equation. By choosing the geometry of the resonator structure, it can be achieved that the separation of the types of waves and their harmonics in terms of the magnitude of the phase velocity, and hence the excitation voltage, will be acceptable for the stable operation of the RM.

Based on the method used to calculate and optimize the parameters of the slowing-down system, the anode blocks with 6, 8, 12 vane-type resonators of direct and inverted geometry and with 12 cavities of the 'slit–hole' type were designed, calculated and manufactured for various applications. A different resonator structure was used to increase the frequency separation for the last magnetron.

The chapter also analyzes the influence of the dimensions of RM elements and power supply parameters on the amplitude and frequency characteristics of the generated microwave pulses and considers the possibility of controlling them. At the same time, the following assumptions used in the calculation of the resonator systems of magnetrons are determined: the absence of space charge, the two-dimensionality of the problem, the limited number of harmonics taken into account, and the accuracy of calculating the roots of the dispersion equation. The 'cold' measurements of the relativistic 6-cavity magnetron have shown that the calculated and measured values of the frequencies of the π-type of oscillations are close, which indicates the applicability of the above assumptions.

The radiation characteristics of a relativistic magnetron were studied at various high-current electron accelerators at voltages of 0.3–1 MV, currents of 3–20 kA, and pulse voltage durations of 60–150 ns. As a result of experiments it was shown that an increase in the power of the electron beam with practically unchanged electronic efficiency of the device leads to a noticeable increase in the output power. The observed electronic frequency shift in a relativistic magnetron is described, i.e., the carrier frequency of the generated microwave pulse approaches the frequency of the 'cold' system. Moreover, in the region of large EFS coefficients, an increase in the frequency with increasing cathode–anode voltage can be explained by a decrease in the capacitive effect produced by a bunch of electrons at the resonator gap. In this case, as the high-frequency voltage on the resonator increases, phase-focusing forces that tend to reduce the angle of shift between the maximum of the microwave field and the position of the electron spoke Θ_1 increase. As the voltage increases and the angle Θ_1 decreases, the effect of the electron cloud charge on the generated frequency increases, which reduces the value of the EFS coefficient.

The influence of the diameter of the cathode on the carrier frequency of the radiation of π-type oscillations is revealed, which is associated with an increase in the value of the cathode-anode capacitance. A noticeable effect on the frequency of microwave radiation and the output power is provided by the type of the resonator system: the end caps installed on both sides, half open (one end cap) and open. A closed system differs from the rest by the presence of a variation of the microwave electric field along the length of the anode block with an antinode at the centre and nodes at the edges, values of the unloaded *Q*-factor, and resonance frequencies of the modes.

The mechanical tuning of the radiation frequency of a relativistic magnetron is conducted by a device based on the use of end caps with various internal radii. It is shown that to change the radiation frequency of the RM within 8%, it is advisable to change the internal radii of both end caps. In this case, the reduced power level does not exceed 10%.

The results of experiments with relativistic magnetrons in the regime of amplification of an external signal indicate the possibility of creating microwave amplifiers with crossed fields using high-current relativistic electron beams. The amplifier studied has a relatively high gain and reproduces the frequency of the input signal, whose power is much less than the power of the high-current beam.

The research of the resonant compression of microwave pulses at the output of a relativistic magnetron is described. In this case, the peak radiation power can be increased several times. In experiments, the relativistic magnetron had a radiation power of up to 180 MW with a pulse duration of ~120 ns and a repetition frequency of 10 Hz. The compression device was created from waveguides of section 7.2×3.4 cm^2 and was a double waveguide tee with symmetrical short-circuited lateral arms. The experiment showed that when the compressor is powered by a relativistic magnetron, as well as from a conventional classical magnetron, it is required to arrange a certain connection between them and to reduce the reflected wave from the resonator in the initial period of its excitation to pull the oscillator frequency by a high-quality storage resonator. When using a waveguide tract of a suitable length or a ferrite valve, the output power of the microwave compressor was 1100 MW for a duration of 5 ns. Thus, for the first time the possibility of the relativistic magnetron generator operating on the resonant load – the microwave compressor – was experimentally demonstrated.

The following abbreviations are used in the tables:

TPU – Tomsk Polytechnic University;
IAP – Institute of Applied Physics of the Academy of Sciences of the USSR;
FIAN – Physical Institute of the Academy of Sciences of the USSR;
MIT – Massachusetts Institute of Technology;
LLNL – Lawrence Livermore National Laboratory (Livermore);
PIC – Physics International Company (San Leandro);
SNL – Sandia National Laboratory (Albuquerque);
Varian – Varian Associaties, Advanced Technology Group;
NRL – Naval Research Laboratory (Washington);
University of St. Andrews – North Haugh, UK;
DERA – Great Malvern, Works, UK;
Technion Research and Development, Haifa, Israel;
N is the number of resonators of the anode block;
d is the value of the cathode-anode gap;
U is the voltage;
I is the current;
τ is the voltage pulse duration;
P_{mw} is the generated microwave power of the device;
f is the radiation frequency;
τ_{mw} is the duration of the pulse of microwave radiation;
F is the repetition rate of the pulses;
P_{av} is the average power of microwave radiation.

Table. Experimental results for RMs

Organization, year research	N	d, cm		U, MV	I, kA	τ, ns	P_{mw}, GW	f, GHz	τ_{mw}, ns	Efficiency, %	Features of design [reference]
MIT 1976	6	0.52	π	0.36	14	40	1.7	3.0	20	35	A type of slowing-down system (SDS) is not specified [8]
IAP 1977	8	0.69	π	0.97	6	70	0.5	9.0	20	15	Diffraction power output [10]
TPU 11978	6	0.9	π	1.2	4	60	2.0	2.4	30	40	Half-open SDS [11]
TPU 1978	6	1.35 0.92	π	0.3	6	800	0.9	2.9	300	43	Closed SDS. The gap of .92 cm takes cathode plasma movement into account [12]
MIT 1979	6	0.53	2π	0.36	12	30	0.9	4.6	-	20	Closed SDS (frequency band 40 MHz) [14]
TPU 1979	6	1.35	π	1	23	60	10	2.9	30	43	Closed SDS [15]
LLNL 1979	6	0.52	2π	0.9	33	–	4,5	4.6	-	15	Symmetric power output from of all RM resonators Washer cathode [16]
Stanford University 1978–1982	12	2.5	π	0.5 0.4	4 3	1000 400	1 0.02	3.0 3.0	1000 35	50 1.9	Oxide thermionic cathode. Power output at RM axis. on wave H_{01}. Upper line – planned, lower - the results of [17, 19, 20]

Table. Continuation

Organization, year of research	N	d, cm	Mode	U, MV	I, kA toA	τ, ns	P_{mw}, GW	f, GHz	τ_{mw}, ns	Efficiency, %	Features of designs [reference]
TPU 1979	6	1.35	π	Beam power 5GW; input microwave power 40kW; gain factor 30dB; adjustment band 300MHz							Amplifier [21, 22]
IAP FIAN 1980	8	0.64	π	1.0	12	–	4	9.2	–	12	Diffraction power output [23]
NRL 1979-1980	54	–	π	0.6	9	55	1.0 0.1	3.2 3.2	40	18	Inverted coaxial RM, without preliminary excitation [32–34]
TPU 1983	12	1.0	$\pi/6$ (–1)	0.7	22	60	8.3	2.4	40	47	Specified electronic efficiency of RM. Half-open different resonator SDS of type slit-hole type [24]
MIT 1983	8 10	1.38 0.84	π π	2.0 1.6	4 9	– –	0.4 0.5	3.7 3.7	30 30	12–3.5	Inverted RM, frequency band 50 MHz. Power output through slits in cathode [35]
PIC 1985	6	0.52	2π	1.2	50	65	6.9	4,5	20–40	10–35	Symmetric power output from all RM resonators. Washer cathode[25]
PIC 1986	6	–	π	1.1	25	80	0.05 0.3	8.5	30	0.2	Waveguide power output. Axial power output[26]

Table. Continuation

Organization, year	N	d, cm	Mode	U, MV	I, kA	τ, ns	P_{mw}, GW	f, GHz	τ_{mw}, ns	Efficiency, %	Design features [reference]
PIC 1986	6		π 2π	1	360	60	3	2.8 3.8			2 RMs at 1.5GW connected by a waveguide ~$\lambda/2$. Phase matching at 5 ns. Washer cathode [27]
SNL 1986	12		π					3.8	5–70	2.0	Inverted RM. Frequency band 25 MHz. Axial power output at E_{01} wave [36]
TPU 1987	6	1.05	π	0.3	4	80	0.36	2.9	80	30	Source LIA with gas discharger $F = 160$ Hz (packet – 3 pulses) [51]
PIC 1987–1988	6	0.84	π	0.7	20	65	3.2	2.8	20	11	Symmetric power output from 3 RM resonators. Washer cathode [28]
North Carolina University 1986	16	0.64	π $7\pi/8$	0.38	12	60	0.07	4.6	40	1.5	Open SDS with different resonators [45]
TPU 1987	8 10 12	1.1 1.0 1.0	π π π	0.5 0.5 0.4	6.0 6.0 8.5	1500 1200 1000	0.1 0.2 0.35	3.3 4.3 2.6	1200 1000 70	4 6 10	Inverted relativistic magnetron. Frequency band 260–600 MHz Axial power output at E_{01} wave [37,38]

Table. Continuation

Organization, year of research	N	d, cm	Mode	U, MV	I, kA	τ, ns	P_{mw}, GW	f, GHz	τ_{mw}, ns	Efficiency, %	Design features [reference]
Varian 1988	16	0.94	$3\pi/8$	0.42	4	60	0.08	3.3	40	4.5	SDS with blade-type resonators [47]
TPU IAP	6 12	1.35 1.0	π 2π	0.64 0.25	6.2 4	60 1000	1.0 0.1	3.0 3.0	30 400	up to 20 up to 10	External injection of an electron beam. The different resonator magnetron [49.50]
IRT Corporation, Cornell Univer 1988	6		2π	0.34-0.4	5.7	250	0.02-0.2	4.4	90	2-10	Superconducting magnet. 20-step pulse volatge generator (PVG), F=1Hz [56]
MIT 1988	6		2π	0.3 0.7	30 0.7	>500 3000	0.005	4.6	200		Superconducting magnet. Power source PVG. Velvet cathode. F=4Hz [59]
PIC 1989			π	1	360	120	0.8	2.86-2.41	40		The output from the RM output came to the region of the virtual cathode and increased the radiated power of the vircator from 100 to 500 MW [60]

Table. Continuation

Organization, year of research	N	d, cm	Mode	U, MV	I, kA	τ, ns	P_{mw}, GW	f, GHz	τ_{mw}, ns	Efficiency, %	Design features [reference]
TPU 1989	6	1.0	π	0.4	4	70	0.25	2.9	50	15	The RM and the LIA are located on a movable platform. F=20Hz
TPU 1989	6	2.0	π	0.4	10	150	0.212 0.045 0.168	1.2	130	6 2.6 5.2	Cylindrical cathode. Closed SDS. Washer cathode. Cathode with symmetric power [95]
MIT 1990	6	1.35	π 2π	0.7	0.7	2000	0.005		250	1–5	Velvet cathode. Parts of RM were subjected to thermal annealing [55]
AAI Corporation 1990	6		2π π	1	17	70	0.7 0.325	4.632 4.27	50	15 7	F=2Hz.Instability of microwave pulses-10%.The lifetime of the anode block is 5000 pulses, of the cathode is 1500 pulses.[57]
General Atomics 1990	6		2π	0.35–0.4			0.300	4.4	100	5–10	Superconducting magnet. PVG.F=1Hz.The instability of the amplitude of microwave pulses is 30-50% [58]

Table. Continuation

Organization, year of research	N	d, cm	Mode	U, MV	I, kA	τ, ns	P_{mw}, GW	f, GHz	τ_{mw}, ns	Efficiency, %	Design features [reference]
PIC 1990	10		π	0.5	11.5	100	0.4-0.6	1.23-2.82	75		RM with 2 outputs.Frequency tuning within 14, 6% for 30-cm and 33.4% for 10-cm RM.F=100Hz (packet 100 pulses) [63].
PIC 1990			π	1 (0.75)	360 (10)	120	0.3	2.8	40		Microwave power from two resonators of the RM is applied to two vircators [64]
North Carolina University, Varian 1990	6		π 2π	0.3 (0.25)	3.1	1000	0.09	2.5 4.17	250	12	Microwave is limited by the appearance of anode plasma [65]
PIC 1992–1993	6	0.95	π	0.75	10	60	1.0 0.7 0.6	1.1	50 50 50		P_{av} = 4.4 kW (F=100 Hz, 50 pulses in packet). P_{av} = 6 kW (F = 200Hz). P_{av} =6,3 kW (F=250Hz) [61,62,66]

Table. Continuation

Organization, year of research	N	d, cm	Mode	U, MV	I, kA	τ, ns	P_{mw}, GW	f, GHz	τ_{mw}, ns	Efficiency, %	Design features [reference]
Varian 1994							0.048 0.046	2,832 2,84	700 900		Thermocathode.Molybdenum coating of anode block. F=10Hz [54]
Rafael 1996	6		2/3 π	0.22 (0.18)	0.7		0.05 (0.1)	2.5	150 (70)		The cathode of RM is grounded.The HCEA delivers a pulse of positive polarity (10 Hz) to it.Metal-dielectric cathode [67]
TPU 1992-1997	6	1.0	π	0.3	4	80	0.2	2.9	50	15	Coherent addition of power from two RMs (powered by 2 synchronized LIAs) [98]
University of St. Andrews, DERA 1998	6	0.95	π	500		150	810 620 450	1.1-1.3	30		Cylindrical velvet-cathode. The disk cathode.PIN cathode [84]
TPU 1999	6	0.9	π	0.33	3	180	0.2 1.1	2.84	120 5	20	Power source - LIA on magnetic elements (F=320Hz) [69].With a microwave compressor (F=10Hz) [118]
TPU 2000	6	0.9	π	0.33	3	180	0.2	2.84	120	20	Mechanical tuning of the RM radiation frequency within 8% [70]

Table. Continuation

TPU 2001	6 8	0.9	π π	0.4	4.5	180	0.35 0.35	2.84 3.03	120 120	20 20	Power supply – LIA on magnetic elements ($F = 200$ Hz)
TPU 2008	6 8	0.9	π	0.4	4.5	180	0.4 0.45	2.84 3.03	120 120	30 30	External resonator communication channel [121-123]. Distributed power output [124].Antenna array [125–128]
Technion 2009	6	1.1	π	0.3	3	150	0.35	3.05	100	20	Ferroelectric plasma cathode [86] cathode

References

1. Hull A.W., Phys. Rev. 1921. V. 18. P. 31–57.
2. Alekseev N.F., Malyarov D.E., Zh. Teor. Fiz. 1940. V. 10. No. 15. P. 1297–1300.
3. Kovalenko V.F., Introduction to super high frequency electronics Moscow, Sov. radio, 1955.
4. Magnetrons of centimeter range: Trans. from Eng. under ed. S. A. Zusmanovsky. M.: Sov. radio, 1950. V. 1.
5. Buneman O., The RF theory of crossed field devices. New York, Academic Press, 1961. V. 1. 209 p.
6. Bychkov S.I., Questions of theory and practical applications of instruments of the magnetron type. Moscow, Sov. radio, 1967.
7. Samsonov D.E., Basics calculating and designing magnetrons. Moscow, Sov. radio, 1974.
8. Bekefi G., Orzechovski T., Phys. Rev. Lett. 1976. V. 37. No. 6. P. 379–382.
9. Bugaev S.P., et al., The phenomenon of explosive electron emission. Discovery. Diploma No. 176, Otkr. Izobr. Promst. Obraztsy, Tov. Znaki, 1976. No. 41. P. 3.
10. Kovalev N.F., et al., Pis'ma Zh. Teor. Fiz., 1977. Vol. 3. No. 20. P. 1048–1051.
11. Didenko A.N., et al., *ibid*, 1978. Vol. 4. No. 3. P. 10–13.
12. Didenko A.N., et al., *ibid*, 1978. Vol. 4. No. 14. P. 823–826.
13. Nechaev V.E., et al., in: Relativistic high-frequency electronics. Gorky: IPF AN USSR, 1979. P. 114–130.
14. Palevsky A., Bekefi G., Phys Fluids. 1979. V. 22. No. 5. 986–996.
15. Didenko A.N., et al., in: Proc. 3 Int. Top. Conf. on High Power Electron and Ion Beam. Novosibirsk, 1979. V. 2. P. 683–691.
16. Craig G., et al., Abstr. IEEE Int. Conf. on Plasma Science. Montreal, 1979. P. 44.
17. Ballard W.P., A relativistic magnetron with thermionic cathode, Supr. Report No. 840. 1981.
18. Ilic D.B., et l., Proc. IEEE Int. Conf. on Plasma Science. Monterey, 1978. P. 286.
19. Ballard W.P., et al., Abstr. IEEE Int. Conf. on Plasma Science. Montreal, 1979. P. 45.
20. Ballard W.P., et al., J. Appl. Phys. 1982. V. 53. No. 11. P. 7580–7591.
21. Zherzlitsyn A.G., et al., Zh. Teor. Fiz., 1979. Vol. 49. No. 11. P. 2480–2481.
22. Zherzhitsyn A.G., et al., Proc. doc. 9 th All-Union. Conf. on Microwave Electronics Kyiv, 1979. P. 90–91.
23. Kovalev N.F., Krastelev EG, Kuznetsov MI and others. Powerful relativistic magnetron from long waves 3 cm. Pis'ma Zh. Teor. Fiz. 1980. T. 6. No. 8. P. 459–462.
24. Vintizenko I.I., et al., Pis'm Zh. Teor. Fiz. 1983. V. 9. No. 8. P. 482–485.
25. Benford J., et al., IEEE Trans. 1985. V. 13. No. 6. P. 538–544.
26. Benford J., et al., In: Proc. 6 Int. Conf. on High-Power Particle Beams. Kobe, 1986. P. 577.
27. Benford J., Phys. Rev. Lett. 1986. V. 62. P. 8. number 969–971.
28. Sze H., et al., IEEE Trans. on Plasma Science. 1987. V. PS-15. No. 3. P. 327–334.
29. Phelps D.A., in: Abstr. 7 Int. Conf on High-Power Particle Beams. Karlsruhe, 1988. P. 321.
30. Benford J., et al., Abstr. 7[t] Int. Conf. on High-Power Particle Beams. Karlsruhe, 1988. P. 327.

31. Didenko A.N., et al., Abstr. 7 Int. Conf. on High-Power Particle Beams. Karlsruhe, 1988. P. 331.
32. Black W.M., et al., In: Proc. Int. Electron Devices Meet. Washington, 1979. P. 175–178.
33. Parker R.K., et al., Abstr. IEEE Int. Conf. on Plasma Science. Montreal, 1979. P. 44.
34. Black W.M., et al., Proc. Int. Electron Devices Meet. Washington, 1980. P. 100.
35. Close R.A., et al., J. Appl. Phys. 1983. V. 54. No. 7. P. 4147–4151.
36. Early LM,et al., Prib. Nauch. Issled., 1986. No. 9. P. 86–96.
37. Vintizenko I.I.,et al., Pis'm Zh. Teor. Fiz. 1987. V. 13. No. 10. P. 620–623.
38. Vintizenko I.I., Sulakshin A.S. Output of microwave-radiation from the converted magnetron on wave E_{01}, Dep. at VINITI. 1987. No. 5983–1987.
39. Vintizenko I.I., et al., in: Proc. doc. of the 7th All-Union. Conf. High-current electronics. Ch. 1. Tomsk, 1988. P. 197–199.
40. Chernogalova L.F., et al., In: Abstr. 7 Int. Conf. on High-Power Particle Beams. Karlsruhe, 1988. p. 329.
41. Didenko A.N., et al., Pis'm Zh. Teor. Fiz. 1981. V. 7. No. 17. P. 1025–1028.
42. Benford J., Microwave Journal. 1987. V. 30. No. 12. P. 97–106.
43. Didenko A.N., DAN CSSR. 1988. V. 300. No. 6. P. 1363–1366.
44. Bekefi G., Proc. Int. Electron Devices Meet. San Francisco, 1984. P. 822.
45. Lemke R.W., Clark M.C., J. Appl. Phys. 1987. V. 62. No. 8. P. 3436–3440.
46. Estrin V., Phelps D.A., Abstr. 7 Int. Conf. on High-Power Particle Beams. Karlsruhe, 1988. P. 322.
47. Treado T.A., et al., IEEE Trans. on Plasma Science. 1988. V. PS-16. No. 2. P. 237–248.
48. Smith R.R., et al., IEEE Trans. on Plasma Science. 1988. V. PS-16. No. 2. P. 234–237.
49. Vintizenko I.I., et al., in: Relativistic high-frequency electronics. IAP AN SSSR. Gorky, 1988. No. 5. P. 125–140.
50. Chernogalova L.F., et al., in: Proc. 6 Int. Conf. on High-Power Particle Beams. Kobe, 1986. P. 573–576.
51. Vasiliev V.V., et al., Pis'ma Zh. Teor. Fiz. 1987. V. 13. No. 12. P. 762–766.
52. Levine J.S., et al., J. Appl. Phys. 1991. V. 70. No. 5. P. 2838–2848.
53. Benford J., in: Proc. Int. Conf. o n High-Power Particle Beams. Karlsruhe, 1988. P. 327.
54. Treado T.A., et al., IEEE Trans. on Plasma Science. 1994. V. 22. P. 616–626.
55. Chen S-S., et al., in: Proc. Intense Microwave and Particle Beams. Los-Angeles, 1990. V. 1226. P. 36–43.
56. Phelps D., et al., In: Proc. 7 Int. Conf. on HighPower Particle Beams. Karlsruhe, 1988. WP112. P. 1347–1352.
57. Spang S.T., et al. IEEE Trans. on Plasma Science. 1990. No. 3. V. 18. P. 586–593.
58. Phelps D.A., IEEE Trans. on Plasma Science. 1990. V. 18. No. 3. 577–579.
59. Chen S-C., et al., Relativistic Magnetron Research , A Reprint from the Proc. SPIE V. 873. Microwave and Particle Beam Sources and Propagation, 1988. P. 18 –22.
60. Price D., Fittingholf D., J. Appl. Phys. 1989. V. 65. No. 12. P. 5185–5189.
61. Aiello N., in: Proc. 9 Int. Conf. on High Power Particle Beams. Washington, 1992. P. 203–210.
62. Sincerny P., et al., NATO Series V. G34, Part A, 1993. 12 p.
63. Smith R.R., et al., in: Proc. Int. Workshop on High Power Microwave and Pulse Schortering. Edinburgh, 1997. P. 1–9.
64. Sze H., et al., J. Appl. Phys. 1990. V. 68. No. 7. P. 3073–3079.

65. Treado T., et al., IEEE Trans. on Plasma Science. 1990. V. 18. No. 3. 594–602..
66. Ashby S., et al., IEEE Trans. on plasma Science. 1992. V. 20. No. 3. 344–350.
67. Schnitzer I.,et al., SPIE. 1995. V. 2843. P. 101–109.
68. Levine J., et al., Frequency agile relativistic magnetron, Preprint PIC No. 23–95, 1995. 6 p.
69. Butakov L.D.,et al., Pis'ma Zh. Teor. Fiz. 2000. V. 25. No. 13. P. 66–71.
70. Vintizenko I.I., Pis'ma Zh. Teor. Fiz. 2000. V. 26. No. 17. P. 67–70.
71. Sulakshin A.S., Zh. Teor. Fiz. 1983. V. 53. No. 11. P. 2266–2268.
72. Glazer I.Z., et al., Pis'ma Zh. Teor. Fiz. 1980. V. 6. No. 1. P. 44–49.
73. Miller R., Introduction to physics of high-current beams of charged particles. Moscow, Mir, 1984..
74. Lovelace R.V., Ott E., Phys. Fluids. 1974. V. 17. P. 1263.
75. Shlifer E.D., Itogi nauki i tekhniki. 1985. V. 17. P. 169–209.
76. Palevsky A., ert al., J. Appl. Phys. 1981. V. 52. P. 8. number 4938–4941.
77. Chan H.-W., et al., Appl. Phys. Lett. 1990. V. 57. No. 12. P. 1271–1273.
78. Gunin A.V., et al., Pis'ma Zh. Teor. Fiz. 1999. V. 25. No. 22. P. 84–94.
79. Kapitsa P.L., Electronics large capacities, Moscow, Publishing house of AN USSR, 1962..
80. Certificate No. 13937 of the Russian Federation, IPC H 01 J 25/00. Magnetron, I.I. Vintizenko (RF). No. 99125713; Declared. 07.12.1999; Publ. BI. 2000. No. 16.
81. Fedosov A.I.,et al., Izv. VUZ, Fizika, 1977. No. 10. P. 134–135.
82. Nechaev V.E., et al., Pis'ma Zh. Teor. Fiz. 1977. Vol. 3. No. 15. P. 763–767.
83. Didenko A.N., et al., Pis'ma Zh. Teor. Fiz. 1983. V. 9. No. 24. P. 1510–1513.
84. Saveliev Y.M., et al., New cathodes for a relativistic magnetron, British crown copyright, DERA, 1988.
85. Vintizenko I.I., et al., Pis'ma Zh. Teor. Fiz., 2012. V. 38. No. 21. P. 56–62.
86. Hadas Y., et al., J. Appl. Phys. 2009. V. 105, 083307-1 to 083307-7.
87. Rosenman G., et al., J. Appl. Phys. 2000. V. 88, 6109.
88. Bosman H., et al., Improvement of Output Characteristics of Magnetrons Using the Transparent Cathode.
89. Lemke R.W., et al., IEEE Trans Plasma Sci. 2000. V. 28. No. 3. P. 887–897.
90. Okress E.Ed., Crossed-Field Microwave Devices, Academic Press, New York & London. 1961. Ch.4. V. 1.
91. Weatherall J.C., IEEE Trans. Plasma Sci. 1990. No. 18. P. 603–610.
92. Dubinov A.E., et al., Inzh.-fiz. zh., 1998. V. 71. No. 5. P. 899–901.
93. Zakharov V.V., et al., Radiotekhnika elektron., 1989. V. 34. No. 5. P. 1095–1100.
94. Himmelblow, D. Applied nonlinear programming. Moscow, Mir, 1975.
95. Zakharov V.V., Krutikov V.N., Kibernetika, 1984. No. 6. P. 94.
96. Aleksandrovich D.V., et al., Pis'ma Zh. Teor. Fiz. 1989. V. 15. No. 22. P. 35–39.
97. David E.E., Phasing with high-frequency signals, in: Electronic ultrahigh frequency devices with crossed fields. Ed. M. M. Fedorov. Moscow, IL, 1961. V. 2. P. 327–348.
98. Novikov S.S., et al., Proc. SPIE in Intense Microwave Pulses IV. San-Diego, 1995. V. 2557. p. 492–498.
99. Novikov S.S., et al., Proc. SPIE in Intense Microwave Pulses V. San-Diego, 1997. V. 3158. p. 271–277.
100. Kanaev G.G,, et al., Pis'ma Zh. Teor. Fiz. 1995. V. 21. No. 20. P. 51–54.
101. Fomenko G.P., et al., 1998. Izv. VUZ, Fizika, No. 4. Addition, P. 146–157.
102. Vladimirov S.N., et al., in: Nonlinear oscillations of multifrequency autooscillatory systems. Tomsk: Publishing house of the Tomsk University, 1993.
103. Novikov S.S., Maidanovskiy A.S., Radiotekhnika elektron., 1983. V. 28. No. 3. P.

513–517.
104. Novikov S.S., Maydanovskiy S.A., Radiotekhnika elektron., 2003. V. 48. No. 5. P. 595–600.
105. Novikov S.S., Proc. In SPIE Intense Microwave Pulses, V., San-Diego, 1997. V. The of P. 3158. 260–264.
106. Novikov S.S., Zarevich A.I., in: Proc. in 12th International symposium on high current electronics and high power microwave sources. Tomsk, 2000. V. 2. P. 466–469.
107. Zarevich A.I., Novikov S.S., Izv. VUZ, Fizika, 2006. No. 9. Appendix. P. 276–282.
108. Marder B.M., J. Appl. Phys. 1988. V. 65. No. 3. P. 1338–1349.
109. Striffler C.D., et al., IEEE Trans. on Nuclear Science. 1983. V. NS-30. No. 4. P. 3429-3431.
110. Bekefi G., et al., Radiation measurements from a rippled-field magnetron (crossed field FEL), Preprint MIT PFC JA-83-3. 1983. 18 p.
111. Vintizenko I.I., et al., Pis'ma Zh. Teor. Fiz. 1987. V. 13. No. 22. P. 1384–1388.
112. Lebedev I.V., Equipment and microwave devices. Moscow, Vysshaya shkola, 1972.
113. Vintizenko I.I., Proc. 2 IEEE Int. Vacuum Electronics Conference IVEC. Noordwijk, 2001. P. 49–53.
114. Vintizenko I.I., Relativistic magnetron. Certificate No. 13036 of the Russian Federation. Publ. BI. 2000. No. 16.
115. Didenko A.N., Yushkov Yu.G. Powerful microwave pulses of nanosecond duration. Moscow, Energoatomizdat, 1984.
116. Novikov S.A., et al., Izv. VUZ, Radiofizika. 1987. V. 30. No. 12. P. 1520–1521.
117. Sretensky V.N. Basics applications devices of superhigh frequencies. Moscow, Sov. radio, 1963.
118. Augustinovich V.A., et al., in: Proc. 4 Int. Work shop 'Strong microwaves in plasmas'. Nizhnyi Novgorod Nizny, IAP, 1999. V. 2. P. 915–929.
119. Didenko A.N., et al., DAN. 1999. V. 366. No. 5. P. 619–621.
120. Novikov S.A., et al., Pis'ma Zh. Teor. Fiz. 1990. V. 16. No. 20. P. 46–48.
121. Artemenko S.N., et al., Zh. Teor. Fiz. 1993. V. 63. No. 2. P. 105–112.
122. Vintizenko I.I., et al., Pis'ma Zh. Teor. Fiz. 2003. V. 29. No. 7. P. 64–70.
123. Vintizenko I.I., et al., Izv. VUZ, Fizika. 2006. No. 9. P. 114–118.
124. Vintizenko I.I., et al., Pis'ma Zh. Teor. Fiz. 2006. V. 32. No. 23. P. 40–47.
125. Vintizenko I.I.,et al., Pis'm Zh. Teor. Fiz. 2005. V. 31. No. 9. P. 63–68.
126. Vintizenko I.I., Novikov S.S., Izv. VUZ, Fizika. 2008. No. 9/2. P. 154–156.
127. Vintizenko I.I., Novikov S.S., Pis'ma Zh. Teor. Fiz. 2009. V. 35. No. 23. P. 88–95.
128. Vintizenko I.I., Novikov S.S., Zh. Teor. Fiz., 2010. No. 11. P. 95–104.

2

Physical Processes in Relativistic Magnetrons

Theoretical works devoted to the study of the properties of the magnetron as an auto-oscillatory system were started in the 1950s of the last century [1–3] under the following simplifying assumptions:

- the non-relativistic case is considered, since the electron velocity in the classical magnetrons is much smaller than the speed of light;
- the interaction of electrons is considered only with the main component of the high-frequency field (π-type of oscillations);
- the influence of the space charge of electrons is not taken into account on the basis of the following considerations. The electrostatic field in the interaction space decreases in proportion to $1/r$. As a result, the radial force of the Coulomb field directed from the cathode to the anode is partially compensated by the force due to the inhomogeneity of the static electric field, which should lead to a decrease in the role of the space charge;
- the cathode current is limited by the space charge, therefore, the cathode potential and the radial field component at the cathode are zero;
- the inhomogeneity of the field along the axis of the magnetron and the inhomogeneity of the axial distribution of the charge in the spoke are neglected. This makes it possible to reduce the problem to a two-dimensional one. In a real magnetron, this is true for regions not too far from the median plane, and at the ends of the magnetron an edge effect will appear. The introduction of the function $\varsigma(z)$, which describes the inconstancy of the radial and tangential components of the field along the anode block, allows one to take into account the edge effects. In the anode blocks of the

open structure, $\varsigma(z) = 1$, for closed $\varsigma(z) \sim \sin(\pi z/L)$, where L is the length of the anode block. The distribution of the microwave field along the block can also be determined experimentally under 'cold' measurements and substituted it in the calculations. However, such an evaluation of the influence of the field inhomogeneity does not take into account the appearance of the electron velocity component along the z axis, and also the edge effect in calculating the static electric field.

As the presentation proceeds, the validity of these simplifications, as well as some additional ones, will be argued.

In the theory of a magnetron, both a cylindrical and a planar model of a generator are considered. The electrostatic field in a cylindrical magnetron can not be constant, but decreases like

$$E_r = -\frac{U}{r \ln R/r_c}, \tag{2.0.1}$$

therefore, the drift velocity of electrons depends on the radius according to the law

$$r\dot{\varphi} \equiv V_\varphi(r) = \frac{\bar{V}_\varphi \cdot r_s}{r}, \tag{2.0.2}$$

where $\bar{V}_\varphi$ is the azimuthal velocity for $r = r_s$ (r_s is the synchronous radius).

The phase velocity of the travelling electromagnetic wave depends on the radius according to a different law:

$$V_p = \bar{V}_p \frac{r}{r_s}, \tag{2.0.3}$$

where $\bar{V}_p$ is the phase velocity at $r = r_s$.

Consequently, the exact synchronism of electrons and waves is possible only at a synchronous radius. As will be shown below, the impossibility of exact synchronism in the entire interaction space leads to an additional azimuthal drift of the electrons, curvature of the spokes of the space charge. In a planar model, the distribution of the potential is linear and the contribution from all horizontal layers of the spoke to the useful power is the same (energy exchange of electrons and waves occurs with the same efficiency over the entire height of the spoke).

The essential difference between the magnetrons and the O-type Cherenkov instruments is that the translational motion of electrons is a drift $\beta_e = [\mathbf{E}_0 \times \mathbf{H}_0]/H^2_0$ in crossed electrostatic $\mathbf{E}_0$ and magnetostatic $\mathbf{H}_0$ fields. The drift of electrons to the anode, during which they receive energy from the electrostatic field and transmit it to the electromagnetic field, occurs under the action of a synchronous wave, and the kinetic energy of the electrons does not change significantly.

If the magnetic field exceeds the value (1.1.6) for a given voltage and interelectrode gap, then the electrons do not hit the anode and the electron cloud rotates at azimuthal speed:

$$r\dot{\varphi} \equiv V_\varphi = \frac{cE_r}{H}. \tag{2.0.4}$$

The form of the trajectories of individual electrons depends on various assumptions about the initial conditions and is still the subject of discussions. There are ideas about the laminar [4] and cycloidal [5] character of the motion. In the model of the Brillouin (parapotential) electron flux, it is assumed that the electrons rotate rapidly with a high cyclotron frequency in a small-radius orbit and drift slowly around the cathode surface along the equipotentials of the electric field with the drift velocity determined by (2.0.4). The model assumes that the initial electron velocities are zero, there are no temporal variations of any fields, and for magnetic fields $H > H_{\text{cr}}$ there will be no current to the anode. In the dual-flux model, it is assumed that electrons move along cycloidal trajectories beginning and ending at the cathode, that is, there are two counterflows – one toward the anode, the other toward the cathode.

Which model better corresponds to reality? It is known [6] that at magnetic fields exceeding the critical value, the flow of an electron current across the cathode–anode gap is experimentally observed. Attempts were made at different times to explain this process by various reasons: the spread of the thermal velocities of the cathode-emitted electrons, the influence of edge effects, the inclination of the magnetic field lines, etc. However, only recognizing the fact that there is no stationary state of the electron flux in crossed electric and magnetic fields, one can explain the violation of the magnetic insulation of the diode. In this case, the motion of electrons to the anode must be accompanied by the excitation of electromagnetic oscillations, which are caused by the development of instabilities in the flow. The speed of electrons depends on the radius on which they are located, hence, we can conclude that there is a free energy source that excites diocotron instability. Since the interaction space of the magnetron is a closed structure in the azimuthal direction, the electron beam has its own resonance frequencies, and at these frequencies it is possible to amplify the microwave oscillations.

The work [7] describes measurements of the parameters of electromagnetic oscillations in a magnetron diode over a wide range of magnetic fields. The experiments were carried out using diodes with a small curvature of the electrodes $R/r_c \sim 1.3$, i.e., the geometry approached the planar case. For magnetic fields smaller than the critical value, the current–voltage characteristic is well defined by the Langmuir current [8], bounded by the space charge $I \sim U^{3/2}/d^2$ with two additional effects. The first effect is associated with the curvature of trajectories under the influence of the magnetic field and the increase in the time of motion of the electrons. Slow electrons shield the plasma surface with their own space charge with respect to the anode potential. Therefore, as the magnetic field increases from 0 to H_{cr}, a slow decay of the current is observed. The second effect is caused by the flow along the surface of the cathode and cathode holder of the axial current I_z, which forms the azimuthal magnetic field:

$$H_\varphi = \frac{\mu_0 I_z}{2\pi d_{cat}}, \tag{2.0.5}$$

where d_{cat} is the distance from the cathode holder.

Approximate calculations [6] show that the current flowing through the gap at $H < H_{cr}$ is calculated as

$$I \approx 2\pi\left(1+\frac{eU}{mc^2}\frac{H_{cr}}{\mu_0}\right)\left[1-\left(\frac{H}{H_{cr}}\right)^2\right]^{1/2}, \tag{2.0.6}$$

where H_{cr} is determined from the relation (1.1.6).

At magnetic fields $H > H_{cr}$ the experimentally measured current to the anode is up to 5% of the maximum current and is accompanied by electromagnetic radiation in a wide frequency range of 7–40 GHz [6]. The radiation is recorded only in the range of magnetic fields exceeding the critical value. This indicates that it is associated with the instability of the Brillouin flow, i.e., with the destruction of the azimuthal homogeneity of the flow and the development of the diocotron instability. This instability is confirmed by the resonance character of the dependences of the power of the microwave radiation on the magnetic field. One, two or even several resonances were observed, which are inherent in a magnetron diode of a certain geometry. At ‘cold’ measurements (in the absence of an electron beam), excitation of diodes by a microwave signal was used to find resonances with identical frequencies and Q values. Thus, the electromagnetic radiation of an unstable rotating electron cloud is amplified at these frequencies. The surface of the Brillouin flow acquires a corrugated shape along the azimuth and excites modes of the type H_{0m} of a circular waveguide, which were recorded experimentally. The oscillations observed in a smooth diode create starting conditions for the development of oscillations in a magnetron generator.

Numerical calculations of electron trajectories in the cloud surrounding the cathode of the relativistic magnetron [8–10], in the pre-generation period, also show the development of azimuthal inhomogeneities of the outer boundary and the presence of electrons moving according to both the Brillouin and two-flux models.

Since the phase velocities of electromagnetic waves in a smooth magnetron diode always exceed the speed of light, the interaction of the wave with the space charge turns out to be weak: the radiation spectrum is wide and the efficiency is low. If the wave is slowed down by a periodic structure – a slowing-down (resonator) system, then the synchronism between the phase velocity of the wave and the speed of the azimuthal drift of the electrons is possible. This can lead to a strong interaction of the wave and space charge and cause an efficient generation of microwave radiation. As shown by numerical calculations [10], at the initial moment of the development of microwave oscillations, electrons perform ‘positive’ work over the field, transferring their energy. As long as the wave increases energy, its field acts on the space charge, forming spatial bunches that increase the wave growth. As a result, space charge spokes are formed, and, ultimately, the amplitude of the microwave field and the anode current of electrons reach their working values.

In the interaction space, there are two types of electrons: correct-phased, giving their energy to the microwave field and reaching the anode, and incorrect-phased, taking energy from the field and falling on the cathode. The drift of the correct-phased electrons to the anode, during which they receive energy from the electrostatic field and transmit it to the electromagnetic field, occurs under the action of a synchronous wave, and the kinetic energy of the electrons does not change significantly. According to the analytical calculations

[11–13] and computer simulation [9, 10], the motion of correct-phased electrons to the anode is a superposition of fast rotation with a cyclotron frequency with small radii, azimuthal rotation in synchronism with the microwave wave and a slow radial motion in the space-charge spoke. The results of observations of the motion of electrons along the first loops of their trajectories in the near-cathode region of a multiresonator nonrelativistic magnetron are presented in [12]. For this purpose, a method of longitudinal probing of the interaction space by a thin electron beam was used. The idea of the method lies in the fact that the injected electrons and electrons of the magnetron itself move along identical trajectories. Getting on the fluorescent screen at the end of the device, the injected electrons give information about the nature of their motion in the interaction space. At voltages lower than the threshold level trajectory, have a loop appearance. This once again confirms the validity of the two-flux state of the electron beam. As the voltage increases, the dimensions of the glowing spot on the screen increase, which indicates an increase in fluctuations in the interaction space and the formation of space charge spokes. The electrons slowly drift to the anode in the bottom of spokes in the operating mode, i.e.. they perform a relatively large number of cyclotron revolutions. As the microwave power was increased due to an increase in the emission current, not only the expansion of the region occupied by the electrons was observed, but also its displacement along the angle. This illustrates the process of bunching of the electrons at the base of the spoke and shows that the electrons move not strictly in sync with the field wave, and their angular velocity increases with increasing power.

2.1. The theoretical model of the relativistic magnetron of planar geometry

A mathematical solution of the problem of the interaction of electrons with fields is based on a joint solution of the field equations and the equations of motion. If the anode of the coaxial diode is constructed as an azimuthal periodic structure, in the condition of Cherenkov phase matching (proximity of the electron drift velocity to the azimuthal wave phase velocity $\beta_e = E_r/H \approx \beta_p = V_p/c$) the energy of the electrons can be converted with high efficiency to electromagnetic radiation. The classical magnetron is an example of such a generator in a non-relativistic region. The best experimental samples of magnetrons achieve the efficiency of up to 82%.

The use of high-current electron accelerators for powering magnetrons opens new prospects for these devices. Moreover, in contrast to the O-type generators, the interaction of a high-current relativistic electron beam with high-frequency fields occurs directly in the diode region, i.e., in the region of creation of an electron beam. In this case, the limitations associated with providing high electron energy at a low velocity spread and mastering of high currents are removed. In the first successful experiment [7] with the use of a relativistic magnetron of a high-current electron accelerator at an anode voltage of 360 kV as a power source, a microwave power of 1 GW in a pulse of 20 ns duration with an efficiency of 35% was achieved at a wavelength of 10 cm. In this case, the traditional magnetron design was used – a 6-cavity anode block with an axial length of $L/\lambda \approx 0.7$ and a power output through a slot in one of the resonators. Obtained at the time record output parameters and ease of construction of the device, known as 'relativistic magnetron', immediately attracted the attention and gave rise to many questions regarding the energy capacity of such devices. It was necessary to find out the wavelength range in which the relativistic magnetron has the best characteristics, where are there limitations on the efficiency at the transition to the relativistic energy range, how long is the pulse length of the microwave radiation, what is the rise time of oscillations in such a generator, which is the traditional design optimal for the region of relativistic energies. These issues were discussed in a theoretical article [13] and the thesis of M.I. Fuchs [14], on the basis of which the theory of the relativistic magnetron of planar geometry is presented below.

Let's consider the simplest – a planar model of the interaction space of a multi-resonator magnetron (Fig. 2.1). Such a model corresponds to a cylindrical design with N resonators (in Fig. 2.2 they have the form of rectangular grooves), the diameters of the anode and cathode in which should be determined as

$$2R_{a,k} = \frac{ND}{\pi} \pm d, \qquad (2.1.1)$$

where D is the period of the slowing-down system; d is the interelectrode gap.

Below is an approximate method for calculating the anode current and power for a planar model of a multiresonator relativistic magnetron. The calculation is based on a preliminary analysis of the problem of the motion of electrons in given fields of the system.

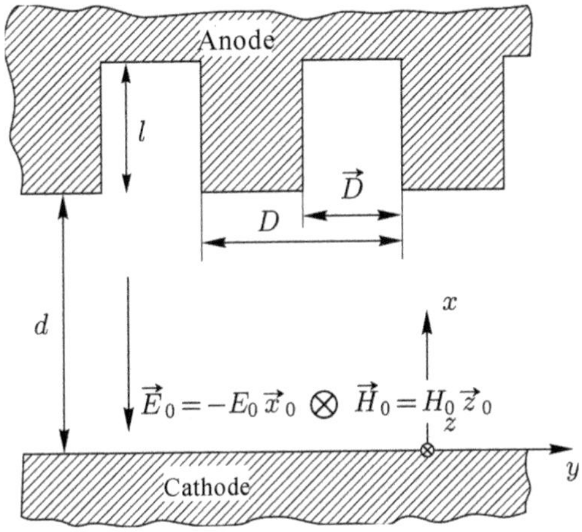

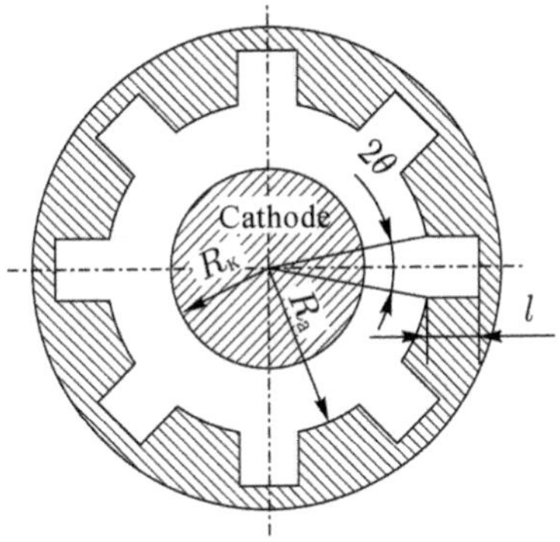

Fig. 2.1. Geometry of the magnetron interaction space.
Fig. 2.2. (right) Magnetron cross section.

Exact solutions of the problem of the motion of electrons in the interaction space and the grouping of the space charge of electrons in spokes can be obtained only by numerical methods. The use of these solutions for further calculations for each particular model is very difficult. The value of approximate analytical solutions is related precisely to the possibility of subsequent analysis. Therefore, it seems expedient to consider an approximate theory that allows one to estimate limiting possibilities on simple physical considerations and to specify the optimal parameters of a magnetron without limiting the relativistic factor.

2.1.1. Motion of electrons in a relativistic magnetron of planar geometry

Disregarding the action of space charge, we consider the motion of electrons in a planar model. The following equations describe the motion of electrons:

$$\frac{m}{e}\frac{d^2y}{dt^2} = \frac{\partial\Phi}{\partial y} - B\frac{dx}{dt},$$
$$\frac{m}{e}\frac{d^2x}{dt^2} = \frac{\partial\Phi}{\partial x} + B\frac{dy}{dt}, \tag{2.1.2}$$

where B is the magnetic induction; Φ is the scalar electric potential. Further in Chapter 2, for compactness of recording, a system of units typical of relativistic electronics will be used, in which $e = m_0 = c = 1$, the unit of potential is $m_0c^2/e = 511$ kV, current $m_0c^3/e = 17.04$ kA, magnetic field strength $m_0c/e = 1.7$ kOe/cm, power $m_0^2c^5/e^2 = 8.7$ GW.

If a cyclicity condition of the type $f(x)\sin(ny - \omega t)$ is imposed, then the synchronous (travelling) harmonic of the high-frequency field, which can be limited when analyzing the motion of electrons, is expressed as

$$E_y = E_{\mathrm{HF}}\frac{\mathrm{sh}(px)}{\mathrm{sh}(pd)}\cdot\cos(h_\vartheta y - \omega t),$$

$$E_x = \frac{h_\vartheta}{p}\cdot E_{\mathrm{HF}}\cdot\frac{\mathrm{ch}(px)}{\mathrm{sh}(pd)}\cdot\sin(h_\vartheta y - \omega t), \tag{2.1.3}$$

$$H_z = \frac{k}{p}E_{\mathrm{HF}}\frac{\mathrm{ch}(px)}{\mathrm{sh}(px)}\sin(h_0 y - \omega t).$$

Here $h_\vartheta = k/\beta_p$ is the 'azimuth' wave number satisfying the condition of cyclicity

$$h_\vartheta = -\frac{2\pi}{ND}(n + mN) \tag{2.1.4}$$

(for oscillations of π-type, providing the best characteristics of magnetrons in the centimeter range $h = \pi/D$); $p = \sqrt{h_\vartheta^2 - k^2}$ is the transverse wave number;

$$E_{\mathrm{HF}} = E_1\frac{\overline{D}}{D}\sin\left(\frac{h_\vartheta\overline{D}}{2}\right)/\frac{h_\vartheta\overline{D}}{2}$$

is the effective amplitude of the synchronous harmonic on the surface of the anode block; E_1 is the amplitude of the high-frequency field at the opening of lamellae; $\overline{D}$ is the distance between the lamellae at $x = d$. If you enter $\lambda_p = (1 - \beta_p^2)^{-1/2}$, then the wave numbers are related by

$$k = \beta_p h = \beta_p\gamma_p p = \sqrt{\gamma_p^2 - 1}\,p. \tag{2.1.5}$$

Let us consider the motion of electrons in crossed electrostatic $-E_0\overrightarrow{x_0}$ and magnetostatic $H_0\overrightarrow{z_0}$ fields and in a high-frequency field without taking into account the intrinsic fields of the electron beam. In the reference frame K' associated with the phase velocity of the wave

β_p, all the fields become static and are expressed, according to the Lorentz transformations, in the form

$$\overrightarrow{E'} = -\nabla\Phi',$$
$$\overrightarrow{H'} = \gamma_p H_0(1-\beta_p\beta_e)\overrightarrow{z_0} = \frac{E_0}{\gamma_p\beta_p}(1-\alpha_0\gamma_p^2)\overrightarrow{z_0}. \tag{2.1.6}$$

Here the potential Φ' is equal to

$$\Phi' = \Phi_0' + \tilde{\Phi}' = \alpha_0\gamma_p E_0 x' - \frac{\tilde{E}}{p}\mathrm{sh}(px')\sin(py'); \tag{2.1.7}$$

$$\alpha_0 = 1 - \frac{\beta_p}{\beta_e} = \frac{U - U_{\text{th}}}{U_{\text{th}}} \tag{2.1.8}$$

is the parameter of the dissynchronism, expressed through the anode voltage U and the threshold voltage:

$$U_{\text{th}} = \beta_p H_0 d. \tag{2.1.9}$$

Value U_{th} corresponds to the exact synchronism:

$$\beta_p \approx \beta_e; \quad y' = \gamma_p(y - \beta_p t); \quad x' = x; \quad p = h_0/\gamma_p; \quad \tilde{E} = E/\mathrm{sh}(pd).$$

The equations of motion in the system K' are thus reduced to the form

$$\frac{dp_y'}{dt} = -\tilde{E}\cdot\mathrm{sh}(px)\cos(py') + \beta_x'\frac{E_0}{\beta_p\gamma_p}(1-\alpha_0\gamma^2),$$
$$\frac{dp_x'}{dt} = \alpha_0\gamma_p E_0 - \tilde{E}\cdot\mathrm{ch}(px)\sin(py') - \beta_y'\frac{E_0}{\beta_p\gamma_p}(1-\alpha_0\gamma_p^2), \tag{2.1.10}$$

where p_x, p_y is the electron impulse. When $\alpha_0 = 0$ ($\beta_e = \beta_p$, $\gamma_e = \gamma_p$) and the absence of high-frequency fields $\tilde{E} = 0$, the motion of electrons occurs in a magnetostatic field

$$H_0' = H_0/\gamma_e, \tag{2.1.11}$$

and in this case, as can be seen from the equation (2.1.10), the electron, starting from the cathode at a rate of $\dot{y} = -\beta_e$, rotates without changing the speed of angular frequency $\omega_H' = \omega_H/\gamma_e$, where $\omega_H = H_0/\gamma_e$ is the relativistic cyclotron frequency. The radius of rotation is

$$R' = \frac{\beta_e}{\omega'_H} = \frac{E_0\gamma_e^2}{H_0^2} = \frac{E_0}{H_0^2 - E_0^2}. \tag{2.1.12}$$

It can be shown that even in the presence of a high-frequency field with a small dissynchronism α_0, the Larmor radius practically does not change if the following condition is fulfilled

$$|E'| << |H'|. \tag{2.1.13}$$

In this case, the motion of the electron (Fig. 2.3) is a superposition of cyclotron rotation along a circle with a radius R' and a relatively slow drift.

We write the equations of motion in a complex form:

$$\dot{p}' = i\omega'_H p' - E', \quad \dot{\xi} = p'/\gamma, \tag{2.1.14}$$

where the impulse $p'=p'_x + ip'_y$; coordinate $\xi' = x'+iy'$; field $E' = E'_x + iE'_y$; $\gamma = (1+|p|^2)^{1/2}$; a dot denotes differentiation with respect to time.

We substitute the coordinate ξ' and the momentum p' in the form

$$\xi' = Z' + R'e^{i\psi}, \quad p' = p'e^{i\psi}. \tag{2.1.15}$$

Since four variables are introduced instead of two, two additional conditions must be imposed: first, let $\dot{\psi} = \omega'_H$ is the instantaneous rotation speed and, thus, R' is the radius and Z' is the centre of rotation, and secondly, let

$$\dot{Z}' + \dot{R}'e^{i\psi} = 0. \tag{2.1.16}$$

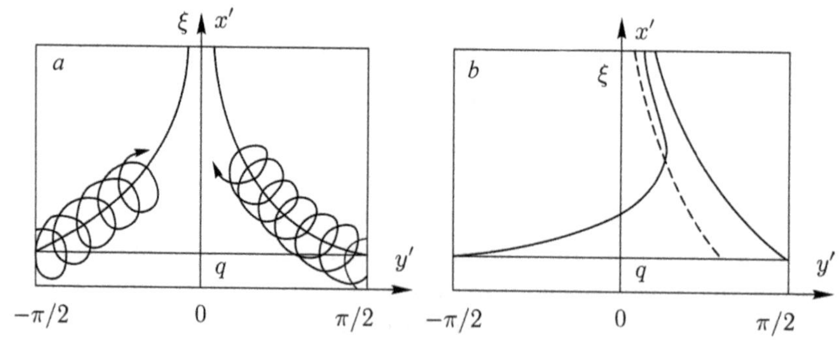

Fig. 2.3. Trajectories of the centres of electron orbits: *a*) in the absence of velocity dissynchronism, $\alpha_0 = 0$; *b*) for $\alpha_0 < \bar{\alpha}_0$ the dotted line is the synchronism line.

For $E' = 0$ all the quantities in (2.1.15) are constant, condition (2.1.16) is usual in the method of variation of constants. The substitution of (2.1.15) into (2.1.14) leads, with allowance for (2.1.16), to

$$\dot{p}' = -E'e^{-i\psi}, \quad R' = -i\frac{p'}{H'}. \tag{2.1.17}$$

This implies:

$$\dot{R}' = i\frac{E'}{H'}e^{-i\psi}, \quad \dot{Z}' = -i\frac{E'}{H'}. \tag{2.1.18}$$

In the absence of space charge fields ($\Delta\Phi' = 0$), the field $E' = F(\xi'^*)$ is an an analytic function. Under the condition (2.1.13), the radius and the centre of rotation should change slowly, so they can be replaced by the mean values ($R' \approx \overline{R}'$, $\dot{\overline{Z}}' \approx \dot{\overline{Z}}'$, the bar denotes averaging over the period of rotation), so that from (2.1.18) we obtain

$$\begin{aligned} \dot{\overline{R}}' \approx \overline{\dot{R}'} &= \frac{i}{H'}\int_0^{2\pi} F(\overline{Z}'^* + \overline{R}'^* e^{-i\psi})\, e^{-i\psi} d\psi = 0, \\ \dot{\overline{Z}}' \approx \overline{\dot{Z}'} &= -\frac{i}{H'}\int_0^{2\pi} F(\overline{Z}'^* + \overline{R}'^* e^{-i\psi})\, d\psi = -i\frac{F(\overline{Z}'^*)}{H'}. \end{aligned} \tag{2.1.19}$$

Thus, in the first approximation, the radius of rotation, and therefore the impulse, energy and frequency remain unchanged. As for the motion of the rotation centres, it follows from (2.1.19) for their 'drift' velocities

$$\beta_e' = [\mathbf{E}' \times \mathbf{H}']/H'^2, \tag{2.1.20}$$

where E' is the field strength at the centre of the cyclotron orbit. Formula (2.1.20) remains valid for any ratio of the radius of the orbit to the size of the inhomogeneity of the weak electrostatic field E'.

From (2.1.19), (2.1.20) for the motion of the centres we have

$$\begin{aligned} \frac{dX'}{d\tau} &= -\frac{d\varphi'}{dY'} = \tilde{\varepsilon}\,\mathrm{sh}\,X'\cos Y', \\ \frac{dY'}{d\tau} &= \frac{d\varphi'}{dX'} = \varepsilon_0 - \tilde{\varepsilon}\,\mathrm{ch}\,X'\sin Y', \end{aligned} \tag{2.1.21}$$

where

$$X' = px', \quad Y' = py', \quad \varepsilon_0 = \alpha_0 p \gamma_p^2 \beta_p, \\ \tilde{\varepsilon} = p\beta_p \gamma_p \tilde{E} / E_0, \quad \tau = \omega'_H t, \quad \varphi' = \varepsilon_0 X' - \tilde{\varepsilon}\,\mathrm{sh}\,X' \sin Y' \tag{2.1.22}$$

is the reduced potential.

Trajectories of centres of orbits, the equation for which, as follows from (2.1.21), has the form

$$\frac{dY'}{dX'} = \frac{\varepsilon_0 - \tilde{\varepsilon}\,\mathrm{ch}\,X' \sin Y'}{\tilde{\varepsilon}\,\mathrm{sh}\,X' \cos Y'}, \tag{2.1.23}$$

coincide with the equipotentials φ' = const – (2.1.22). Here, the constant φ' is determined by the coordinates of departure of the 'centre' from the 'power' plane $q = pR'$. Depending on the departure coordinate Y', the electrons can fall into the region of correct phases:

$$2\pi\left(s - \frac{1}{4}\right) \leqslant Y' \leqslant 2\pi\left(s + \frac{1}{4}\right) \tag{2.1.24}$$

(s is any integer) in which the motion is from the cathode to the anode and the potential energy of the electrons is given to the high-frequency field or to the region of incorrect phases where the electrons bombard the surface of the cathode with the energy they take away from the high-frequency field. In view of the periodicity of the system, it suffices to consider the motion in the phase region $-\pi \le Y' \le \pi$. The lines $Y' = \pm\pi/2$ are isoclines on which $dX'/d\tau = 0$. They divide the fluxes into regions of correct and incorrect phases. Other isoclines $dY'/d\tau = 0$, called synchronism lines, divide the interelectrode space into regions in which electrons overtake the wave of the field and lag behind it. The equation of the synchronism lines has the form

$$\varepsilon_0 - \tilde{\varepsilon}\,\mathrm{ch}\,X' \sin Y' = 0. \tag{2.1.25}$$

In Fig. 2.3 to the left of the synchronism line (indicated by dashed lines) $dY'/d\tau > 0$, to the right $-dY'/d\tau < 0$.

From the equation (2.1.23) it follows that when $\varepsilon_0 > \tilde{\varepsilon}\,\mathrm{ch}(q)$, i.e. at dissynchronism $\alpha_0 > \bar{\alpha}_0 = \tilde{E}\,\mathrm{ch}(q) / \gamma_p E_0$, there is a special 'saddle' point with the coordinates

$$X'^{*} = \mathrm{arcch}\frac{\varepsilon_0}{\tilde{\varepsilon}}, \quad Y' = \pi / 2. \tag{2.1.26}$$

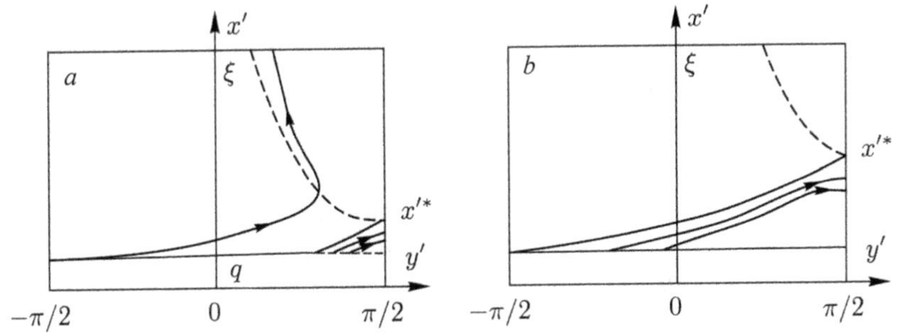

Fig. 2.4. Trajectories of the centres of electronic orbits: a – for $\alpha < \alpha_0 < \alpha_{0кр}$; b – for $\alpha_0 = \alpha_{0cr}$

The integral curve $\varphi' =$ const, passing through the singular point (2.1.26), may cross the power plane $X' = q$ at a certain coordinate Y' in the region of correct phases. All averaged trajectories on one side of this asymptote turn to the anode (electrons are captured by the high-frequency field), and on the other side (Fig. 2.4 a) – to the cathode (after crossing the boundaries of the phase regions $Y' = \pi/2$). The asymptote equation has the form

$$\sin Y' = \frac{\frac{\varepsilon_0}{\bar{\varepsilon}}(X' - X'^*) - \operatorname{sh} X'^*}{\operatorname{sh} X'}. \tag{2.1.27}$$

It can be seen from (2.1.27) that with increasing $\varepsilon_0 / \bar{\varepsilon}$ the coordinate of the singular point X'^* can so increase that the asymptote stops crossing the power plane $X' = q$ in the region of correct phases (Fig. 2.4 b). This means that there is a critical ratio $(\varepsilon_0 / \bar{\varepsilon})_{cr}$ (or critical dissynchronism α_{0cr}) corresponding to the instant when the electrons cease to fall on the anode. Thus, if in dissynchronism $\alpha_0 \neq 0$ the electrons are captured by the RF field only from a certain amplitude $\tilde{\varepsilon}$.

2.1.2. Output characteristics of a relativistic magnetron. Anode current

In the region of incorrect phases, the centres of the electron orbits shift to the cathode, which leads to bombardment of its surface and the appearance of intense secondary emission. The latter plays an extremely important role in the work of the non-relativistic magnetron, since the secondary electrons constitute the main part of the anode current. Under conditions when there is an electric field in the magnetron with a strength of ~10^6 V/cm, the emission mechanism is not yet clear. In a coaxial diode with magnetic insulation, where a high-frequency field is absent, the question of the mechanism is solved unambiguously in favour of 'explosive emission'. Concerning the same magnetron, there are a number of considerations in favour of the traditional secondary emission mechanism. Each selected point on the surface of the cathode is in the region of correct phases for a time $t_0 = Dh_\vartheta/\omega$, which usually does not exceed 0.3 ns. The electron bombardment of the cathode surface in this region is absent, there is no secondary reproduction. However, for the 'explosive emission' the time interval 0.3 ns is also insufficient. The delay time of 'explosive emission' at a field strength of ~10^6 V/cm on the smooth surface of the cathode is much greater than 0.3 ns, even taking into account the fact that the concentration of the electric field on the micro-projections of the cathode surface leads to an increase in the field strength by a factor of 10–100. On a small number of microprojections, field amplification is possible by more than 10^2 times, and with an increase in the strength E_0, the number of microprojections that can decay in time t_0 increases, but specify exactly the boundary from which the mechanism of 'explosive emission' becomes predominant until does not seem possible.

In the region of incorrect phases, explosive emission can also be difficult if the cyclotron rotation period is much less than t_0. Under these conditions, intensive bombardment of the cathode surface by the electrons leads to a rapid multiplication of secondary electrons (the secondary emission coefficient of the cathode is greater than unity) and, as the space charge increases, to decrease the electrostatic field intensity at the cathode. To estimate the anode current, it will be assumed that it consists of secondary electrons that multiply in the region of incorrect phases. The secondary electrons accumulating in these regions drift, as seen from (2.1.21), through the boundaries $py' = \pm\pi/2$ to the region of correct phases (Fig. 2.3) with average velocities

$$\beta_y' = \beta_p\gamma_p(\alpha_0\gamma_p \pm \frac{E_{\mathrm{HF}}}{E_0}\frac{\mathrm{ch}(q)}{\mathrm{sh}(\xi)}), \tag{2.1.28}$$

where $q = pR' = 2\beta_p\gamma_p/\lambda E_0 = \omega/\omega_H'$; $\xi = pd$. The charge (per unit area) of the cathode layer of secondary electrons in the regime of current limitation by the space charge is

$$\sigma' = \frac{\sigma}{\gamma_p} = \frac{E_0}{4\pi\gamma_p}. \tag{2.1.29}$$

Thus, for the electron fluxes flowing into the region of correct phases through both boundaries ($py' = \pm\pi/2$), when $\alpha_0 < \bar{\alpha}_0$ (see Fig. 2.3), the current is

$$I' = \frac{\beta_p L}{4\pi}\frac{2E_{\text{HF}}\text{ch}(q)}{\text{sh}(\xi)}, \tag{2.1.30}$$

or through one left boundary $py' = -\pi/2$, when $\alpha_0 > \bar{\alpha}_0$ (see Fig. 2.4, *a*):

$$I' = \frac{E_0\beta_p L}{4\pi}\left(\alpha_0\gamma_p + \frac{E_{\text{HF}}}{E_0}\frac{\text{ch}(q)}{\text{sh}(\xi)}\right). \tag{2.1.31}$$

Here L is the axial (in z) length of the device. The highest currents are attained in the second case. From (2.1.31) follows an explanation of the known current growth and radiation power of the magnetron almost proportional to dissynchronism α_0. Such an increase, however, can not be unlimited, since the electrons can get to the anode only in the event that the dissynchronism is not too large $\alpha_0 < \alpha_{0\text{cr}}$. It is seen from (2.1.26), (2.1.27) that the maximum currents are reached with the following connection of the parameters:

$$M \cdot \text{arch}(M) - \sqrt{M^2 - 1} = Mq + \text{sh}(q). \tag{2.1.32}$$

Here $M = \varepsilon_0 / \tilde{\varepsilon}$. The dependence of M on q is shown in Fig. 2.5. The

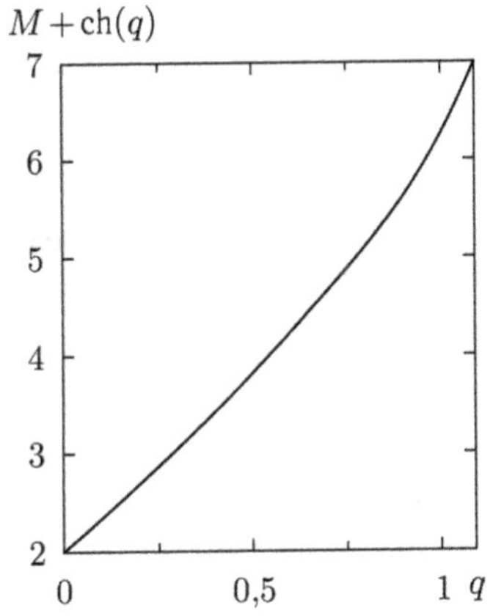

Fig. 2.5. The dependence of the parameter M, which is associated with the optimal α_0, on $q = \gamma\omega/\omega_H$.

most interesting from the practical point of view of the parameters is $q < 1$ (region of high efficiency). In this region, the relation (2.1.32) is well approximated by the dependence

$$M + \mathrm{ch}(q) \approx 2 + 3.5q. \qquad (2.1.33)$$

Thus, for the current of one spoke in the laboratory system of coordinates, $I_{\mathrm{sp}} = I'/\gamma_p$ (the invariant with respect to the coordinate transformation is the x component of the current density), we have

$$I' = \frac{\beta_p L}{4\pi\gamma_p} E_{\mathrm{HF}} \frac{2 + 3.5q}{\mathrm{sh}(\xi)}. \qquad (2.1.34)$$

Efficiency. The electrons, leaving the cathode at zero speed, along with the translational drift velocity β_e, also acquire an oscillatory motion, which in the reference frame K' is rotation with a radius $R' = \beta_e/\omega'_H$. Under the conditions of the Cherenkov synchronism and in the absence of cyclotron resonance, the action of the wave on electrons does not perturb their cyclotron rotation (2.1.19), i.e., $R' =$ const. Therefore, the electron gets to the anode approximately with the same kinetic energy $E_0 \cdot 2R'$, which it had at the top of the first loop of his trajectory (here it is taken into account that the transverse spatial scales are invariant with respect to the reference frame). The rest of the energy $E_0 d$, taken by the electron near the electrostatic field, turns out to be transformed into wave energy. Thus, the RF field is given the energy $E_0(d - 2R')$ and the efficiency is described by formula

$$\eta = 1 - \frac{2R'}{d} = 1 - \frac{2}{U}\frac{\beta_{\mathrm{e}}^2}{1 - \beta_{\mathrm{e}}^2}. \qquad (2.1.35)$$

From this it is clear that the efficiency is expressed in terms of geometric parameters in the same way as in the non-relativistic case, and a high efficiency is achieved for small R'/d, regardless of the energy of the motion of the electrons. At a fixed efficiency, an increase in the anode voltage leads, according to (2.1.35), to a decrease in the slowing-down ($\beta_{\mathrm{e}} \to 1$).

Radiation power. We express the voltage $U = E_0 d$ through the parameters ξ and $r_p = \beta_p \gamma_p$:

$$U = E_0 \lambda r_p \xi / 2\pi. \tag{2.1.36}$$

Taking into account (2.1.34)–(2.1.36), the radiation power per one spoke, $P_{sp} = \eta U I_{sp}$, is equal to

$$P_{sp} = \left(1 - \frac{2q}{\xi}\right) \xi \frac{r_p^2}{1 + r_p^2} \frac{L \lambda E_0 E_{HF}}{8\pi^2 \text{ch}\xi} (2 + 3.5q). \tag{2.1.37}$$

Taking into account that $q = 2\pi r_p / \lambda E_0$, it is easy to see that the power is expressed in terms of the quantities E_0, E_{HF}, λ, which are advantageous to increase, and the parameters r_p, ξ, over which the quantity P_{sp} should be optimized.

The choice of the length L of the device, as well as the number of resonators N, is determined by the method for solving the problem of selection of the oscillation modes for the purpose of seprtion of one working mode. Since the selection issues are solved differently depending on the design of the instrument, but there are no restrictions on r_p, ξ, E_0, E_{HF}, we will optimize the power value (2.1.37) given by one electron spoke per unit electric length L/λ.

The dependence of the output power of the magnetron on the parameters r_p and ξ is rather complicated. With increasing ξ, the efficiency and the required anode voltage U increase, but it is increasingly difficult to ensure the capture of electrons in the spoke, so the allowable dissynchronism α_0 and current I decrease. On the other hand, an increase in the parameter r_p for a fixed ξ decreases the efficiency (the Larmor radius R' grows in proportion to r_p^2), but it is possible to enhance U by increasing d. Therefore, for each wavelength λ it is possible to specify the optimal r_p and ξ, which make it possible to reach limiting powers. Denoting $a = E_0\lambda/4\pi$, and taking into account the above restrictions we present the equation for the power in a convenient form for maximization:

$$\frac{P}{nL/\lambda} = \left(1 - \frac{r_p}{a\xi}\right) \frac{a^2 r_p^2}{1 + r_p^2} \frac{\xi}{sh\xi} \frac{E_{HF}}{E_0} \left(2 + 1.75\frac{r_p}{a}\right). \tag{2.1.38}$$

Figure 2.6 shows the optimum values of the phase velocity β_p and parameter ξ characterizing the degree of decay of the RF field from the anode to the cathode, and the efficiency values corresponding to these parameters.

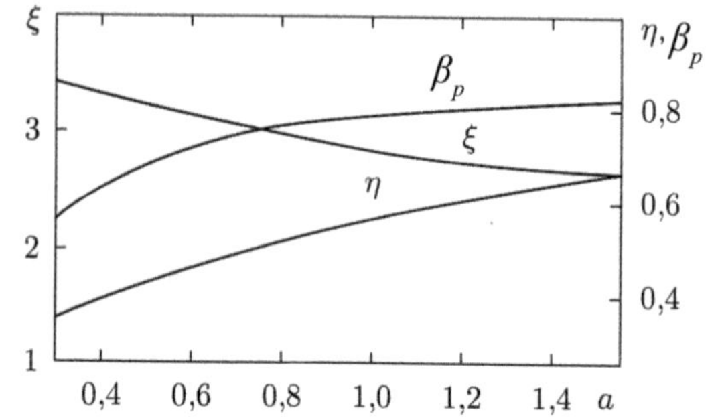

Fig. 2.6. Dependence of optimum parameters β_p, ξ and efficiency on $a = E_0\lambda/4\pi$ when $E_{HF}/E_0 = 0.025$.

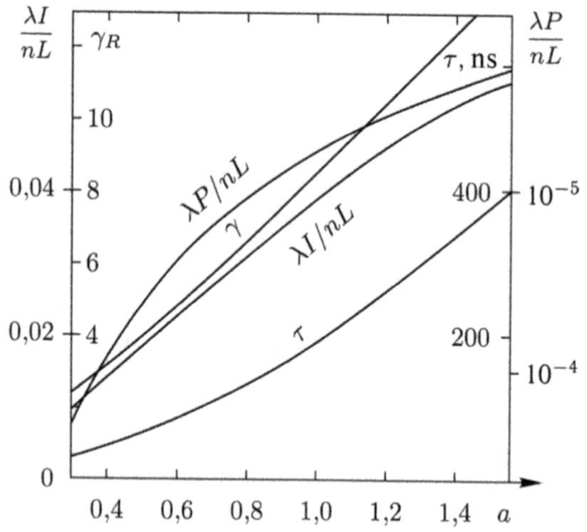

Fig. 2.7. Dependence of specific power $\lambda P/nL$ and current $\lambda I/nL$, anode voltage γ and the permissible pulse duration τ in the optimum regime on $a = E_0\lambda/4\pi$ when $E_{HF}/E_0 = 0.025$.

The anode current and the radiation power associated with one spoke, reduced to the unit of electrical length L/λ, and the anode voltage U for optimal power modes depending on the parameter a at $E_{HF}/E_0 = 0.025$ are shown in Fig. 2.7.

For a fixed value of E_0, these curves describe the dependence on the wavelength λ. It is seen that with shortening of λ the energy potential of magnetrons at restrictions on the field intensities, received above, fall sharply, but in the short-wave part of the centimeter range one can still get a high enough power level.

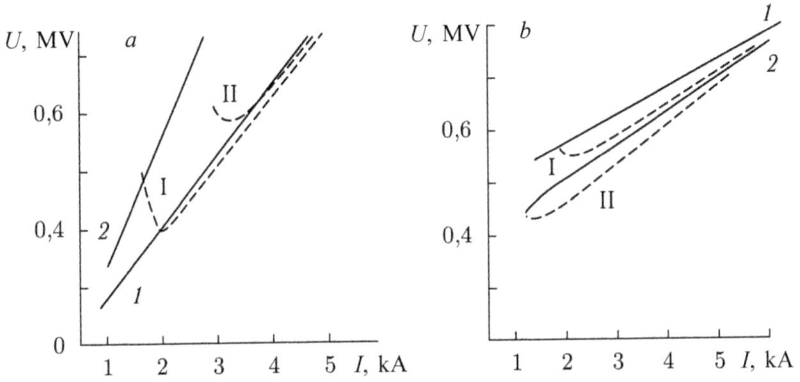

Fig. 2.8. The operational characteristics of the relativistic magnetron: the line of constant efficiency taking into account the relativistic factor (*1* and *2*) and without taking into account the relativistic factor (I and II); *1* and I – $\eta = 0.33$, *2* and II – $\eta = 0.35$ (*a*). Lines of constant generated power, taking into account the relativistic factor (*1* and *2*) and without taking into account the relativistic factor (I and II); *1* and I – $P = 1.0$ GW, *2* and II – $P = 0.6$ GW (*b*).

2.1.3. Performance characteristics of a relativistic magnetron

The parameters of the magnetron performance $U = f(I)$ are the generated power, magnetic induction and efficiency. Figure 2.8 shows the current–voltage characteristics (lines of constant power and constant efficiency) calculated for a 6-cavity relativistic magnetron of a 10-cm wavelength range, taking into account the relativistic corrections (curves *1* and *2*) and without taking them into account (curves *I* and *II*). As shown by Fig. 2.8, theoretical estimates show a significant influence of the relativistic effects on the microwave generation processes.

Calculations of the dependence of the microwave power on the magnitude of the magnetic field (Fig. 2.9, curve *1*), taking into account the actual shape of the voltage pulse, carried out using formula (2.1.38) for a six-resonator magnetron of the 10-cm range show satisfactory agreement with a similar dependence obtained experimentally.

For comparison, Fig. 2.9 shows the results of calculating the same dependences obtained for various constant electrostatic field strengths. The sharp increase in power in these curves corresponds to the increase of a current to the anode with increasing dissynchronism α (2.1.31) – electrons fall to the region of the correct phases only through the left border (see Fig. 2.4 *a*).

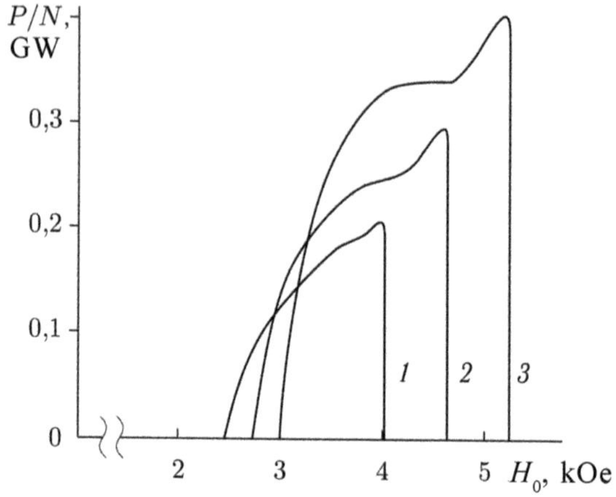

Fig. 2.9. Dependence of microwave power at various electrostatic field strengths: E_0 = 0.36 (*1*), 0.41 (*2*) and 0.47 (*3*) MV/cm.

If we limit the intensity of electric fields to the value 10^6 V/cm, then we get $\alpha_0 = \lambda/2\pi$. For each wavelength of the generated wave λ and respectively α_0 one can find the optimal parameters r_p and ξ. Figure 2.6 shows the optimum values of the phase velocity β_p and parameter ξ characterizing the degree of decay of the RF field from the anode to the cathode, and the corresponding parametric value efficiency – η and the power relating to a spoke – $P\lambda/NL$, reduced per unit electrical length L/λ.

We note that the studies [15–17] describe a program that makes it possible to numerically simulate relativistic magnetrons. The program is two-dimensional, completely relativistic ad uses the particle-in-cell method. The resulting self-consistent solutions of Maxwell's equations, taking into account the effect of space charge, confirm the possibility of the existence of a quasi-Brillouin layer of electrons around the cathode and the correctness of the concepts of the subsequent development of space-charge spokes that move synchronously with the working high-frequency mode of the magnetron.

2.2. The theoretical model of the relativistic magnetron of cylindrical geometry

2.2.1. High-frequency fields of the anode block

For the mathematical solution of the problem of the interaction of the electrons with the fields it is necessary to jointly solve the field equations and the equations of motion written in a cylindrical coordinate system (transition to a planar coordinate system is carried out at $R/r_c \to 1$). It will be assumed that the component of the high-frequency field $\tilde{E}_\varphi = \text{const}$ along an arc connecting the walls of neighbouring resonators and separating the resonator space from the interaction space. Suppose that the field in adjacent slits differs only in the phase of oscillations. On the surface of the cathode and anode segments $\tilde{E}_\varphi = 0$. Thus, the boundary conditions have the form

$$\begin{aligned} &\tilde{E}_\varphi(r_c,\varphi) = 0, \\ &\tilde{E}_\varphi(R,\varphi) = \tilde{E}_1 \exp(i2\pi nq/N) \text{ at } \left(\vartheta_q - \vartheta\right) < \varphi < \left(\vartheta_q + \vartheta\right), \\ &\tilde{E}_\varphi(R,\varphi) = 0, \end{aligned} \tag{2.2.1}$$

where $\tilde{E}_1$ is the microwave field strength at the anode; 2ϑ is the central angle per one slit; $\vartheta_q = 2\pi nq/N$ is the period of the slowing system with respect to the angle ϕ (the angle at which one period of the anode block is seen from the centre of symmetry), $q = 0, 1, ..., N - 1$ is the number of the gap, $n = 1, 2, ..., N - 1$ is the number of the mode of oscillation.

The components of the high-frequency field strength are found from the solution of the wave equation, which in the cylindrical coordinates has the form

$$r\frac{\partial}{\partial r}\left(r\frac{\partial \Phi}{\partial r}\right) + \frac{\partial^2 \Phi}{\partial \varphi^2} + k^2\Phi = 0, \tag{2.2.2}$$

where Φ is the scalar potential; $k = 2\pi/\lambda$ is the wave number. The solution of the wave equation allows us to represent the electromagnetic field in the interaction space in the form of a Fourier series, each term of which corresponds to a rotating wave:

$$\tilde{E}_r(r,\varphi) = -\frac{iN\vartheta\tilde{E}_1}{\pi kr} \sum_{p=-\infty}^{\infty} \chi \left(\frac{\sin \chi\vartheta}{\chi\vartheta}\right) \frac{Z_\chi(kr)}{Z'_\chi(kR)} e^{i(\chi\varphi+\omega t)}; \tag{2.2.3}$$

$$\tilde{E}_\varphi(r,\varphi) = \frac{N\vartheta\tilde{E}_1}{\pi} \sum_{p=-\infty}^{\infty} \left(\frac{\sin \chi\vartheta}{\chi\vartheta}\right) \frac{Z'_\chi(kr)}{Z'_\chi(kR)} e^{i(\chi\varphi+\omega t)}; \tag{2.2.4}$$

$$\tilde{H}_z(r,\varphi) = -\frac{iN\vartheta\tilde{E}_1}{\pi}\sum_{p=-\infty}^{\infty}\left(\frac{\sin\chi\vartheta}{\chi\vartheta}\right)\frac{Z_\chi(kr)}{Z'_\chi(kR)}e^{i(\chi\varphi+\omega t)}, \tag{2.2.5}$$

where $\chi = n + pN$, p is the number of the spatial harmonic (any integer, including 0),

$$Z_\chi(kr) = J_\chi(kr) - \frac{J'_\chi(kr_c)}{N'_\chi(kr_c)}N_\chi(kr); \tag{2.2.6}$$

$$Z'_\chi(kr) = J'_\chi(kr) - \frac{J'_\chi(kr_c)}{N'_\chi(kr_c)}N'_\chi(kr), \tag{2.2.7}$$

$J'_\chi(kr)$, $N'_\chi(kr)$ are the derivatives with respect to the argument of the Bessel and Neumann functions.

The number of possible modes of oscillation is limited due to the fact that the oscillation system is closed and the total phase shift around the system must be a multiple of $2\pi n$. Oscillations with $n > N$, for which the phase difference between the resonators is a multiple of 2π, correspond to the harmonics of the individual resonators. These types of oscillations do not play a significant role in the work of magnetrons and are observed very rarely.

Without taking into account the space-charge fields, the high-frequency potential under the condition $kr \ll \chi$ can be represented in the form

$$\Phi = \frac{iNr\vartheta\tilde{E}_1}{\pi}\sum_{p=-\infty}^{\infty}\left(\frac{\sin\chi\vartheta}{\chi^2\vartheta}\right)\frac{Z'_\chi(kr)}{Z'_\chi(kR)}e^{i(\chi\varphi+\omega t)}. \tag{2.2.8}$$

Although the number of harmonics p is infinite, the practical excitation of oscillation modes is observed until $p = -1$. The role of higher values of p is insignifcant due to high values of χ and the intensity decreases proportionally to the field components $(r/R)^\chi$. In addition, the product $\chi \cdot \lambda$, entering the relation

$$\beta_p = 2\pi R/\chi\lambda \approx E_r/H,$$

requires substantially different operating conditions.

For the particular case of interest to us $n = N/2$ (π-type of oscillations), i.e., when the electron moves with an angular velocity close to the velocity of the slowest wave of the slowing-down system, the action of the remaining waves on it is reduced to the creation of rapidly varying perturbations which can be neglected. The possibility to take into account only one of the waves allows us to introduce a coordinate system rotating together with a wave in which the electromagnetic field does not depend on time. Since the π-type oscillations are non-degenerate (the field represents a pure standing wave), the phase shift between the electric and magnetic fields is 90°, then (2.2.3)–(2.2.5) show that the action on the part of the RF magnetic field is $(V_p/c)^2$ times less than from the side $\tilde{E}_r(r,\varphi)$. Therefore, for non-relativistic and weakly relativistic electron velocities the action of the variable magnetic field is neglected.

Under the above condition $kR \ll \chi$, that is, when the wavelength of the oscillations is much larger than the distance between the resonators the following expression can be written for the potential of the zero harmonic of the high-frequency electric field

$$\Phi_1 = \frac{\tilde{E}r_c}{2n}\left[\left(\frac{r}{r_c}\right)^n - \left(\frac{r_c}{r}\right)^n\right]\sin(n\varphi + \omega t), \qquad (2.2.9)$$

where $\tilde{E} = \frac{N\tilde{E}_1}{\pi}\left(\frac{\sin\chi\vartheta}{\chi}\right)$ is the effective amplitude of the synchronous harmonic of the high-frequency electric field at the anode. The condition $kR \ll \chi$ is well satisfied for relativistic magnetrons with a high slowing-down of the electromagnetic wave. We will use this simplification in the future, since this will allow us to obtain an analytical solution of the problem in a form convenient for analysis.

The potential Φ_1 in the interaction space creates the following components of forces acting on the electrons:

$$\tilde{F}_r = -\frac{\partial}{\partial r}\left(\frac{e}{m}\Phi_1\right) = -\frac{e}{m}\tilde{E}\frac{r_c}{2r}\left[\left(\frac{r}{r_c}\right)^n + \left(\frac{r_c}{r}\right)^n\right]\sin(n\varphi + \omega t); \qquad (2.2.10)$$

$$\tilde{F}_\varphi = -\frac{1}{r}\frac{\partial}{\partial \varphi}\left(\frac{e}{m}\Phi_1\right) = -\frac{e}{m}\tilde{E}\frac{r_c}{2r}\left[\left(\frac{r}{r_c}\right)^n - \left(\frac{r_c}{r}\right)^n\right]\cos(n\varphi + \omega t). \qquad (2.2.11)$$

dt

2.2.2. Equations of motion of electrons in a relativistic magnetron of cylindrical geometry

We write down the equation of motion of the electrons:

$$\frac{d}{dt}(m\mathbf{V}) = e\left(\mathbf{E} + \frac{1}{c}[\mathbf{V}\times\mathbf{H}]\right). \tag{2.2.12}$$

We decompose the vector equation for the two equations in a cylindrical coordinate system, corresponding to the projections of vectors on the axis r and φ:

$$\frac{d}{dt}(m\dot{r}) - mr\dot{\varphi}^2 = F_r; \tag{2.2.13}$$

$$\frac{1}{r}\frac{d}{dt}(mr^2\dot{\varphi}) = F_\varphi, \tag{2.2.14}$$

where

$$m = \frac{m_0}{\sqrt{1-(V/c)^2}}; \tag{2.2.15}$$

$$F_r = e(\tilde{E}_r + E_r + r\dot{\varphi}(H+\tilde{H})/c); \tag{2.2.16}$$

$$F_\varphi = e(\tilde{E}_\varphi - \dot{r}(H+\tilde{H})/c), \tag{2.2.17}$$

F_r, F_φ are the components of the force acting on the electron from the static electric (E_r) and magnetic (H) fields, high-frequency electric ($\tilde{E}$) and magnetic ($\tilde{H}$) fields.

Integration of the equations of motion meets great mathematical difficulties. The main obstacle is the ambiguity of the electron velocity in a rotating coordinate system, connected with the intersection of electron trajectories. The method of solving the equations of motion is based on the periodicity of the processes occurring, due to a constant magnetic field and high-frequency oscillations. This periodicity is eliminated by applying the mathematical averaging operation [1–3], which makes it possible to obtain expressions that are convenient for analysis. The motion of an electron is considered to consist of a motion along a circular orbit and the motion of the centre of this circular orbit. For large values of the cyclotron frequency, in comparison with the frequency of the resonator system, the shift of the centre of the orbit will be small during the time of the full period of revolution of the electron and the influence of the orbital motion of electrons on the motion of the centre of their orbit can be considered as 'average' for a short period of time.

2.2.3. A simplifying assumption for the analytical solution of the equations of motion

We first need to simplify the equations of motion (2.2.13) and (2.2.14), and then use the averaging method.

First we consider the simplest case – the absence of microwave fields under the condition

$$E_{\text{kin}} = \frac{m_0 V^2}{2} = eU \frac{\ln r/r_c}{\ln R/r_c}; \tag{2.2.18}$$

$$E_{\text{kin}} = \frac{m_0 c^2}{\sqrt{1 - V^2/c^2}} - m_0 c^2 = eU \frac{\ln r/r_c}{\ln R/r_c}, \tag{2.2.19}$$

where the expressions (2.2.18) and (2.2.19) are the relations for the kinetic energy of electrons at non-relativistic and relativistic velocities of motion. Let us find the trajectories of the motion of electrons in these cases.

From the last expressions we obtain the following relations for the velocity:

$$V = \sqrt{\frac{2eU \ln r/r_c}{m_0 \ln R/r_c}}; \tag{2.2.20}$$

$$V = c\sqrt{1 - \frac{m_0 c^2}{m_0 c^2 + eU \dfrac{\ln r/r_c}{\ln R/r_c}}}. \tag{2.2.21}$$

Since $V^2 = r^2\dot{\varphi}^2 + \left(\frac{dr}{d\varphi}\frac{d\varphi}{dt}\right)^2$, we can find the expression for the angular rotation speed:

$$\dot{\varphi} = \frac{V}{\sqrt{r^2 + (dr/d\varphi)^2}}, \tag{2.2.22}$$

which we substitute into the equation of motion (2.2.14).

After a single integration

$$\frac{m_0 r^2 V}{\sqrt{r^2 + \left(\dfrac{dr}{d\varphi}\right)^2}} = eB\left(\frac{r^2}{2} - \frac{r_c^2}{2}\right); \tag{2.2.23}$$

$$\frac{m_0 r^2 V}{\sqrt{1 - \dfrac{V^2}{c^2}}\sqrt{r^2 + \left(\dfrac{dr}{d\varphi}\right)^2}} = eB\left(\frac{r^2}{2} - \frac{r_c^2}{2}\right) \tag{2.2.24}$$

provided that the rate of emission of electrons from the cathode surface is zero. In high-voltage diodes, the height of the potential barrier is negligibly small in comparison with the applied voltage. In this case, the minimum of the potential practically coincides with the surface of the cathode. Thus, it can be assumed that the electric field vanishes at the cathode surface and the initial velocity of the electrons emitted by the cathode is close to zero.

With allowance for (2.2.20) and (2.2.21), the equations for the electron trajectory in the non-relativistic and relativistic cases take the following form:

$$\varphi = \int_{r_c}^{r} \frac{dr}{r\sqrt{\dfrac{m_0}{e}\dfrac{8r^2 U \ln r/r_c}{H^2(r^2 - r_c^2)^2 \ln R/r_c} - 1}}; \tag{2.2.25}$$

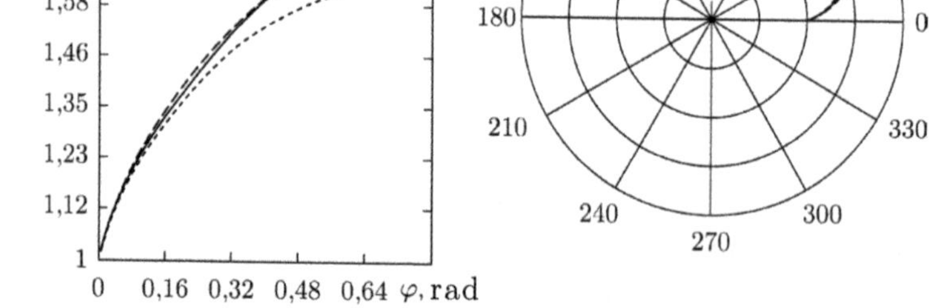

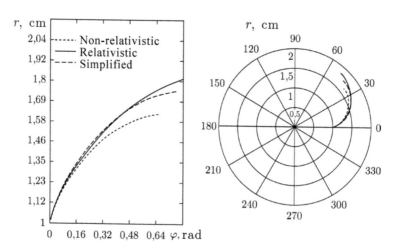

Fig. 2.10. The electron trajectories in the magnetron diode for non-relativistic (dotted curve) and relativistic (solid curve) energy values and using the proposed simplification (dashed curve); $\beta_p = 0.45$; $\gamma_p = 1.09$.

$$\varphi = \int_{r_c}^{r} \frac{dr}{r\sqrt{\dfrac{4r^2\left[\left[1+U\dfrac{\ln r/r_c}{\ln R/r_c}\dfrac{e}{m_0c^2}\right]^2-1\right]}{\dfrac{e^2}{m_0^2c^2}H^2(r^2-r_c^2)^2}-1}} \tag{2.2.26}$$

Figure 2.10 shows the trajectories of electrons obtained by numerical integration of equations (2.2.25) and (2.2.26) for the non-relativistic and relativistic values of the electron energy. To estimate the influence of relativistic effects, the same ratio of the electric field intensities to the magnetic field was chosen for these cases. Since the time equations are excluded from the trajectory equations, and the ratio r/r_c is used as an independent variable, then the analysis is limited to the motion of the electron only to the top of the cycloid. It follows from the calculations that the trajectories of the electrons with energies not exceeding 50 keV coincide, i.e., relativistic effects do not appear. A noticeable distinction of the trajectories begins to be observed at energies greater than 100 keV. Naturally, the difference between trajectories is related to the increase in the mass of an electron in proportion: $m = m_0 / \sqrt{1-V^2/c^2}$. The motion at non-relativistic and relativistic energies with the same ratio E_r/H differs in the latter case by larger radii of rotation of the electrons.

Suppose that the relativistic motion of an electron is equivalent to the non-relativistic motion of an electron by a mass

$$m = m_0 / \sqrt{1-V^2/c^2},$$

and the radicand will be considered a constant value, defined as follows:

$$\sqrt{1-V^2/c^2} \approx \sqrt{1-(r\dot{\varphi}/c)^2} \approx \sqrt{1-(E_r/H)^2}. \tag{2.2.27}$$

In fact, we neglect the radial velocity of electrons in comparison with the azimuthal velocity. Indeed, in the absence of microwave fields, the motion of electrons is a slow azimuthal drift of the centre of the orbit and a rapid rotation with a cyclotron frequency. In the microwave fields of the anode block, the electrons automatically acquire such a rotational speed that they are always in synchronism with the field.

The radial velocity of the electrons in this case is small, since during the period of revolution the electron changes its value from zero under the middle of the lamella to the maximum in the middle plane of the resonator, then the velocity again drops to zero. If the electron acquires a high radial velocity, this is accompanied by an increase in the Lorentz force acting on it, which wraps the electrons to the cathode. We also note that it is the magnitude of the radial velocity of electrons that determines the energy loss at the anode, i.e., the efficiency of the device. The desire to minimize its magnitude lies at the heart of the design of any magnetron generator. Experimental data [12] and computer simulation [9, 10] of electron dynamics in a relativistic magnetron also indicate a low value of the radial velocity.

Let us compare the trajectories of the electrons using the non-relativistic formula (2.2.25), substituting in it the expression (2.2.27), with the calculation for the relativistic case by the formula (2.2.2 6). It follows from Fig. 2.10 that the electron trajectories in the above cases are very close (the discrepancy does not exceed 3.5%, while the difference between the non-relativistic and relativistic cases reaches 11%). Of course, such a simplification method has applicability limits – specific range of ratios between the electric and magnetic fields (the value of $\beta_p < 0.5$). It should be noted that the real value of existing devices $\beta_p = 0.1–0.45$, or the efficiency of the device is low.

2.2.4. Analytical solution of the equations of motion

We use the proposed simplification for solving the equations of motion (2.2.13) and (2.2.14), which can be transformed to the form

$$m_0\gamma(r\ddot{\varphi}+2\dot{r}\dot{\varphi})+m_0 r\dot{\varphi}\frac{d\gamma}{dt}=e\tilde{E}_\varphi-\frac{e}{c}\dot{r}(H+\tilde{H}); \qquad (2.2.28)$$

$$m_0\gamma(\ddot{r}-r\dot{\varphi}^2)+m_0\dot{r}\frac{d\gamma}{dt}=e(E_r+\tilde{E}_r)+\frac{e}{c}r\dot{\varphi}(H+\tilde{H}), \qquad (2.2.29)$$

where

$$\frac{d\gamma}{dt}=\frac{e}{m_0 c^2}(\dot{r}(E_r+\tilde{E}_r)+r\dot{\varphi}\tilde{E}_\varphi). \qquad (2.2.30)$$

Taking into account that $\tilde{H}=\tilde{E}_r rV_p/c$ (it follows from (2.2.3) and (2.2.5)), $rV_\Phi \approx r\dot{\varphi}$ and $\dot{r} \ll c$, the equations of motion can be transformed to the following form

$$\ddot{r} - r\dot{\varphi}^2 = \frac{e}{m_0\gamma_p}\left(E_r + \frac{1}{c}(r\dot{\varphi}\cdot H) - \frac{\tilde{E}_\varphi}{\gamma_p^2}\right); \tag{2.2.31}$$

$$r\ddot{\varphi} + 2\dot{r}\dot{\varphi} = -\frac{e}{m_0\gamma_p}\left(\frac{1}{c}(\dot{r}\cdot H) + \frac{\tilde{E}_r}{\gamma_p^2}\right), \tag{2.2.32}$$

where $\gamma_p = (1 - \beta_p)^{-1/2}$. Thus, as seen from (2.2.31) and (2.2.32), in the relativistic magnetron compared to the non-relativistic one the effect on the electrons by the static magnetic and electric fields is weakened γ_p times, and by the high-frequency electric field is weakened γ_p^3 times. Since the magnetrons effectively operate at slow electromagnetic waves $\beta_p \sim 0.2–0.45$, the relativistic correction is $\gamma_p \sim 1.02–1.09$, $\gamma_p^3 \sim 1.06–1.3$.

In the general case, the intensities of high-frequency electric fields are determined by the relations (2.2.3) and (2.2.4). However, if we use the approximate expression for the high-frequency potential (2.2.9), the equations of motion of the electrons in a cylindrical relativistic magnetron can be written in a complex form for the variable $z = r \cdot e^{i\varphi}$:

$$\ddot{z} + \frac{i\Omega\dot{z}}{\gamma_p} = \frac{f_0}{\gamma_p} + \frac{\tilde{F}}{\gamma_p^3}, \tag{2.2.33}$$

where $\Omega = \frac{eH}{m_0 c}$ is the cyclotron frequency; $f_0 = -\frac{e}{m_0}\frac{U}{\ln(R/r_c)z^*}$ is the acceleration produced by the static electric field;

$$\tilde{F} = -i\frac{e\tilde{E}r_c}{2m_0 z^*}\left[\left(\frac{z^*}{r_c}\right)^n e^{-i\omega t} - \left(\frac{r_c}{z^*}\right)^n e^{i\omega t}\right]$$

is the acceleration created by the synchronous high-frequency field (the equations (2.2.10) and (2.2.11) are used for the components of the high-frequency field).

To exclude the time factors $e^{-i\omega t}$, it is necessary to move to a new complex coordinate:

$$z' = ze^{i\frac{\omega}{n}t} = re^{i\varphi'}; \quad \varphi' = \varphi + \frac{\omega}{n}t. \tag{2.2.34}$$

In this case the equation of motion becomes

$$\ddot{z}' + i\Omega_p'\dot{z}' = \frac{f'}{\gamma_p}, \tag{2.2.35}$$

where

$$f' = -\frac{eU}{m_0 \ln(R/r_c)z'^*} - \Omega_p''\gamma_p\frac{\omega}{n}z' - \\ -i\frac{e\tilde{E}r_c}{2m_0\gamma_p^2 z'^*}\left[\left(\frac{z'^*}{r_c}\right)^n - \left(\frac{r_c}{z'^*}\right)^n\right]; \tag{2.2.36}$$

$$\Omega_p' = \Omega_p - 2\omega/n; \tag{2.2.37}$$

$$\Omega_p'' = \Omega_p - \omega/n; \tag{2.2.38}$$

$$\Omega_p = \Omega/\gamma_p. \tag{2.2.39}$$

We are looking for a solution in the form

$$z' = \alpha' + \beta' e^{-i\Omega_p' t}. \tag{2.2.40}$$

We use the averaging method for the solution (2.2.35) (the averaging is carried out over the period of rotation of the electron similarly to [14]) and, as a result, we obtain

$$\dot{\alpha}' = -\frac{i}{\Omega_p'\gamma_p}f'(\alpha',\alpha'^*), \qquad \dot{\beta}' = -i\frac{\Omega_p''}{\Omega_p'\gamma_p}\frac{\omega}{n}\beta'. \tag{2.2.41}$$

From the last equation it follows that β' varies according to the law

$$\beta' = \beta_0' \exp\left(-i\frac{\Omega_p''}{\Omega_p'\gamma_p}\frac{\omega}{n}t\right), \quad \beta_0' = \text{const}; \tag{2.2.42}$$

$$\beta' \exp\left(-i\Omega_p' t\right) = \beta_0' \exp\left(-i\Omega_{p*}' t\right), \\ \Omega_{p*}' = \frac{\left(\Omega_p''\right)^2}{\Omega_p'} - \frac{\omega}{n} + \frac{\omega\left(\Omega_p'' - \Omega_p\gamma_p\right)}{\Omega_p'\gamma_p n}. \tag{2.2.43}$$

From this it is clear that in the rotating coordinate system the orbital motion of electrons occurs on the average with an angular velocity Ω_{p*}' with a constant radius β_0'.

If we put:

$$\alpha' = re^{i\varphi'}, \qquad \dot{\alpha}' = \left(\dot{r} + ir\dot{\varphi}'\right)e^{i\varphi'}, \tag{2.2.44}$$

then this gives two equations for the motion of the centres of the orbital trajectories:

$$\dot{r} = -\frac{e\tilde{E}r_c}{2m_0\Omega_p'\gamma_p^3 r}\left[\left(\frac{r}{r_c}\right)^n - \left(\frac{r_c}{r}\right)^n\right]\cos(n\varphi'); \tag{2.2.45}$$

$$\begin{gathered} n\dot{\varphi}' = \frac{neUr_c}{m_0\Omega_p'\gamma_p r^2 \ln(R/r_c)} + \frac{\omega\Omega_p'}{\Omega_p'} + \\ + \frac{ne\tilde{E}r_c}{2m_0\Omega_p'\gamma_p^3 r^2}\left[\left(\frac{r}{r_c}\right)^n + \left(\frac{r_c}{r}\right)^n\right]\sin(n\varphi'). \end{gathered} \tag{2.2.46}$$

2.2.5. Electron trajectories in a cylindrical relativistic magnetron

Let us analyze the solution found. From the equation (2.2.45), it follows that depending on the coordinates of the departure the electrons can travel to the correct phase region ($2\pi(s-1/4) \le \varphi' \le 2\pi(s + 1/4)$, where s – any integer) and reach the anode surface, giving energy to the microwave field. In the region of the incorrect phases, electrons take energy from the high-frequency field and fall on the cathode. The lines $n\varphi' = \pm\pi/2$ are isoclines on which $\dot{r} = 0$. They divide the fluxes into the regions of the correct and incorrect phases. Other isoclines $n\varphi' = 0$, called the matching lines, share the space in the interaction region in which the electrons overtake the wave of the microwave field or lag behind it.

The differential equation for the electron trajectory is obtained by dividing (2.2.45) by (2.2.46). Integrating the expression obtained in quadratures, we can get the equation of the trajectory:

$$\begin{gathered} \frac{eU\ln(r/r_c)}{m_0\Omega_p'\gamma_p \ln(R/r_c)} + \frac{\omega}{2n}\frac{\Omega_p'' r^2}{\Omega_p'} + \\ + \frac{e\tilde{E}r_c}{2nm_0\Omega_p'\gamma_p^3}\left[\left(\frac{r}{r_c}\right)^n - \left(\frac{r_c}{r}\right)^n\right]\sin(n\varphi') = \text{const}. \end{gathered} \tag{2.2.47}$$

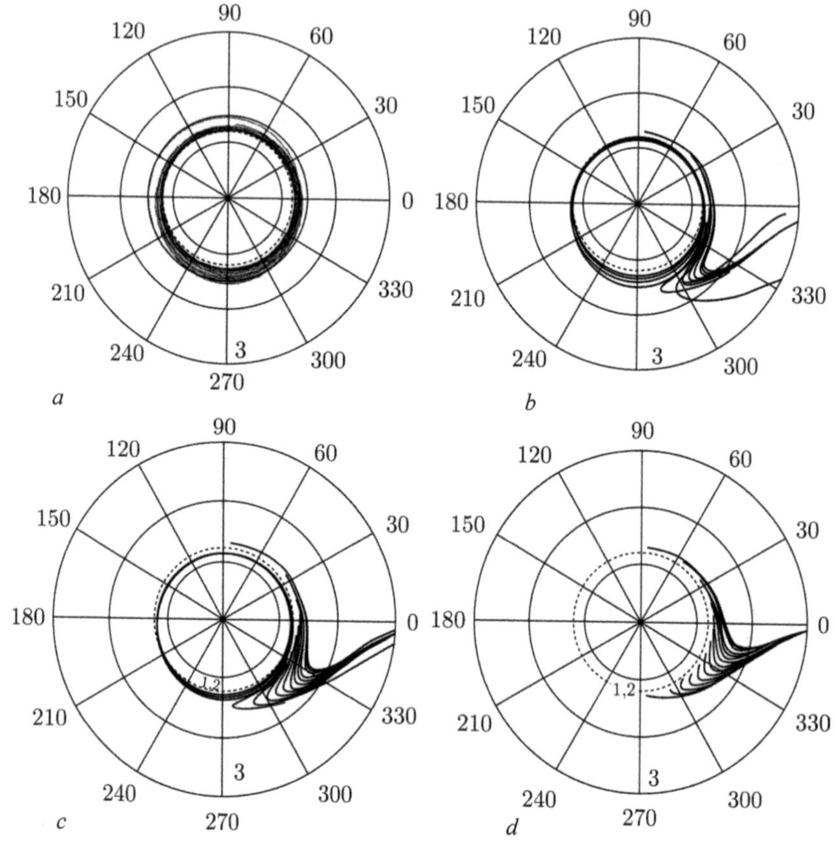

Fig. 2.11. The trajectories of electrons in the relativistic magnetron with respect to a different ratio of the strength of the high-frequency electric field to the strength of the static electric field $\varepsilon = 0.08$ (*a*), 0.11 (*b*), 0.14 (*c*) 0.23 (*d*). $\beta_p = 0.45$.

We note that the expression obtained coincides with the expression for the reduced potential in the rotating reference system (the equation for the potential can be obtained from (2.2.36)). Thus, the trajectories of the electrons will coincide with the reduced equipotential. The trajectories of electrons in the relativistic magnetron for different initial phases $n\varphi'$ and at a different ratio of the strength of the high-frequency electric field to the static electric field $\varepsilon = \tilde{E}r_c\ln(R/r_c)/U$, calculated using (2.2.45) and (2.2.46), are shown in Fig. 2.11.

Figure 2.11 *a* shows that for ε less than a certain critical value, the electrons can not enter to the anode. Indeed, for small amplitudes of the microwave field the radial velocity of the electrons is low, the azimuthal drift of electrons caused by the speed difference (V_φ)

and (V_p) outputs the electrons from the region of the correct phase and the bunching effect of the microwave field ceases. In this case the near-cathode electron layer acquires a rippled structure. By increasing the microwave field the radial drift is accelerated and electron spokes form. When $\varepsilon > \varepsilon_{crit}$ the average trajectories of the electron that left the anode at zero velocity are limited by the extreme trajectories. These trajectories near the anode approach so closely that the width of the space-charge spoke is determined mainly by the radius of cyclotron rotation of the electrons. The upper spokes extreme trajectory starts from the point with the coordinate $n\varphi' = \pi/2$, the lower trajectory coincides with the asymptote of a singular point (singular points are those in which rT and $n\phi T'$ are zero, and they are located on the lateral boundaries of the correct phase regions). All trajectories of the electrons on one side of this asymptote rotate to the anode, and on the other side to the cathode after crossing the boundary of the phase regions $n\varphi' = -\pi/2$ (Fig. 2.11 *b*). The electron trajectories for high ratios of the high-frequency field strengths to the static electric field are shown in Fig. 2.11 *c* and *d*. As can be seen, the processes of formation of the trajectories are analogous to those typical for the planar geometry of a relativistic magnetron.

2.2.6. Output characteristics of a relativistic magnetron. Anode current

In the region of the incorrect phases, the electrons shift to the cathode and bombard its surface, which leads to the appearance of secondary emission electrons (the secondary emission coefficient of cathodes is greater than one). It is known that such electrons for a conventional magnetron are the main part of the anode current. By analogy, it is assumed that in the relativistic magnetron the anode current is formed by the secondary emission electrons that multiply in the region of incorrect phases. The electrons accumulated in these areas drift across the borders $n\varphi' = \pm\pi/2$ into the region of the correct phases, determined by the equation (2.2.46).

Near the boundaries of the regions of correct phases, the charge of the near-cathode layer completely shields the cathode, and its value per unit surface in the space charge limitation regime is

$$\sigma' = \sigma / \gamma_p = E_r / 4\pi\gamma_p. \qquad (2.2.48)$$

If the electrons flow into the region of correct phases through both boundaries (there are no singular points), the value of the anode current will be

$$I_{\text{anode1}} = \sigma' L\left[n\dot{\varphi}'(-\pi/2) + n\dot{\varphi}'(\pi/2)\right] = \\ = -\sigma' L \frac{ne\tilde{E}r_c}{\gamma_p^3 m_0 \Omega_p'^{\,2} r^2}\left[\left(\frac{r}{r_c}\right)^n + \left(\frac{r_c}{r}\right)^n\right]. \tag{2.2.49}$$

For the electrons to flow through only one boundary ($n\varphi' = -\pi/2$), i.e., in the presence of a singular point, the magnitude of anode current is

$$I_{\text{anode2}} = -\sigma' L \times \\ \times\left[\frac{neUr_c}{\gamma_p m_0 \Omega_p' \ln(R/r_c) r^2} + \omega\frac{\Omega_p''}{\Omega_p'} + \frac{ne\tilde{E}r_c}{2\gamma_p^3 m_0 \Omega_p' r^2}\left[\left(\frac{r}{r_c}\right)^n + \left(\frac{r_c}{r}\right)^n\right]\right]. \tag{2.2.50}$$

As follows from the expressions obtained, the larger the value of γ_p, the smaller is the anode current of the relativistic magnetron, which naturally reflects on the value of the generated power.

Electronic and general efficiency of a relativistic magnetron. The electronic efficiency can be calculated by estimating the energy losses of electrons at the anode (we neglect the energy losses of electrons on the cathode):

$$\eta_e \approx \frac{E_c - E_a}{eU}, \tag{2.2.51}$$

where

$$E_a = \frac{1}{2}m(R\omega)^2 + \frac{1}{2}m(\dot{r})^2_{\text{anode}} \tag{2.2.52}$$

is the energy of the electrons reaching the anode, consisting of the energy of rotational motion in the crossed static electric and magnetic fields and the radial motion energy. The value ($\dot{r}$) of the anode is determined by the formula (2.2.45) taking into account $n\varphi' = \pi$,

$$E_c = -e\Phi_b + \frac{1}{2}m(r_s\omega)^2 \tag{2.2.53}$$

is the electron energy at the cathode, which can be converted into microwave radiation, includes the potential energy of the Brillouin flow Φ_b and the energy of the azimuthal rotation. The potential of a rotating Brillouin electron flow is

$$\Phi_b = -U + \frac{eB^2}{8mc^2}\left(r_s^2 - r_c^2\right). \tag{2.2.54}$$

Thus, the electronic efficiency of the device is

$$\eta_e = 1 - \frac{eB^2(r_s^2 - r_c^2)/4mc^2 + m\omega^2(R^2 - r_s^2) + m(\dot{r})^2_{\text{anode}}}{2eU}. \tag{2.2.55}$$

As follows from the above expression, for relativistic magnetrons, a decrease in the electronic efficiency is associated with an increase in the electron mass (the second term of the difference increases in proportion to γ_p).

For the classical magnetrons the overall efficiency of the device is expressed as a product of the electron efficiency and the circuit efficiency – η_{cir}, with the the latter is expressed as follows:

$$\eta_{\text{cir}} = \frac{Q_{\text{load}}}{Q_{\text{ext}}} = \frac{Q_{\text{o}}}{Q_{\text{o}} + Q_{\text{ext}}}, \tag{2.2.56}$$

where Q_{o}, Q_{ext} and Q_{load} is the intrinsic, external and loaded Q of the resonator system.

In addition, the relativistic magnetrons are characterized by current losses associated with 'parasitic' emission from the surfaces of high-voltage electrodes and the presence of an end current I_{end}. If we enter a value

$$\eta_T = \frac{I_{\text{anode}}}{I_{\text{total}}} = \frac{I_{\text{anode}}}{I_{\text{end}} + I_{\text{anode}}}, \tag{2.2.57}$$

then the overall efficiency of the relativistic magnetron can be determined as follows:

$$\eta = \eta_e \eta_{\text{cir}} \eta_T. \tag{2.2.58}$$

Thus, setting the parameters of the relativistic magnetron (R, r_c, N, β_p, λ, the radius of the drift tube), we can calculate the distribution in the interaction space of the components of the microwave field (2.2.3)–(2.2.5), the anode current (2.2.49) or (2.2.50), the full efficiency (2.2.58) and the generated power according to the formula

$$P = \eta U I_{\text{anode}}. \tag{2.2.59}$$

It can be concluded that to obtain high values of electronic efficiency and power, it is advisable to produce the RMs with the greatest slowing-down of the electromagnetic wave, and to increase the overall efficiency of the device, measures must be taken to eliminate the end-current losses. The possibilities and ways to implement these recommendations are discussed below.

From (2.2.59), if we substitute the expressions for the factors, we can see that the power is expressed in terms of the quantities that it is advantageous to increase: U, $\tilde{E}$, L, $N = 2n$ for the π-type of oscillations and in terms of R, r_c, for which P should be optimized. However, the parameters N and L are limited by the need to ensure a sufficient separation of the modes by the frequency and magnitude of the electromagnetic wave slowing-down. The level U depends on the energy capabilities of high voltage sources and the magnetic field, the value of $\tilde{E}$ should not exceed the breakdown values in the resonators and power output devices. As for R, r_c, the estimates show that for each magnetic field there is an optimal cathode radius and for a given cathode radius there is an interval of working magnetic fields. It should be noted that for these estimates the value ε was equated to a constant whose choice, generally speaking, is unrelated. The change in the effective amplitude of the fundamental spatial harmonic during the duration of the voltage pulse should be determined from the joint LIA–relativistic magnetron computer model (see Chapter 3).

Calculations show that the influence of relativism is manifested in an increase in the radius of the synchronous layer due to an increase in γ_p of the radius of cyclotron rotation and the weakening of the effect on electrons by a factor of γ_p^3 times of the high-frequency field. The first effect leads to an improvement in the capture of electrons in space-charge spokes, since the intensity of the high-frequency electric field on the surface of the near-cathode layer increases, and the second effect, on the contrary, makes it difficult to form the spokes. In general, a smaller number of electrons are captured in the space-charge spokes in RM with all other conditions being equal, i.e., the region of correct phases narrows because of a decrease in E_r.

The formulas and performance characteristics allow us to determine the areas of the generator's optimal power and efficiency parameters, to estimate the current consumed, and to reveal the requirements for the design of the device. These data were used in the calculation and design of RMs, as well as for interpreting the experimental results.

2.2.7. Limits of applicability of the averaging method

An important question in this theoretical model of the relativistic magnetron is the determination of the limits of applicability of the averaging method. In other words, it is necessary to find the conditions under which the averaging carried out does not distort the character of the motion of the electrons. As shown in [11], a sufficient condition for the applicability of the averaged equations is the smallness of the distance δ (δ is the distance over which the electron moves for a time equal to the averaging time $(2\pi/\Omega'_p)$, compared to the characteristic geometric dimensions of the slowing-down system or spatial variation of the high-frequency field):

$$k\delta << 1, \tag{2.2.60}$$

where $\delta = f'/(\Omega'_p)^2$. Substituting the expression for f' and simplifying, we obtain the following relation:

$$\frac{e\tilde{E}}{m_0\lambda\gamma_p^3(\Omega'_p)^2}\left[\left(\frac{R}{r_c}\right)^{N/2}-\left(\frac{r_c}{R}\right)^{N/2}\right]<<1. \tag{2.2.61}$$

It follows from the expression that the conditions for the applicability of the averaging method are well satisfied for magnetrons: 1) operating at high magnetic fields, 2) low-frequency, 3) having a small ratio of the radii of the anode and cathode, and 4) containing a small number of anode block resonators. The last three conditions are natural for the RMs, since, as a rule, they are made for a wavelength of ~10 cm, they tend to develop the interaction space of the device to reduce thermal loads, the number of resonators is chosen small to ensure a satisfactory separation of frequencies between modes. Substituting these 6-resonator magnetrons intended for repetitively pulsed mode with a small slowing-down of the electromagnetic wave $\beta_p = 0.45$, it can be noted that the conditions for the applicability of the averaging method are well satisfied to the values of the high-frequency electric field intensity at the cathode ~200 kV/cm.

Taking into account the space charge, the condition of applicability of the method acquires the form [11]

$$\omega_p << \Omega_p, \quad (2.2.62)$$

where $\omega_p = \sqrt{4\pi e\rho / m}$, ρ is the density of the space charge.

Determining the charge density of a 6-cavity relativistic magnetron (voltage ~400 kV, total current ~3.6 kA, output power ~300 MW) through the anode current, which, according to estimates from formula (2.2.49), is about 2000 A, we note that the above relation is much worse (the difference in frequencies is 4–5 times). This circumstance for the RMs is completely natural, since from the classical magnetrons they are much more different in the magnitude of the anode current than in such parameters as geometric dimensions, magnetic field strength, and so on.

2.2.8. The effect of space charge

Let us analyze the effect of space charge on the characteristics of the instrument with the cylindrical geometry of the electrodes. As noted earlier, in a cylindrical magnetron, the exact synchronism of the wave and the angular velocity of rotation of the electron is possible only at a synchronous radius. The dissynchronism of the velocities causes an additional azimuthal drift distorting the spokes of the space charge. Moreover, the value of the electron displacement can be estimated by the following expression:

$$\Delta(r\varphi) \approx (V_\varphi - V_p)T_{el}, \quad (2.2.63)$$

where T_{el} is the time of flight of the electrons. We note that above and below the synchronous radius, the velocity dissynchronism changes the sign. Thus, for electrons that are located at radii below the synchronous one, velocity dissynchronism causes a dephasing effect, that is, it counteracts the phasing action of the electromagnetic wave. At radii above the synchronous effect of the azimuthal component of the HF field and the velocity dissynchronism are directed in one direction. It follows from the above considerations that the closer the synchronous radius to the feed radius (the outer boundary of the electron cloud rotating around the cathode), the more electrons can be captured into the electron spoke.

To take into account the effect of the space charge, we use the following condition: the spokes move at the same speed as the space charge. The additional electron drift caused by the action of the space charge is especially dangerous near the feed radius where the phasing field of the electromagnetic wave is small and, in addition, there is an additional drift associated with the dissynchronism of the velocities. For a relativistic magnetron, this situation is exacerbated by the fact that the action from the high-frequency field is attenuated γ_p^3 times. Therefore, for efficient capture of electrons in the spoke we require sufficiently large values of the radius of rotation of the electron cyclotron r_c, which determine the power range. Naturally, this condition requires low values of working magnetic fields, that is, high β_p. (For small r_c even the classical magnetron generators operate poorly – in special experiments proportionally increased voltage elevated magnetic field to maintain matching conditions at constant emission current In this case, there was a decrease. $r_c \sim 1/H^2$, and there was a reduction of efficiency.) Above the synchronous radius, the pushing action of the space charge is compensated by the dissynchronism of the electron and electromagnetic wave velocities.

To estimate the influence of the space charge on the instrument›s limiting parameters on the anode current, and hence on the power, it is necessary to determine its perturbing action near the feed radius. The largest perturbation in the motion of electrons will be the radial component of the space-charge field, since its combined action with a constant magnetic field causes the electrons to drift in the azimuthal direction and the electrons to leave the regions of the correct phases.

Let us consider the case in which along with the electrostatic field E_r a small perturbing field operates, also directed along the radius by the intensity ΔE_r. (Since the field of the space charge moves in synchronism with the spokes, it will have the same effect on the electrons as the electrostatic field.) This field corresponds to the acceleration

$$\Delta F_r = -\frac{e}{m}\Delta E_r. \qquad (2.2.64)$$

We add this acceleration to the equation (2.2.36), and average (2.2.35) and, as a result, we obtain

$$\dot{r} = -\frac{e\tilde{E}r_c}{2\gamma_p^3 m_0 \Omega_p' r}\left[\left(\frac{r}{r_c}\right)^n - \left(\frac{r_c}{r}\right)^n\right]\cos n\varphi', \qquad (2.2.65)$$

$$n\dot{\varphi}' = \frac{neUr_c}{\gamma_p m_0 \Omega_p' \ln(R/r_c) r^2} + \omega \frac{\Omega_p''}{\Omega_p'} + \\ + \frac{ne\tilde{E}r_c}{2\gamma_p^3 m_0 \Omega_p' r^2}\left[\left(\frac{r}{r_c}\right)^n + \left(\frac{r_c}{r}\right)^n\right]\sin(n\varphi') + \frac{ne}{\gamma_p m_0 r}\frac{\Delta E_r}{\Omega_p'}. \tag{2.2.66}$$

It follows from the last equation that the space charge increases the angular velocity of electrons by $ne\Delta E_r/\gamma_p m_0 \Omega_p$. It should be noted that the perturbed motion retains the same potential character as the unperturbed motion.

We shall determine the limiting values of the additional acceleration caused by the space charge, at which the generation of microwave radiation is conserved, i.e., the electrons are captured in the spokes. For this, we find the equation of the trajectory of the motion of electrons for the given case. We divide (2.2.65) by (2.2.66) and integrate the resulting differential equation in quadratures. The equation of the electron trajectory has the form

$$\frac{neU\ln(r/r_c)}{\gamma_p m_0 \Omega_p' \ln(R/r_c)} + \frac{\omega}{2}\frac{\Omega_p'' r^2}{\Omega_p'} + \\ + \frac{e\tilde{E}r_c}{2\gamma_p^3 m_0 \Omega_p'}\left[\left(\frac{r}{r_c}\right)^n - \left(\frac{r_c}{r}\right)^n\right]\sin(n\varphi') + \frac{ne\Delta E_r r}{\gamma_p m_0 \Omega_p'} = \text{const}. \tag{2.2.67}$$

An interesting point is the singular point with the coordinates $n\varphi' = -\pi/2$ and $r = r_s$ (we assume that the feed radius coincides with the synchronous radius). The main distinguishing feature of this point is the derivative of the trajectory along the radius is zero for $r = r_s$, which leads to the expression

$$\frac{neUr_c}{\gamma_p m_0 \Omega_p' \ln(R/r_c) r_s} + \frac{\omega \Omega_p'' r_s}{\Omega_p'} + \\ + \frac{ne\tilde{E}r_c}{2\gamma_p^3 m_0 \Omega_p' r_s}\left[\left(\frac{r_s}{r_c}\right)^n + \left(\frac{r_c}{r_s}\right)^n\right] + \frac{ne}{\gamma_p m_0}\frac{\Delta E_r}{\Omega_p'} = 0. \tag{2.2.68}$$

From the last equation, one can obtain the condition for the conservation of the spoke-like motion of the electrons:

$$\frac{eUr_c}{\gamma_p m_0 \ln(R/r_c)r_s} + \frac{\omega\Omega_p'' r_s}{n} + \frac{e\tilde{E}r_c}{2\gamma_p^3 m_0 r_s}\left[\left(\frac{r_s}{r_c}\right)^n + \left(\frac{r_c}{r_s}\right)^n\right] \geqslant -\frac{e}{\gamma_p m_0}\Delta E_r. \tag{2.2.69}$$

When the condition (2.2.69) is satisfied, the effect on the electrons from the microwave field exceeds the dephasing action from the side of the space charge, and all electrons in the region of correct phases are captured into the spokes. If ΔE_r does not satisfy this condition, the formation of the spokes is broken, the anode current decreases and can even fall to zero.

Let us determine the relationship of ΔE_r with the space charge parameters. Knowing the shape of the spokes and the density of the space charge distribution in them, it is possible to calculate the field at a point where the space charge action is maximal. For this it is necessary to find the field of space charges by integrating the fields created by each element of the area. To simplify the calculations, it is proposed in [15] to approximate the shape of the space-charge spoke by a triangle, which leads to the following expression in the cylindrical case:

$$\Delta E_r = \rho\frac{2\pi r_s}{n}. \tag{2.2.70}$$

The electron current density J to form a single spokes is related to the space charge ρ by the ratio

$$J\frac{2\pi r_s}{n} = \int_{-\pi/2}^{\pi/2} \rho \dot{r} d\varphi. \tag{2.2.71}$$

Using the expressions for the current density (2.2.71), the stability condition for the conservation of motion in the spokes (2.2.69) and equation (2.2.70), we can obtain the value of the limiting electron charge density:

$$\rho=\frac{n}{2\pi r_s}\frac{m_0\gamma_p}{e}\times$$
$$\times\left[\frac{eUr_c}{\gamma_p m_0\ln(R/r_c)r_s}+\frac{\omega\Omega_p''r_s}{n}+\frac{e\tilde{E}r_c}{2\gamma_p^3 m_0 r_s}\left[\left(\frac{r_s}{r_c}\right)^n+\left(\frac{r_c}{r_s}\right)^n\right]\right] \tag{2.2.72}$$

and the critical current density of a relativistic magnetron when working on an oscillation with the number n:

$$J=\frac{n\tilde{E}}{4\pi^2 r_s\Omega_p'}\left[\left(\frac{r_s}{r_c}\right)^n-\left(\frac{r_c}{r_s}\right)^n\right]\times$$
$$\times\left[\frac{eUr_c}{\gamma_p m_0\ln(R/r_c)r_s}+\frac{\omega\Omega_p''r_s}{n}+\frac{e\tilde{E}r_c}{2\gamma_p^3 m_0 r_s}\left[\left(\frac{r_s}{r_c}\right)^n+\left(\frac{r_c}{r_s}\right)^n\right]\right]. \tag{2.2.73}$$

Despite the fact that the expression obtained has a rather complex form, some useful conclusions can be drawn for the design of devices. 1) The limiting current is proportional to the value of $\tilde{E}^2$, which, in turn, is proportional to the energy of the high-frequency field in the resonators. Therefore, the electromagnetic energy in the resonators must be limited to a certain value so that the anode current generated by it does not exceed the limiting value (2.2.73). This circumstance imposes certain requirements on the choice of the magnitude of the circuit efficiency of the device, that is, to the design of the output of the relativistic magnetron power. 2) The limiting current of the device is proportional to the number of resonators, which seems quite natural. 3) When choosing the relationship between the radius of the cathode and the feed radius, the following should be considered. With increasing r_c, on the one hand, the conditions for the capture of electrons in the spoke are facilitated, and on the other hand, the electronic efficiency is reduced. The radius of the cathode is actually given when constructing a relativistic magnetron, and the synchronous radius is chosen experimentally with a change in the strength of the magnetic field.

2.2.9. *Electronic frequency shift in a relativistic magnetron*

It was established above that the RM operate in the presence of a strong space charge. The spatial charge in the cathode–anode gap of the magnetron increases the phase velocity of the electrons, which leads to the appearance of a phase shift between the first harmonic of the induced current and the voltage and, as a consequence, causes the frequency of the oscillations to shift relative to the frequency of the resonator system. For the classical magnetrons the concept of the EFS is introduced, defined as [18]

$$\chi_{\text{EFS}} = df / dI_{\text{anode}} \tag{2.2.74}$$

and calculated with constant magnetic field strength, constant external load and temperature of the anode block. In the case of an RM, the value of the anode current can be varied only by changing the voltage, since the cathode of the device operates in the regime of current limitation by the space charge. In this experiment, it is necessary to adjust the magnetic field in order to fulfill the synchronism conditions. The electronic frequency shift is an important parameter of the RMs, since voltage pulses can have significant differences from the rectangular shape and the associated frequency changes need to be estimated quantitatively.

The problem of determining the value of the EFS is reduced to finding the angle of error between the anode current and the voltage across the resonator:

$$\text{tg}\ \Theta_1 = -\tilde{P}_e / P_e = 2Q_{\text{load}}\left(\omega - \omega'_n\right) / \omega'_n, \tag{2.2.75}$$

where ω is the generation frequency; ω'_n is the circular frequency, with which the frequency field of n-th oscillation mode varies; $\tilde{P}_e$ is the reactive power in the resonator; P_e is the power given by the electrons of one spoke to the high-frequency field.

The values of active and reactive power are determined in accordance with the expressions [19]

$$P_e = -\overline{\int\limits_{V_{\text{res}}} j(t)E(t)\,dV_{\text{res}}}; \tag{2.2.76}$$

$$\tilde{P}_e = -\int\limits_{V_{\text{res}}} j(t)\frac{\partial E(t)}{\partial(\omega t)}\,dV_{\text{res}}, \tag{2.2.77}$$

where $j(t) = \rho V_{\text{res}}$ is the exciting current; $E(t)$ is the field in the interaction space excited by the current; V_{res} is the volume of the resonator.

Consider a two-dimensional model and take into account that the density of the space charge of electrons in any section of the spoke remains unchanged. In this case it is possible to obtain the following relations for the active and reactive power of the relativistic magnetron:

$$P_e = -\rho L \iint [\dot{r}(\tilde{E}_r + E_r) + r\dot{\varphi}\tilde{E}_\varphi] r dr d\varphi; \qquad (2.2.78)$$

$$\tilde{P}_e = -\rho L \iint [\dot{r}\frac{\partial \tilde{E}_r}{\partial(\omega t)} + r\dot{\varphi}\frac{\partial \tilde{E}_\varphi}{\partial(\omega t)}] r dr d\varphi. \qquad (2.2.79)$$

We use the following circumstance: the azimuthal rotation speed of an electron depends little on the intensity of a high-frequency electric field, so for an average speed of rotation we take the n-th mode of oscillation $r\dot{\varphi} = \frac{\omega_n' r}{n}$. The limits of integration over the radius are chosen from r_c to R, and over the angle from $-\pi/2$ to $\pi/2$, taking into account the following. The maximum anode current, and hence the magnitude of the space charge, is observed for large values of the ratio of the high-frequency field intensity to the static field strength, i.e., when all the electrons from the regions of correct phases are captured by the microwave field.

From (2.2.9), (2.2.14), (2.2.45), (2.2.78), (2.2.79) we can obtain the following expressions for the active and reactive power in the resonator:

$$P_e \approx \rho L \frac{\tilde{E} r_c}{n} \times \left[\frac{\omega_n' \left(\frac{R^{n+2}}{r_c^n} + \frac{r_c^n}{R^{n+2}} - 2 \right)}{n+2} - \frac{eU \left[\left(\frac{R}{r_c} \right)^n + \left(\frac{r_c}{R} \right)^n - 2 \right]}{n \Omega_p' m_0 \gamma_p \ln(R/r_c)} \right]; \qquad (2.2.80)$$

$$\tilde{P}_e \approx -\rho L \frac{e\tilde{E}^2 \pi r_c^2}{16 n^2 m_0 \Omega_p' \gamma_p^3} \left[\left(\frac{R}{r_c} \right)^{2n} + \left(\frac{r_c}{R} \right)^{2n} - 2 \right]. \qquad (2.2.81)$$

We note that the difference in the products $r\dot{\varphi}\tilde{E}_\varphi - \dot{r}E_r$ gives the contribution to the active power, where the first term determines the energy of the rotating electron spokes, and the second term – the energy loss by the electrons at the anode. The magnitude of the reactive power given by the product $\dot{r}\dfrac{\partial \tilde{E}_r}{\partial t}$, i.e. $\tilde{P}_e$ is created by the radial component of the induced current. This component of the current is shifted by $\pi/2$ relative to the azimuthal component of the induced current, which, in turn, differs in phase by π in relation to the high-frequency voltage. Therefore, in the case of small angles of phase mismatch, the radial induced current is purely reactive. Substituting (2.2.80) and (2.2.81) into (2.2.75), we obtain the following expression:

$$\mathrm{tg}\,\Theta_1 = -\frac{e\tilde{E}\pi r_c\left[\left(\dfrac{R}{r_c}\right)^{2n}+\left(\dfrac{r_c}{R}\right)^{2n}-2\right]}{16\left[\dfrac{nm_0\Omega'_p\gamma_p^3\omega'_n\left(\dfrac{R^{n+2}}{r_c^n}+\dfrac{r_c^n}{R^{n+2}}-2\right)}{n+2}-\dfrac{\gamma_p^2 eU\left(\left(\dfrac{R}{r_c}\right)^n+\left(\dfrac{r_c}{R}\right)^n-2\right)}{\ln(R/r_c)}\right]}. \qquad (2.2.82)$$

Since the ratio of reactive power to active has a negative sign, this means that the induced current of the electron spoke lags behind the high-frequency voltage on the resonators and the reactive power is capacitive in nature. In this case, the radial component of the microwave field increases the intensity of the static field acting on the electron charge of the spoke, and thereby ‘compensates’ the deviation of the electrostatic field strength from the synchronous value.

It may appear that the magnitude of the angle of phase mismatch increases with reaching the intensity of the high-frequency electric field. However, this is not the case, because the expression in the denominator is proportional to E. Indeed, the $U \sim \tilde{E}/\eta_e$ and $\omega'_n = \beta_p \dfrac{c\chi}{R} \sim \dfrac{\tilde{E}}{\eta_e}$. Thus, with increasing voltage, the angle of phase mismatch decreases and the frequency of generation of the relativistic magnetron approaches the frequency of the resonator system without the space charge. With the use of (2.2.82), it is possible to estimate the frequency modulation of the device radiation caused by variations in the accelerating voltage. Calculations show that in the area of operational modes the RMs have Θ_1 in the range $-5°$ ... $-20°$, which, with loaded Q-factors at the level of 50–100 will cause a frequency shift of not more than 100 MHz when the value of the cathode-anode voltage is changed by a factor of two. The estimates given are made in the approximation that when the voltage is varied, the power generated by the device is preserved ($\tilde{E}$ = const). Nevertheless, such insignificant frequency shifts with significant deviations of the amplitude of the voltage pulse from the maximum value are undoubtedly a positive quality of the device, which opens up prospects of applied nature.

2.3. Thermal processes in a relativistic magnetron

In contrast to O-type generators in magnetrons, the collector of the anode current of electrons is the surface of the resonator system, so the thermal processes in such devices are given considerable attention. In particular, the effect of thermal heating of anode blocks on the stability of the performance characteristics of the device and their durability is shown. The anode block of the magnetron is heated from several sources of heat: from radiation from the surface of the cathode, from electron bombardment of lamellae and from the circulation of high-frequency currents along the surface of the resonators. For approximate calculations, the heating of lamellae due to radiation and microwave currents can be neglected because of their small influence [19]. An increase in the temperature of the resonator system leads to deformation of the lamellae and straps (in classical instruments), hence, to a change in the wavelength. Besides, when the temperature increases, the active losses increase, i.e., the intrinsic Q-factor of the oscillatory system and the circuit efficiency decrease. Therefore, in order to eliminate the above-mentioned effects for devices, the concept of a safe temperature at the level of 350–400°C is introduced.

In the RM, the power density of the electron beam deposited on the surface of the anode block becomes so significant that it is possible to develop evaporation, erosion and mechanical deformation processes under the action of thermal shocks. In this case, the pulse duration is so small that the dissipated power at the anode does not have time to not only be discharged outside, but even spread to the entire thickness of the anode, causing instantaneous pulse heating. The surface layer of the metal is heated during each pulse, and cools during pauses between the pulses. The expansion and contraction of the surface layer with respect to the neighbouring non-heated layer causes internal stresses under the influence of which inhomogeneities are formed. With a large number of impulses, inhomogeneities unite and microcracks appear. The microcracks prevent the process of heat transfer from the surface layer into the interior of the metal, so the temperature of the layer increases with each subsequent impulse, which leads to melting and evaporation of the material. The main channels for dissipation of the electron beam energy from the surface layer are thermal conductivity, the propagation of thermomechanical stresses, phase transitions, and evaporation of the material. Depending on the specific density of the energy released and the pulse duration, it is possible to develop certain processes. In the calculations presented below, for obvious reasons, the influence on heating of such factors as electric or microwave breakdown, the formation of a gas discharge is not taken into account.Experimental studies of relativistic magnetrons with the use of HCEA in the single-pulse mode showed a low longevity of the anode blocks – the number of pulses often did not exceed one hundred. Fractures took place in separate parts of the lamellae in the region of sharp edges. For the pulse-periodic mode, the lifetime of the anode block must be at least 10^6 and more impulses. Therefore, it is necessary to evaluate the limiting modes of RM operation that do not lead to destruction of the anode blocks depending on the electron beam energy flux density, repetition frequency, pulse duration, electron beam parameters, and suggest measures to increase the longevity of the anode device blocks. To do this, it is required to determine the areas of deposition of electrons, the depth of their penetration and the uniformity of heating, to consider the process of heat dissipation during the supply pulse and in a pause between pulses.

2.3.1. Areas of electron deposition on the surface of the anode block

In both classical and relativistic magnetrons, the electrons move along trajectories close to cycloidal and fall on the surface of the anode at different angles, determined by the ratio of the radial and tangential component of the velocity near the anode, $\mathrm{tg}\alpha = \dot{r}/\dot{\varphi}$, and the phase of cyclotron rotation [18]. Qualitatively, the process of entering of the electrons is shown in Fig. 2.12. It can be seen from the figure that the deposition of the electrons occurs on the cylindrical and lateral parts of the anode block lamellae. The shown trajectories allow us to conclude that, depending on the ratio of velocities, the

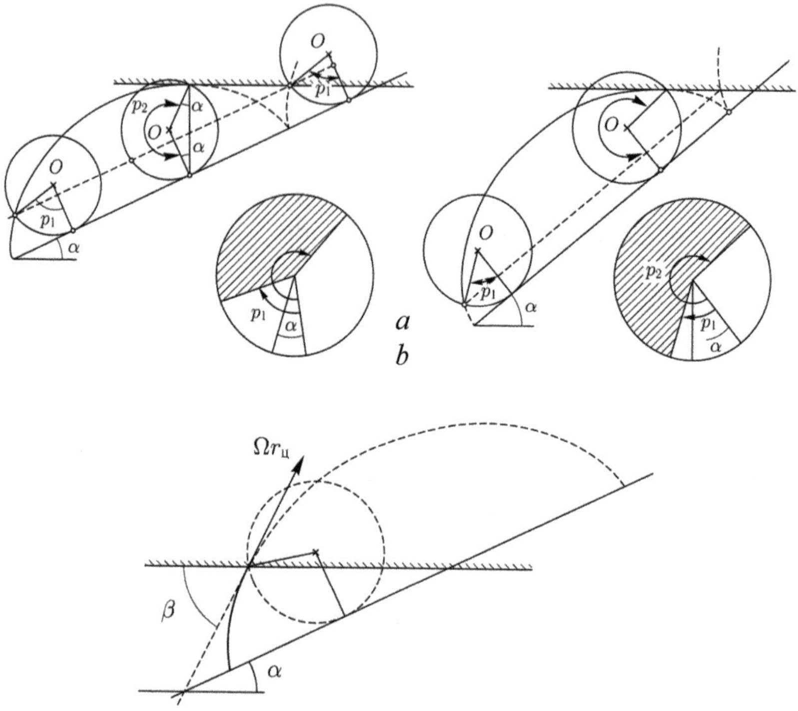

Fig. 2.12. Dependence of the region of possible phases $p_2 - p_1$ of electrons falling on the surface of the anode block of the RM on the angle α (a). The angle of incidence of electrons β on the surface of the anode block as a function of the phase of cyclotron rotation p (b).

electrons can reach the cylindrical part of the lamella in a certain range of phases $p_2 - p_1$ (the angles of rotation of the circle forming the cycloidal trajectory). From the geometrical construction shown in the figure, one can determine the angle $p_2 = \pi + 2\alpha$. The angle p_1 is calculated from the transcendental equation [19]:

$$\cos p_1 + (\sin p_1 - p_1)\operatorname{tg}\alpha = (\pi - 2\alpha - \sin 2\alpha)\operatorname{tg}\alpha - \cos 2\alpha.$$

The higher the radial velocity of the electrons, caused by the action of the azimuthal component of the high-frequency field, the wider the range of possible angles. Naturally, with the change in the angle α, the kinetic energy of the electrons bombarding the anode will also change; this energy is composed of the energy of cyclotron rotation and the radial motion of electrons under the action of the microwave field:

$$E_{\text{inc}} = m \cdot \left[\left(\Omega_p r_c \right)^2 + \left(\dot{r} \right)^2_{\text{anode}} \right] / 2, \tag{2.3.1}$$

where the quantity $(\dot{r})_{\text{an}}$ can be determined using formula (2.2.45) for the radial velocity of electrons at $n\varphi' = \pi$, the cyclotron rotation radius is determined according to formula

$$r_c = mc^2 E_r / eH^2. \tag{2.3.2}$$

The rate of cyclotron rotation of the electrons can vary from $2\Omega_{\text{p}} \cdot r_c$ at the apex of the cycloid to a value close to zero at the origin of the cycloid. Therefore, the average value of the kinetic energy of an electron associated with cyclotron rotation will be taken to be half the sum of the extreme values. Taking into account the additional dissipation of electrons in matter, it will not be a big mistake to assume that heating occurs uniformly over the entire depth of their penetration. The maximum depth of penetration of the electrons is estimated on the basis of the following considerations. At the apex of the cycloid, the direction of the velocity vector of the electron coincides with the angle α and its energy is equal to $2m(\Omega_{\text{p}} r_c)^2$. The kinetic energy components normal to a cylindrical surface are equal to

$$E_c = 2m \left(\Omega_p r_c \right)^2 \sin \alpha + m \cdot \left(\dot{r} \right)^2_{\text{anode}} / 2$$

and for the lateral surface

$$E_{\text{lat}} = 2m \left(\Omega_p r_c \right)^2 \cdot \cos \alpha.$$

The penetration depth of the electrons in the substance δ (cm) depends on the electron energy E (keV) and the density of the material ρ_{m} (kg/cm^3) as follows [20]:

$$\delta = 10^{-5} E_{\text{inc}}^{3/2} / \rho_m. \tag{2.3.3}$$

Consequently, the depth of the maximum range of electrons for a cylindrical surface can be calculated from

$$\delta_2 = 10^{-5} m^{3/2} \left[\left(2\Omega_p r_c \right)^2 \sin \alpha + (\dot{r})_{\text{anode}} \right]^{3/2} / 2\sqrt{2}\, \rho_m. \tag{2.3.4}$$

On the lateral surface of the resonators, the electrons come with phases from 0 to π, so our assumption regarding the uniformity of warming up the depth of the end surface of the lamellae is all the more justified. The maximum depth of penetration of electrons in this case

$$\delta_2 = 10^{-5} m^{3/2} [(2\Omega_p r_c)^2 \cos\alpha]^{3/2} / 2\sqrt{2}\rho_m. \qquad (2.3.5)$$

The height of the region of deposition of the electrons on the lateral surface of the lamellae h also depends on the value of the angle α, the cyclotron rotation radius and can be determined by geometric constructions for the anode block of a particular geometry. Provided $r_c \ll R - r$ the value $h = \bar{D}\,\mathrm{tg}\,\alpha$ (where $\bar{D}$ is the azimuth size of the gap of the resonator), the angle $\alpha \sim 20°$, the phase of cyclotron rotation of the incident electrons vary from 50 to 220° and angles of incidence $\beta = \alpha + (\pi - p)/2$ (Fig. 2.12) on the surface of the lamellae are in the range from 0 to 85°. From the relation $\mathrm{tg}\,\alpha = \dot{r} / r\dot{\varphi}$, one can determine $\dot{r}$ using (2.2.45), then calculate the high-frequency electric field intensity at the cathode $\tilde{E}$ and calculate the generated power.

For a more accurate determination of the region of electron deposition, it is necessary to take into account the effect on the space charge electrons, the shape of the axial distribution of the microwave field in the anode block, the inconsistency of the tangential component of the microwave field over the width of the spoke for a fixed time and its variation in magnitude during the period of oscillations. Taking into account all the factors listed above is hardly possible even with computer simulation of processes in the interaction space of the RM, therefore it is more reliable to estimate the regions of electron deposition in the wake of the erosion of the lamellae obtained in the high power modes of operation of the relativistic magnetron. Such regimes were realized during experiments on the Tonus-1 HCEA, and an erosion of the angles of the lamellae was fairly uniform throughout the entire length of the anode block. This fact confirms the previously stated assumption about the uniformity of the deposition of electrons along the length of the anode block, and on the other hand, indicates the presence of stressed sections in terms of heat release. Indeed, heat fluxes from electrons incident on the cylindrical and lateral surfaces of the lamellae are summed at the lamella angles. Therefore it is sufficient to analyze the thermal state of the lamellae in the volume with the size $\delta_1 \cdot \delta_2 \cdot L$ to determine the maximum operating conditions of the device. It is noted that at the edges of the lamellae the strength of the static and high-frequency electric fields increase and the anode current density also increases. To reduce this effect, a radius of 0.5 mm was used to round the edges of the lamellae of the anode blocks of the RMs, which is 3% of the width of the lamella. A further increase in the radius of curvature leads to a redistribution of the amplitudes of the spatial harmonics of various types of oscillations in the interaction space and, as a consequence, to a decrease in the electronic efficiency [21].

The total anode current of an RM is divided by the current incident on the lateral surface I_{bA} and the current incident on the cylindrical surface I_{cA}, as $\bar{D}/(D-\bar{D})$, where D is the period of the anode block, $\bar{D}$ is the azimuthal dimension of the resonator gap, and $D-\bar{D}$ is the azimuthal dimension of the lamellae. Knowing the area, depth, distribution along the length and magnitude of the current of the anode current incident on certain parts of the lamellae, one can proceed to determine the thermal state of the isolated volume of the anode block during the pulse, in the pause between pulses and in a series of pulses at different repetition frequencies.

2.3.2. Formulation of the problem of thermal calculation and its solution

For pulsed heating with a duty cycle of more than 10^3, the rise in surface temperature over time is described by expression

$$T(t,\tau)=T_{av}(t)+T_{imp}(\tau), \tag{2.3.6}$$

where $T_{av}(t)$ is the average value of the surface temperature heated by the pulse packet; $T_{imp}(\tau)$ is the temperature jump due to heating by a single pulse.

To determine the impulse component of the surface temperature, it is necessary to take into account the features of the RM operation. First, for an impulse the duration of which is from several hundredths to several tenths of a microsecond, an energy with a density of the order of 10^7–10^8 W/cm^2 is released on the working surface of the RM anode, warming up the latter. For such a short time, the heat does not have time to not only go outside, but even spread throughout the thickness of the anode, causing instantaneous pulse heating. Heating, at which the heat released does not have time to spread over the entire thickness of the anode, is called the heat shock. Secondly, the thermal state of the anode is affected not only by the power density of the incident electron beam, but also by the energy of the electrons with which they reach its surface. In the RM, the electron energy reaches hundreds of kiloelectronvolts, so that the electrons penetrate into the anode material to a depth comparable to the depth of penetration of heat, thereby reducing the temperature of its surface. This phenomenon is accounted for by the correction factor $G(d_e / 2\sqrt{a\tau})$, which is called a function of penetration and depends on the ratio of electron penetration depth d_e and the thermal field $\sqrt{a\tau}$ inside the anode material [22], a is the thermal diffusivity, τ is the pulse duration.

The greatest idealization in solving this thermal problem is that the power density of the electron beam is taken to be uniformly distributed over the working surface of the anode. Thus, the impulse component of the temperature is calculated by the formula for the surface thermal shock taking into account the penetration depth of the electrons:

$$T_{\text{imp}}(\tau) = \frac{2\, P_0 \sqrt{a\tau}}{\lambda_m \sqrt{\pi}} G\left(\frac{d_e}{2\sqrt{a\tau}}\right). \tag{2.3.7}$$

Here P_0 is the surface density of the pulsed power; λ_m is the coefficient of thermal conductivity of the anode material.

The thermal shock is the cause of significant temperature stresses on the working surface of the anode. Fatigue processes in the metal result in the formation of micro- and macrocracks which increase the thermal resistance of the anode surface, and with further development they can lead to its complete destruction. To prevent premature destruction of the anode by an electron beam, the amplitude of the impulse temperature on the surface should not exceed the safe value ΔT_s,

$$T_{\text{imp}}(\tau) < \Delta T_s. \tag{2.3.8}$$

The value of ΔT_s is calculated by the formula [23]:

$$\Delta T_s = \frac{2\sigma_T}{\alpha_t E_m}, \tag{2.3.9}$$

where σ_T is the yield strength; α_t is the coefficient of linear expansion; E_m is the modulus of elasticity of the material.

Figure 2.13 shows the dependence of the surface density of the pulsed power of the electron beam on the pulse duration for copper,

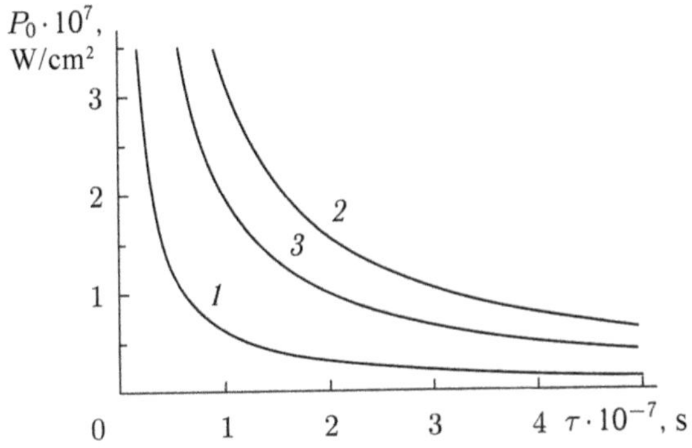

Fig. 2.13. The dependence of the electron beam power density on the pulse duration at $T_{imp}(\tau) = \Delta T_s$: *1* – copper, *2* – molybdenum, *3* – stainless steel.

molybdenum and stainless steel under the condition $T_{imp}(\tau) = \Delta T_s$. The safe temperature values calculated from formula (2.3.9) are 74, 485 and 192°C, respectively.

When working with the beams whose parameters are higher than those shown by the curves in Fig. 2.13, the surface of the anode block of the RM is destroyed in the single-pulse mode.

The component $T_{av}(t)$ in Eq. (2.3.6) can be found by solving the problem of heating an unbounded plate of finite thickness one of whose surfaces is heated by a heat flux whose power is equal to the average pulsed power of the electron beam, while the other one is forced to cool down according to Newton's law. This solution for the heated surface has the form

$$T_{av}(t) = T_0 + \left\{1 + \frac{1}{Bi} - \sum_{n=1}^{\infty} A_n \exp\left(-\mu_n^2 \frac{at}{h^2}\right)\right\} \frac{P_0 h}{\nu \lambda_1}, \qquad (2.3.10)$$

where T_0 is the initial temperature of the anode; t is the packet pulse duration; $Bi = \frac{\alpha}{\lambda} h$ is the Biot criterion which depends on the thickness of the anode and the cooling conditions (for water cooling $\alpha \sim 0.05$–1 W/(cm² · deg), for air $\alpha \sim 10^{-3}$–10^{-2} W/(cm² · deg)); α is the heat transfer coefficient; λ_m is the coefficient of thermal conductivity of the anode material; h is the thickness of the plate; ν is the pulse ratio;

$$A_n = \frac{2\left(\mu_n^2 + Bi^2\right)}{\mu_n^2\left(\mu_n^2 + Bi^2 + Bi\right)};$$

μ_n are the roots of the characteristic equation $\operatorname{ctg}\mu_n = \frac{1}{Bi}\mu_n$. As follows from expression (2.3.10), $T_{av}(t)$ is determined by the average value of the power of the electron flux $\frac{P_0}{\nu}$ and depends on the intensity of the heat sink from the anode working surface. If $0.1 < Bi < 100$, then the heat dissipation rate is determined by the heat transfer both inside the material by means of thermal conductivity and by the transfer of heat from the material surface to the cooling liquid by heat transfer. At $Bi < 0.1$, the heat removal intensity is determined only by heat transfer.

Thus, the equation for calculating the temperature of the anode working surface with a high pulse ratio will be:

$$T(t,\tau) = T_0 + \\ + \left\{1 + \frac{1}{Bi} - \sum_{n=1}^{\infty} A_n \exp\left(-\mu_n^2 \frac{at}{h^2}\right)\right\} \frac{P_0 h}{\nu \lambda_i} + \frac{2 P_0 \sqrt{a\tau}}{\lambda_i \sqrt{\pi}} G\left(\frac{d_e}{2\sqrt{a\tau}}\right). \quad (2.3.11)$$

It is obvious that heating in the pulse-periodic mode is aggravated by the presence of a constant component $T_{av}(t)$, relative to which temperature jumps occur. This component can be reduced by the intensification of cooling and by reducing the thickness of the anode at the sites of deposition of electrons. The temperature value in this case, the excess of which leads to the destruction of the anode surface, can be determined with sufficient accuracy by the melting point T_m [23]:

$$[T] = \frac{1}{3} T_m. \quad (2.3.12)$$

2.3.3. Results of thermal calculations

For the material under consideration, the temperature $[T]$ calculated by formula (2.3.12) is: copper 360°C, molybdenum 875°C and stainless steel 475°C. These are below the temperatures recommended for the use of these materials under continuous heating (copper 500°C [24], molybdenum 1700°C [24], steel 600°C [25]).

Since the impulse component of the temperature, limited by the condition (2.3.12), does not affect the life of the anode, it is obvious that the main contribution to the disruption of the surface is made by the constant component of the temperature $T_{av}(t)$, which depends on the heat sink intensity, as indicated above. In our case at $\alpha = 0.15$

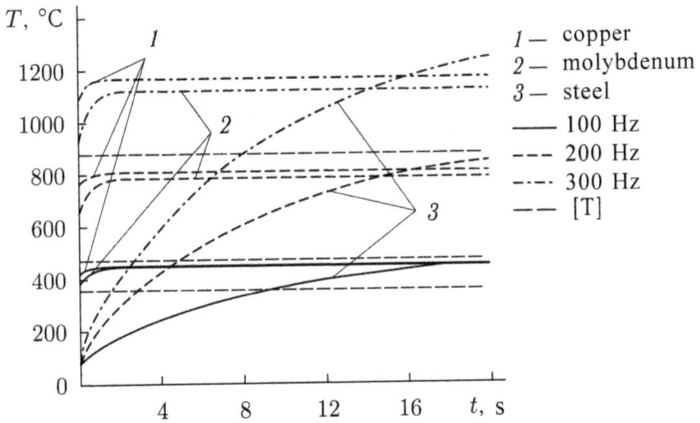

Fig. 2.14. Temperature dependence of the anode surface on the duration of a packet of pulses when $\tau = 10^{-7}$ s and $P_0 = 0.5 \cdot 10^7$ W/cm²: *1* – copper, *2* – molybdenum, *3* – stainless steel. The solid curve is 100 Hz, the dashed curve is 200 Hz, the dot-dashed curve is 300 Hz, the long dashed curve is [*T*].

W/(cm² · deg) the *Bi* criteria calculated for copper, molybdenum and steel criteria are 0.01, 0.03 and 0.3, respectively, and indicate that the fraction of heat, taken from the cooling surface, from the heat supplied through the wall, for steel is several tens of times higher than for copper or molybdenum.

Figure 2.14 shows the dependence of the surface temperature of the anode of the duration of the packet of pulses at the characteristic values for $\tau = 10^{-7}$ s and $p_0 = 0.5 \cdot 10^7$ W/cm² for pulse repetition frequencies of 100, 200 and 300 Hz. It follows from the data given that copper can not be used for pulsed-periodic operation of RM, the use of molybdenum is limited by a repetition rate of slightly more than 200 Hz.

Based on the calculations carried out, and also taking into account the high processability, availability and low cost, the use of stainless steel for the manufacture of anode blocks of the RMs seems to be a fairly successful solution. Figure 2.15 shows the appearance and design of the RM block of the 10-cm wavelength range made of stainless steel.

The presented anode block is designed to work with linear induction accelerators of the Tomsk Polytechnical University. The resonator system *1* of the anode block has a water cooling circuit formed by the outer casing *2*, an outer channel *3*, special grooves *4* made in the body of the lamellae and separated by the regulating

a

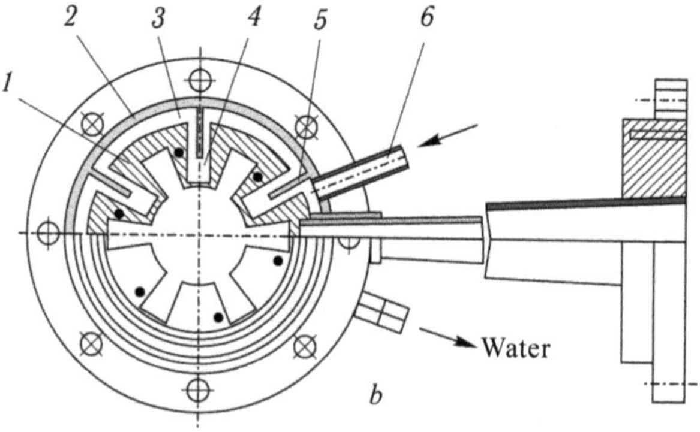

b

Fig. 2.15. Appearance (*a*) and construction (*b*) of the anode block of the 10-cm wavelength range with water cooling from stainless steel.

diaphragms *5*. The water in the cooling circuit is fed through the inlet *6*. When used as a power source LIA 04/6 [26], this anode block allows the generation of pulses of 2.5 s duration with a frequency of 320 Hz (pulse duration of the electron beam 100 ns) in a pulse-periodic mode. The generation interval for packets is about 3 min.

The above calculations of the thermal state of the surface of the anode blocks of relativistic magnetrons are of an evaluative nature. It is known that in a relativistic magnetron a considerable part of the anode current does not reach the surface of the resonator system due to drift in the axial direction. This leads to a decrease in the efficiency of the device, but also to a decrease in the thermal load of the anode. On the other hand, a number of simplifications were introduced in the calculations, such as the rectangular shape of the pulse, the uniform distribution of the power flow, additional heating sources were not taken into account. Also, the depth penetration function of electrons is not accurately determined. Nevertheless, the calculations are in satisfactory agreement with the results of experimental studies on the resource of anode blocks and allow us to work out the following recommendations for the construction of RM anode blocks [27].

Preference should be given to RM anodes with vane-type resonators having a large lateral surface in comparison with slit- or slit–hole resonators. The edges of the lamellae need to be rounded to reduce the strength of the electric fields.

To manufacture RM anode blocks, materials should be selected whose mechanical properties make it possible to withstand sufficiently large jumps in the pulse temperature.

The melting point of the material must be high to ensure that the maximum permissible operating temperatures can be reached.

The thermophysical properties of the material should ensure the value of the Biot criterion in the range 0.1–100.

It is necessary to apply cooling of the anode block lamellae. Cooling should be carried as close as possible to the most heat-loaded parts of the anode, so the thickness of the lamellae in the region of electron beam deposition should be minimum possible.

The material should be accessible and technologically.

All of the above requirements are best satisfied by stainless steel, although it is not a traditional material for microwave devices.

To obtain high average microwave power, it is advisable to use power supplies with an increased voltage pulse duration, since the rise in the surface temperature of the anode block is proportional to the square root of the duration. However, in this case it is necessary to take the original RM designs, for example, described the one in Chapter 5 of this book.

For the devices operating in the pulse-periodic mode, it is permissible to use anode blocks with a low slowing-down of the electromagnetic wave (in spite of a decrease in the electronic efficiency). This leads to the appearance of high values of the energy of cyclotron rotation of the electrons. As a consequence, the electrons penetrate deep into the metal by a large amount, reducing the temperature of pulsed heating.

The chapter presents the theory of a magnetron of planar and cylindrical geometries without taking into account the space charge fields for the relativistic stress range. It is shown that in the RM the effect on the electrons of the static electric and magnetic fields is weakened γ_p times, and that of the high-frequency electric field γ_p^3 times. Trajectories of the motion of electrons in the interaction space are constructed for different emission phases from the cathode surface and as a function of the ratio between the static and high-frequency electric field strengths. Relations are obtained that allow calculating the output parameters of the device and estimating the current consumed by the device. Accounting relativity shows that in the relativistic magnetron compared to the non-relativistic electron efficiency analog value decreases approximately γ_p times, and the magnitude of the anode current falls approximately γ_p^3 times. A relation has been derived for calculating the limiting value of the additional acceleration caused by the space charge at which microwave generation is conserved, and the magnitude of the limiting current of the relativistic magnetron.

The thermal processes on the surface of the anode blocks of the RMs were calculated and it is shown that for the operation of the instrument without breaking the anode block in the single-pulse mode it is necessary to reduce the specific thermal load to values of 10^8 W/cm^2 at a pulse length of less than 10^{-7} s and 10^7 W/cm^2 for a duration of $\sim 10^{-6}$ s. The pulse-periodic operation of the device is possible when the electron beam energy density is not greater than 10^7 W/cm^2, a duration of 10^{-7} s, with a frequency of not more than 100 Hz and is limited by high values of both average and impulse temperature. Exceeding any of the above parameters causes the packet mode to operate with a limited number of pulses.

An experimentally observed phenomenon of the electron frequency shift of an RM is described and explained. Moreover, in the region of high EFS coefficients, the increase in the generated frequency with increasing cathode–anode voltage is associated with a decrease in the capacitive effect produced by a bunch of electrons at the cavity gap. In this case, increasing high frequency voltage to the resonator increases the phase focusing force tending to reduce the maximum shift angle between the microwave field and the position of the electronic spokes Θ_1. As the voltage increases and the angle Θ_1 decreases, the influence of the electron cloud charge on the generated frequency increases, which reduces the value of the EFS coefficient.

References

1. Kapitsa P.L., High-power electronics, Moscow: Publishing House of the USSR Academy of Sciences, 1962.
2. Nechaev V.E., Izv. VUZ, Radiofizika, 1962. Vol. 5. No. 3. P. 534–548.
3. Nechaev V.E., *ibid,* 1964. V. 7. No. 1. P. 146–159.
4. Brillouin L., Phys. Rev. Lett. 1942. V. 42. P. 166.
5. Lovelace R.V., Ott E., Phys. Fluids. 1974. V. 17. P. 1263.
6. Orzechowski T.J., Bekefi G., Phys. Fluids. 1979. V. 22. No. 5. P. 979–985.
7. Bekefi G., Orzechovski T., Phys. Rev. Lett. 1976. V. 37. No. 6. P. 379-382.
8. Langmuir I., Phys. Rev. 1921. V. 21. P. 419.
9. Palevsky A., Bekefi G., Drobot A.T., J. Appl. Phys. 1981. V. 52. No. 8. P. 4938–4941.

10. Chan H.-W., Chen C., Davidson R.C., Appl. Phys. Lett. 1990. V. 57. No. 12. P. 1271–1273.
11. Vainshtein L.A., Solntsev V.A., Lectures on ultrahigh-frequency electronics, Moscow, Sov. radio, 1973.
12. Groshkov L.M., Nechaev V.E., Izv. VUZ, Radiofizika, 1965. V. 8. No. 3. P. 413–416.
13. Nechaev V.E., et al., Pis'ma v Zh. Teor. Fiz. 1977. V. 3. No. 15. P. 763–767.
14. Fuchs M.I., Investigation of electron currents in systems with magnetic insulation: Dissertation, Gor'kii, 1983.
15. Palevsky A., Bekefi G., Phys. Fluids. 1979. V. 22. P. 926.
16. Palevsky A., et al., in: Proc. of the 3 Int. Top. Conf. On High-power Electron and Ion Beam Research and Technology. Novosibirsk, 1979. V. 2. P. 759.
17. Palevsky A., et al., IEEE Int. Conf. On Plasma Science. Montreal, Canada, 1979. P. 43.
18. Bychkov S.I., Questions of the theory and practical application of devices of the magnetron type, Moscow, Sov. radio, 1967.
19. Samsonov D.E. Basics of calculating and constructing multiresonator magnetrons, Moscow: Sov. radio, 1966.
20. Kovalenko V.F., Elektronnaya tekhnika. Ser. 1, Elektronija SVCh, 1972. No. 1. P. 3–11.
21. Shlifer E.D., Calculation of multi-cavity magnetrons, Moscow, MEI, 1966.
22. Khmara A.V., Elektronnaya tekhnika. Ser. 1, 1971. Issue 1. P. 77–82.
23. Kovalenko V.F., Thermophysical processes and electrovacuum devices, Moscow, Sov. radio, 1975.
24. Properties and application of metals and alloys for electrovacuum devices: A reference book, ed. R.A. Nylender, Moscow, Energiya, 1973.
25. Anuriev V.I., Handbook of the designer-machine builder: in 3 volumes. V. 1. 6th ed., Moscow, Mashinostroenie, 1982.
26. Butakov LD,, et al., Pis'ma v Zh. Teor. Fiz. 2000. V. 25. No. 13. P. 66–71.
27. Vintizenko I.I., Pis'ma v Zh. Tekh. Fiz. 2012. V. 38. No. 12. P. 37–44.

3

Power Sources and Components of Experimental Installations

In the work [1] devoted to the development of science, Academician M.A. Markov noted that along with the continuous development of high-energy physics it is necessary to develop the physics of beams of relatively low energy but high intensity. Moreover, such beams can have almost unlimited possibilities of practical application in engineering, medicine, and the national economy. In particular, high-current relativistic electron beams can be used to heat the plasma to the thermonuclear temperature, to develop new methods for collective ion acceleration, to study phase transformations in a solid and the properties of materials, and to generate and amplify microwave radiation. For these purposes, electron beams should have the following parameters: electron energy 0.3–30 MeV, beam current 1–100 kA, pulse duration 10–500 ns. Relativistic high-frequency electronics deals with high-current electron beams with the above parameters. Such beams are formed by cold explosive-emission cathodes, for the creation of which high electric field strengths of hundreds and more kilovolts per centimeter are required.

To produce such beams, high-current electron accelerators are used. The development of these systems began in the 1970s of the last century. The first HCEA worked exclusively in the single-pulse mode – one 'shot' for several minutes. In addition, they had low efficiency. However, as studies began to move into the stage of practical application, high-efficiency electron accelerators began to demand high efficiency, high reproducibility of the parameters, the ability to work with the repetition rate of pulses.

3.1. High-current electron accelerators

In the first experiments and up to the present time, high-current electron accelerators are used to power the RMs [2–4]. Such an accelerator contains a pulse voltage generator (PVG), which is discharged directly from the magnetron diode or through the forming line (FL) to the diode. A block diagram of an installation based on a high-current electron accelerator of nanosecond duration is shown in Fig. 3.1. The HCEA consists of a power source (PS), designed to charge the capacitors of the PVG.

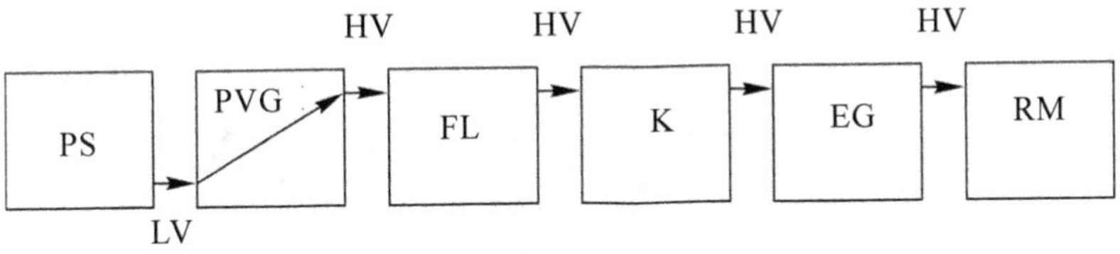

Fig. 3.1. Block diagram of HCEA.

The discharge of the PVG capacitors connected in series allows increasing the voltage from the low (LV) value set by the PS to the high (HV) value of the FL charge. Switching on switch K (usually a gas spark gap) connects the FL via an electron gun (EG) to a load, for example, a relativistic magnetron in our case. The electron gun is designed to separate the oil-filled volume of the accelerator and the vacuum volume of the microwave generator and transfer the high voltage of the FL to the load. As can be seen from Fig. 3.1, FL and K are under high voltage of PVG.

The wave resistance of FL of various HCEAs is from 2.8 to 40 Ohm. The range of voltages is 0.3–2 MV, respectively, the pulse power of the accelerator can reach 60 GW. Naturally, such installations can operate exclusively in a single-pulse mode (single pulse for several minutes). Reducing the output parameters allows realizing a pulse-periodic mode of operation of the installation with a repetition frequency of 2–4 Hz and parameters: voltage 0,3–1 MV, current 17–30 kA.

The direct discharge of the PVG to the load makes it possible to generate voltage and current pulses of microsecond duration (1–1.5 μs) with the following parameters: voltage 0.25–0.5 MV, current 4–8.5 kA (Fig. 3.2).

Since the designs of the HCEA are of the same type, we shall briefly consider them on the basis of the 'Tonus-1' [4] and 'LUCH'

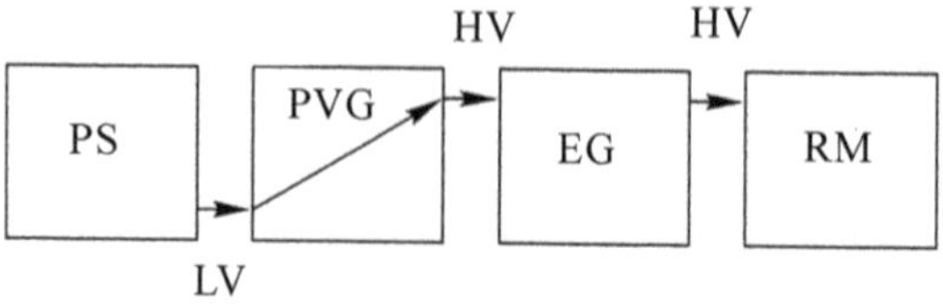

Fig. 3.2. Block diagram of the HCEA of microsecond duration.

Fig. 3.3. External view of the Tonus-1 HCEA with a relativistic magnetron.

[5] devices created in the Tomsk Polytechnic University for various experiments, including relativistic magnetrons. These accelerators were one of the first in the USSR and for their time had record characteristics. Initial RM studies were carried out on a high-current electron accelerator Tonus-1. The accelerator contained an air pulse voltage generator, a coaxial double shaping line with a wave resistance of 24 Ohm, filled with transformer oil. Later, the accelerator was upgraded: the DFL charge device was made on the basis of the Tesla transformer. HCEA 'Tonus-1' makes it possible to form electron beams with energy up to 1.2 MeV, current up to 50 kA and duration 60 ns. An external vies of the HCEA 'Tonus-1' with a relativistic magnetron loaded on a microwave calorimeter is shown in Fig.3.3. In the background, an air pulse voltage generator is seen.

Relativistic magnetrons in the microsecond voltage pulse duration were investigated on a high-current electron accelerator LUCH. The installation includes a PVG that is connected via a high-voltage switch directly to the electron gun to generate microsecond electron beams (electron energy up to 500 keV at a current of ~10 kA) or to the forming line for obtaining nanosecond beams (electron energy up

Fig. 3.4. External view of the LUCH accelerator complex.

to 1 MeV at a current of ~100 kA with the duration of 100 ns). The appearance of the accelerator complex is shown in Fig. 3.4. At the top of the structure there is a pulse voltage generator, on the right below – the DFL case, to the left – an electron gun for the formation of microsecond electron beams.

The primary energy storage device is a pulse voltage generator, assembled according to the Arkadiev–Marks scheme. In order to reduce the volume and level of electromagnetic interference, the PVG is placed in a metal tank measuring 1.8 × 2.6 × 4.2 m, filled with transformer oil. To reduce the number of spark gaps and, consequently, to reduce losses and internal resistance of the PVG, bipolar charging of capacitor stages was used. Each of the 16 stages consists of three parallel-connected capacitors (100 kV, 0.4 μF). Three-electrode gas dischargers are used as commutators. The first discharger of the trigatron type is ignited by a voltage pulse with an amplitude of 30 kV from an external generator. Steady start of the PGV is ensured at a voltage less than the breakdown voltage by no more than 30%. Charging and decoupling resistances of the generator are made of vinyl plastic tubes with a diameter of 50 mm, filled with an aqueous solution of KCl, the value of each resistance is 10 kOhm. The PVG is charged with sources of constant voltage.

Here are the characteristics of the PVG:

- shock capacitance $8 \cdot 10^{-8}$ F;
- the stored energy at a charging voltage of 60 kV 34.6 kJ;
- maximum voltage ~1 MV;
- inductance ~17 μH;
- internal resistance ~16 Ohm.

Electron guns for the formation of beams contain vacuum coaxial diodes and various solenoids with a power system. The vacuum diode is a sectioned cylindrical insulator made of plexiglass, inside of which is placed a high-voltage electrode of smaller diameter. For a more even distribution of the potential on the surface of the insulator, active division of the voltage is used by means of resistances located between the gradient rings. The electron gun is placed in a steel tank with a diameter of 1.2 m and a length of 1.5 m, filled with transformer oil.

Nanosecond voltage pulses are generated according to the following scheme: double forming line with a commutator – a separating discharger – a transmission line – a vacuum diode. Deionized water is used as a dielectric.

The magnetic field is formed in RMs by the pulsed Helmholtz pairs and solenoids of different design and dimensions. The magnetic field in them is created by the discharge current of the capacitor bank. The automation unit provides charging and maintaining the voltage on the PVG and the capacitor bank at a given level and their synchronized start.

The parameters of the accelerator and the beam were measured by a standard technique: capacitive and active voltage dividers, reverse current shunts, Rogowski coils and Faraday cylinders.

The accelerators described above make it possible to form electron beams in a single mode (approximately one pulse per minute).

3.2. Linear induction accelerators

Linear induction accelerators (LIA), like HCEA, can generate electron beams of considerable energy, in which currents reach unity and tens of kiloamperes. However, LIAs have certain advantages over the accelerators of other types: relatively low voltage on the structural elements, not exceeding the excitation voltage of one inductor;

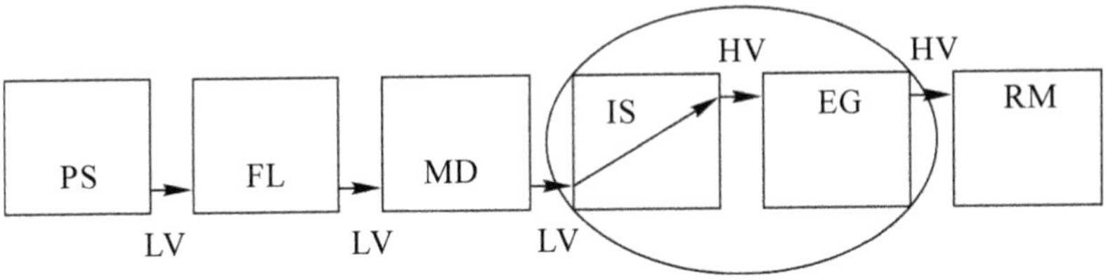

Fig. 3.5. LIA block diagram.

acceleration rate up to 3 MeV/m; high efficiency of conversion of electrical energy into beam energy; the kinetic energy of the electrons is increased by using the same type of accelerating modules. Another extremely important feature of the LIA for the supply of RMs is that when working on a mismatched load, the energy of repeated pulses is scattered not in the magnetron diode, but is spent on the magnetization reversal of the cores of the induction system.

The block diagram of the LIA is shown in Fig. 3.5; it includes the following elements: a power source for charge of the forming lines (FL), a multichannel discharger (MD) connecting the FL to an induction system (IS) increasing the voltage from a low voltage (LV)) to high (HV) values. Note that the induction system can be structurally integrated with the electron gun.

The principle of the induction system of LIA is that in a system of a series of successively installed toroidal ferromagnetic cores a magnetic flux that varies in time is excited, which creates a vortex electric field accelerating the electrons. The electric field strength on the axis of the induction system is defined as [6]

$$E(t) = N_c U(t)/L_{\text{in}}. \tag{3.2.1}$$

where N_c is the number of ferromagnetic cores; $U(t)$ is the excitation voltage of the core; L_{in} is the length of the induction system. Thus, the energy acquired by the electrons corresponds to the sum of the voltages on the primary windings of the inductors, and for a sufficiently short system length the rate of acceleration can reach high values.

When the cores are excited with rectangular pulses $U(t) = U_0$, the increment of the kinetic energy of the beam with current I_n is

$$\Delta P = \int_0^{\tau} I_b(t) U(t) N_c dt = I_b S \Delta B N_c, \tag{3.2.2}$$

where S is the cross-section of the ferromagnetic core; ΔB is the induction increment the core, τ is the pulse duration. It can be seen from the formula that the energy is proportional to the induction increment. To fully utilize the core material, it is necessary to transfer the ferromagnetic core to the region of negative saturation before applying the working voltage pulse to the magnetizing coil, i.e. carry out the process of demagnetization (magnetization reversal).

Since the value of ΔB in steel 50 NP is not more than 3 T, the increase of S is limited by the increase of the losses in reversal of the core, so the increase of the energy of the beam can take place by the accelerated current or electron kinetic energy by selecting a large number of cores N_c.

Table 3.1 shows the parameters, applications and some design features of linear induction accelerators developed by various organizations of the USSR, Russia, the USA and other countries. One can note a wide range of changes in the energy of electron beams and current, the pulse durations formed by LIA. In this connection, a wide range of accelerator applications is observed.

3.2.1. Layout scheme of the LIA developed at the Tomsk Polytechnic University

LIAs are developed in various scientific centres of Russia, the United States, Japan, France, and others (Table 3.1). An important contribution to the development of accelerator technology has been made by specialists from the Tomsk Polytechnic University. To create accelerators, the original element base was used: low-impedance strip forming lines [7]; multi-channel spark dischargers with forced current common between channels [8]; non-linear saturation chokes [9].

There are injector modules of the LIA intended for feeding a high voltage pulse to the cathode of the electron gun (injector modules) and accelerating modules in which the electrons are accelerated. Figure 3.6 shows the construction of the injector modules, which combine a ferromagnetic induction system, low impedance forming lines, a commutator, and a system for demagnetizing the cores in a common case. On the axis of the induction system of the injector module is a high-voltage electrode connected to the cathode. The accelerating module differs from the injector by the presence of a vacuum tract. The accelerator tracts are made of a dielectric tube with a weakly conducting coating, which serves to drain the charge of electrons that enter the inner surface and equalize the potential along the length of the tract. A solenoid is laid on the top of the tube, one terminal of which is connected to the case and the other to the source. The solenoid current demagnetizes the cores of the induction system and creates a longitudinal focusing magnetic field to transport the electron beam.

The ferromagnetic cores *4* are covered by magnetizing coils *5*, which are connected to the terminals of the electrodes *6* of the forming line, made in the form of a strip-double (DFL) or a single (SFL) forming line. The DFL consists of four plates, laid around along the Archimedes spiral around the cores. Figure 3.6 (top) shows a system of six parallel-connected DFLs. The potential plates of each line are connected to anodes S_0 of a multi-channel discharger. The ground and free plates are connected to the magnetizing turns of the cores, and the ground electrodes are grounded to the accelerator case by the other end, i.e. are connected to the cathode *2* of the discharger.

Depending on the required value of the internal wave resistance of the LIA, 4 to 12 parallel-connected DPLs are used. The system of arrangement of the SFL electrodes consists of several pairs of potential and ground electrodes (Fig. 3.6, below).

The pulse duration τ, the stored energy Q, the wave impedance Z of the strip DFL, the beam current I_b for N_c cores at the maximum voltage U_0 of the charge of the DFL, are related to the size of the line and the insulation thickness d_i and between the electrodes by the following dependences:

$$\tau = 2l\sqrt{\varepsilon}\,/\,c; \tag{3.2.3}$$

$$Q = 2\varepsilon_0 \varepsilon U_0^2 m h_e l_e \,/\, d_i; \tag{3.2.4}$$

$$I_n = K_{en} Q / U_0 N_c \tau; \tag{3.2.5}$$

$$Z = 400\, N_c^2 d_i \big/ K_{en} m h_e \sqrt{\varepsilon}\,, \tag{3.2.6}$$

where ε_0 is the absolute permittivity of vacuum; ε is the relative permittivity of insulation; m is the number of lines connected in parallel; K_{en} is the coefficient of transformation of the energy of the DFL into the kinetic energy of the electrons; h_e, l_e is the width and length of the electrodes.

The efficiency of LIA K_{LIA} is mainly determined by the ratio of the magnetizing current I_μ of the cores and the beam current:

$$K_{\text{LIA}} \approx I_b \big/ (I_b + I_\mu). \tag{3.2.7}$$

The linear size of the section depends on the number of cores N_c,

Table 3.1 Parameters of LIA (electron energy *E*, beam current *I*, pulse duration *τ*, pulse repetition frequency *F*), applications and constructive accelerator features (*FEL = Free-electron laser)

LIA, Organization	*E*, MeV	*I*, kA	*τ*, ns	*F*, Hz	Features
ASTRON Lawrence Livermore National Laboratory, USA	4.2	0.8	300	60	The energy spread is less than 2%. Emittance 25 cm · mrad
ETA-2 Lawrence Livermore National Laboratory, USA	6-7	2	50	5000	For FEL* 140 GHz for heating the plasma in a tokamak. Injector 1.5 MeV and 60 accelerating inductors
ATA Lawrence Livermore National Laboratory, USA	47.5	10	70	1000 to 10 imp.	For FEL 140 GHz for plasma heating in a tokamak
ERA Berkeley, USA	4.25	0.5	45	1	The energy spread is less than 0.5%. Emittance 70 cm · mrad
LIA 3000 JINR, Dubna, NIIEFA, Leningrad, the USSR	3	0.2	350	25	
LIA 30/250 JINR, Dubna, NIIEFA, Leningrad, the USSR	30	0.25	500	50	
LIA-30 IEF, Arzamas-16, the USSR	40	100	20		LIA on radial lines; 288 lines with water insulation, 2432 trigatron for switching lines
RHEPP-1 RHEPP-2 Sandia National Laboratories, Albuquerque, USA	1 2.5	25 25	60	120	The average power of the electron beam is 100 kW The average power of the electron beam is 300 kW
LAX-1 National Laboratory of KEK, Japan	2	2	120		For FEL. Forming lines with adjustable capacity for the motion of dielectric plates between the electrodes of lines

Table 3.1 Completion

LIA, Organization	*E*, MeV	*I*, kA	τ, ns	*F*, Hz	Features
AIRIX France	20 – project	0.35	60		For radiographic measurement. LIA includes 4-MeV injector, 8 accelerating modules of 8 inductors and the power system from 32 PVG with 250 kV output voltage
I-3000 VNIIEF, Russia	3	10	16		'Non-iron' LIA
LIA for industry application, China	5 – project	0.2	1000 100		Mobile LIA for radiotechnology investigations.
SNOMAD-1 SNOMAD-4 Science Research Laboratory, USA	0.6 1.5	0.6 0.6	60 50	5000 5000	For a vapor laser Thermionic cathode. Average power of the electron beam 500 kW
LELIA Center d'Etudes Scientifiques et techniques d'Aquitaine Comissariat Energie Atomique, France	3 – project 2.1	3 2.5	50	1000	For FEL. Dispenser cathode from osmium
Compact LIA Physics International Company, Olin Corporation Aerospace Division, San Leandro, United States	0.75	10	60	200	For relativistic magnetrons and relativistic klystrons.
3.4 MeV LIA China Academy of Engineering Physics, Chengdu, China	3.4	2	90		For FEL. Injector from 4 cores, 8 accelerated sections of 0.3 MeV. Power supply - two PVGs, 12 water DFL

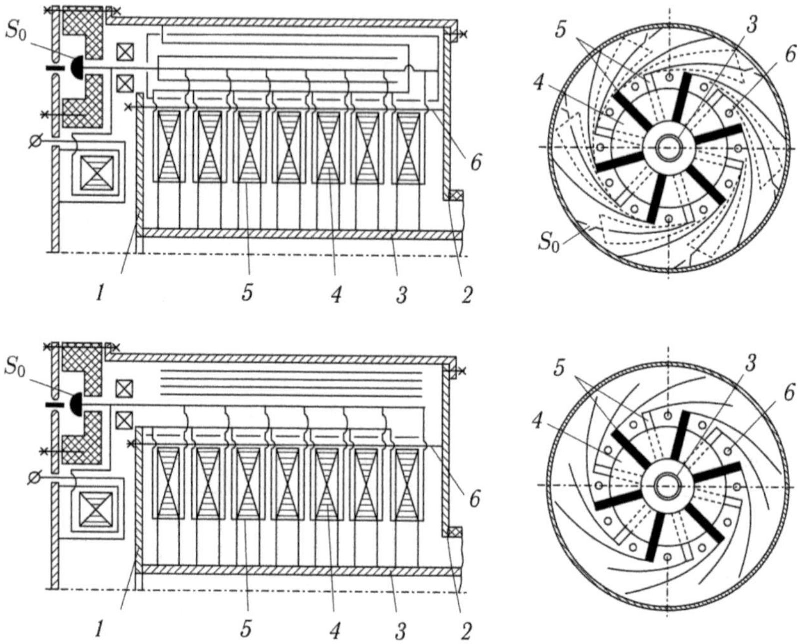

Fig. 3.6. The construction of the injector modules with double forming lines (top), with single forming lines (bottom): *1* , *2* – flanges of the induction system; *3* – high-voltage electrode; *4* – cores of the induction system; *5* – magnetizing turns; *6* – electrodes of strip forming lines; S_0 – multichannel spark discharger.

while the cross section of the steel of one core must be at least

$$S \geqslant U_0 \tau / \Delta B. \tag{3.2.8}$$

The outer diameter of the winding of the DFL is defined as

$$D_m = \left[D_c^2 + 16(\delta_e + d_i) ml / \pi \right]^{1/2}, \tag{3.2.9}$$

where D_c is the outer diameter of the induction system; δ_e is the thickness of the electrodes of the DFL or SFL.

The above relations (3.2.3)–(3.2.9) allow us to evaluate the output parameters (U_0, I_b, τ) and the geometric dimensions of the LIA.

The main disadvantage of the strip DFL (SFL) is an increase in the electric field strength at the edges of the electrodes. The influence of the edge effect can be reduced by applying a weakly conducting layer with a volume resistance along the contour of all the electrodes, which is 10^4–10^5 times smaller than that of the line insulation

[7]. The electric field strength in this case is determined not by the configuration of the edge of the electrode but by the voltage drop along the layer. The material with a bulk resistance of $\sim 10^6$ Ohm · m at a temperature of 20°C, $\varepsilon_c = 4.5$, possessing good adhesion to metals, as well as the dimensions of a weakly conducting layer, was calculated experimentally, and a technique for its application was developed.

3.2.2. Commutators of the forming lines of LIA

One of the most important elements of the LIAs is the commutator of the forming lines, which determines to a large extent the rate of current rise, the amplitude and duration of output pulses on the load, the operating life, and the limiting frequency of repetition of pulses. For switching of lines one requires to switch to a current of 60–420 kA with a duration of 10^{-7} s from having an inductance of not more than 10^{-8} H. For the pulse–perodic regime, this can be done in two ways: by using multichannel spark switching or magnetic commutation based on saturation chokes. Figure 3.7 shows the circuitry of connecting the discharger and the magnetic switch to the accelerator forming line. Located at the end of the section, the switches have almost the same dimensions and successfully fit into its design, forming a low-inductance connection with the line.

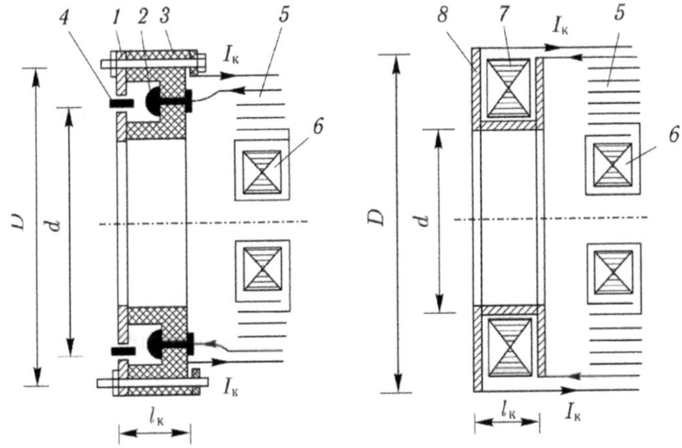

Fig. 3.7. Schemes for connecting a multichannel spark discharger (on the left), magnetic commutator (on the right) to the electrodes of the forming lines: *1* – 'ground' electrode of the discharger; *2* – anodes; *3* – discharger isolator; *4* – discharger cathodes; *5* – electrodes of forming lines; *6* – ferromagnetic cores of the induction system; *7* – magnetic commutator; *8* – magnetizing turn of the magnetic commutator; I_k – current through commutators.

3.2.3. LIAs with multichannel dischargers

The production of kiloampere electron beams in the LIA is possible only if a multichannel commutator is used when all channels are synchronously switched on and the current is divided evenly between the channels.

Indeed, the pulse rise time of at switching of the DFL is determined by the switching time t_{com}, the discharge circuit inductance L_d and the line characteristic impedance ρ_l:

$$t_{fr} = t_{com} + 2.2L_d/\rho_l. \tag{3.2.10}$$

In the case of a multi-channel switching both t_{com} decreases by reducing the current flowing through the channel, and also the value L_d due to the parallel operation of many spark gaps. When choosing the number of channels it must be assumed that the charge commutated by a single channel does not exceed $8 \cdot 10^{-4}$ K. This charge is the threshold value for the operation of the discharger without erosion of the anode material [8]. As a rule, the discharger simultaneously shorts the forming line and the charging power source. Therefore, in order to exclude the flow of additional current through the dischargers, it is necessary to commutate at the end of the line charge. These requirements are met by the low-inductance ring multichannel dischargers developed at the TPU. They have an upper limit on the frequency of operation on the order of 50 Hz in continuous mode and up to 200 Hz in the short packet pulse mode, determined by the recovery time of the electric strength of the gap [10].

Figure 3.8 shows the functional circuit of a multichannel discharger in the expanded form. Structurally, it is located at the end of the accelerator. Electrode *1* and cathode *2* are connected with strip DFL, and spark gaps formed by anodes *4*, starting electrodes *3* with common cathode *2*, are placed in a toroidal dielectric chamber. The electrodes *3* are introduced deep into the spark gap of length h by an amount Δ. The anodes *4* are covered by ferrite rings *5* with a shorted turn *6*.

Let us briefly consider the principle of the discharger, for more details see [8]. At the time when the charging current of the DFL becomes zero the saturation three-winding choke connected in series with the secondary winding of the pulse transformer *PTR* remagnetized because diode *D2* is locked and a high-voltage pulse for switching on the starting discharger *P* is generated. (It should be noted that in order to simplify the circuit of the starting discharger it is possible to turn it on from an external pulse generator. The output pulse of the generator is formed with an adjustable delay, which makes it possible to adjust the switching on of the starting discharger at the end of the charge of the LIA forming lines.)

When the starting discharger is switched on, the voltage across the resistors *R* and therefore on the starting electrodes *3*, is inverted. Initially, the development of the discharge in the spark gaps of the anode *4* – starting electrode *3* begins. With a further change in the voltage at the starting electrode, when the voltage becomes larger than the breakdown voltage: the starting electrode *3* – cathode *2*, there is a breakdown of the latter. Since the formation time of the breakdown is proportional to the length of the spark gap, then, by providing the appropriate speed of voltage inverting on the start electrodes, it is possible to arrange uniform switching of the channels. However, the breakdown of even the same spark gaps is of a statistical nature. With the advanced switching of one of the channels, the conditions for starting the nearest channels deteriorate since a zero voltage wave propagates between the common electrodes *1* and *2* of the spark gap. To prevent shunting of the interelectrode capacitances of the multichannel discharger, equalize the currents

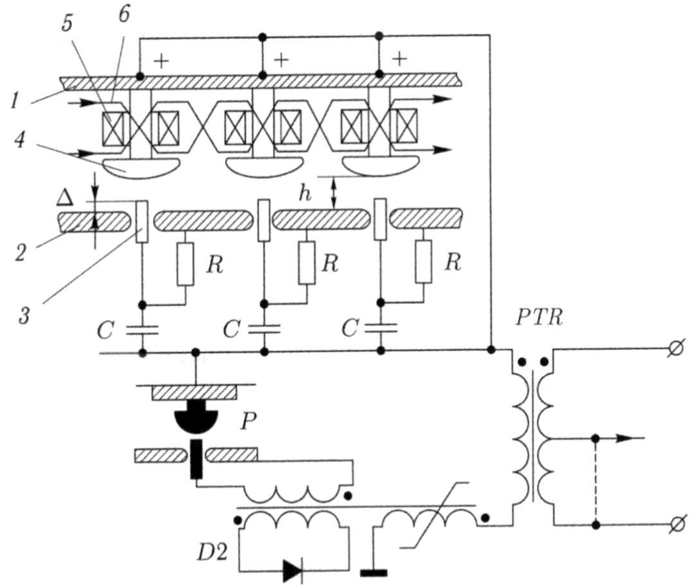

Fig. 3.8. Design of a multichannel discharger: *1* – high-voltage electrode; *2* – cathode; *3* – starting electrodes; *4* – anodes; *5* – ferrite rings; *6* – short circuited turn.

in the individual channels, and reduce the spread of the switching time it is necessary to use an anode divider with the common short-circuited coil *6*. In order to reduce inductance, the coil is made in the form of two counter-parallel turns. With the advancing activation of a number of spark gaps, the current in them is limited at the level of the magnetization reversal of the ferrite core *5*. In this case, emf is induced in the sections of short-circuited turns encompassing the switched-on channels, applied to the sections of turns that cover the channels that have not been turned on. This leads to an additional charge of the interelectrode capacitances of the non-activated channels and their switching on.

A computer simulation program has been developed to optimize such quantities as h and Δ, the values of the elements of the starting circuit R, C, the construction of the three-winding choke, and the electrode profiles determining the work of the discharger, The operational experience of accelerators based on the optimized multichannel dischargers showed high reliability and long service life (> 10^6 activations) of the dischargers with a number of channels from 12 to 24.

Power supplies for LIA with multichannel dischargers. Simple and reliable power circuits have been developed to charge the LIA forming lines with spark gaps that can simultaneously generate a current in the magnetic system of the microwave device. Power circuits use capacitive storage devices based on capacitors, a primary commutator–ignitron or a block of thyristors, a pulse transformer installed in a separate case or located inside a section, as well as a charge circuit of the storage (charging choke, mains transformer, rectifier). In the most efficient model of the LIA with a pulsed electron beam power of ~4 GW, the dimensions of the installation are as follows: diameter 700 mm, length 900 mm, weight 1000 kg [11]. Such accelerators were used in experiments on the generation of microwave radiation by relativistic magnetrons, and for the first time the possibility of a pulse–periodic operation of such a microwave generator was demonstrated. However, the repetition rate of pulses of the LIA with a multichannel discharger in the periodic mode is limited to 50 Hz. Higher frequencies (up to 200 Hz) use the packet mode of the accelerator with the number of pulses not exceeding 5. At the same time, the results of the thermal calculation of the heating temperature of the anode block show that the parameters of the electron beams formed by the LIA enable the RM to work at much higher repetition rates. Therefore, LIAs with magnetic commutation of forming lines are of particular interest.

3.2.4. LIAs on magnetic elements

A switch that is capable of commutating a current of hundreds of kiloamperes with a frequency of almost a kilohertz, with practically unlimited resources, is a magnetic commutator (MC), which is a saturation choke.

To implement such a switch with minimal dimensions and, correspondingly, with a minimum inductance, it is required to charge the forming lines in a time of hundreds of nanoseconds from magnetic pulse generators (MPGs) [12–14]. A block diagram of such an installation is shown in Fig. 3.9.

The MPG is a sequence of *LC*-circuits with increasing natural frequency. The circuit is formed by a capacitor and a saturation choke with a core of a ferromagnetic material. Such systems have reliability, high efficiency of energy compression, the possibility of generating pulses of high power. The MPGs have been used in a significant number of LIAs, primarily in the USSR, Russia, including the TPU [15–17], and in the USA [18, 19].

Layout and electrical circuits of LIA on magnetic elements and the principle of operation. The layout and principal electrical diagram of the accelerator LIA 04/4000 [17] are shown in Fig. 3.10. In a cylindrical case with a diameter of 700 mm and a length of 1600 mm, there are: an induction system of fifteen cores, a magnetic commutator L_1, which is a single-turn saturation choke, and an MPG.

Over the cores of the induction system there are the electrodes of the strip single-forming line C_1 of capacitance C_1 and the capacitor C_2 of the last MPG compression link of C_2 capacitance (made using the single-forming line technology). The high-voltage electrodes C_1 and C_2 are common and connected to the last stage of MPG compression. The other two electrodes are interconnected by the winding of the magnetic switch L_1. A high-voltage electrode and

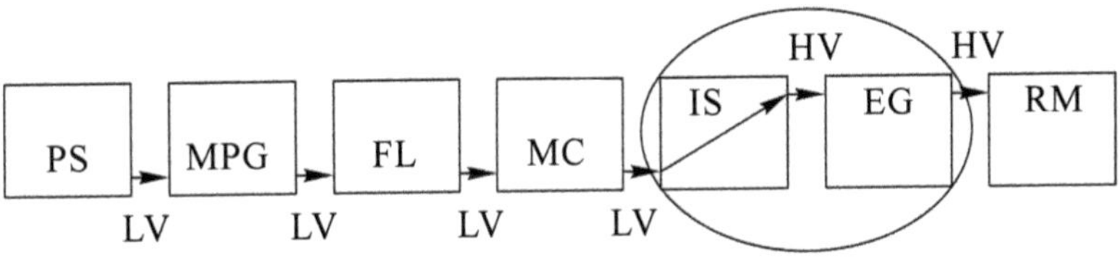

Fig. 3.9. Block diagram of LIA on magnetic elements.

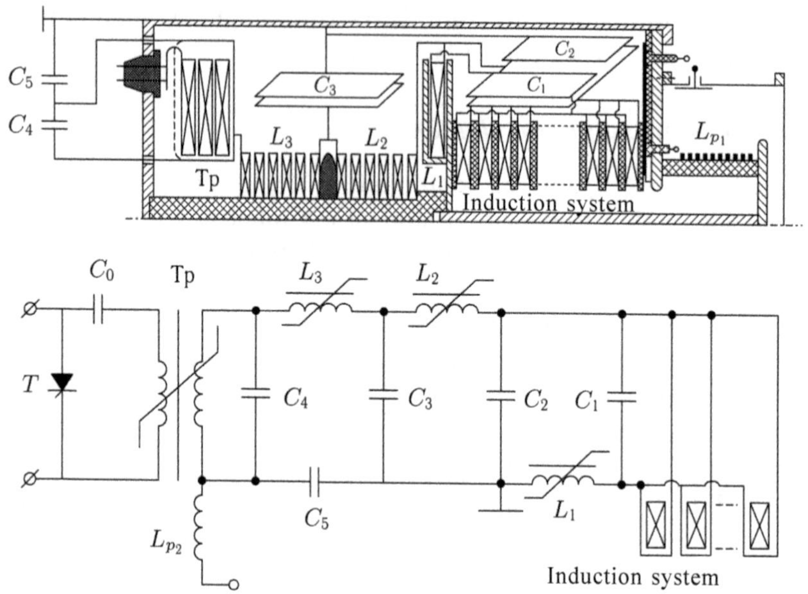

Fig. 3.10. Composition and principal electrical diagram of the accelerator on magnetic elements of the LIA.

a cylindrical insulator are installed along the axis of the induction system of the module. Thus, most of the elements of the LIA are located in a common case, which makes it possible to reduce the inductance of the connection of the elements, to increase reliability, and to reduce the weight-and-size parameters.

The magnetic pulse generator has three stages (links) of compression on the saturation chokes L_3, L_2, pulse transformer Tp and capacitors C_3, C_4, C_5. To reduce the inductance, the capacitor C_3 is also based on the SFL technology. The high-voltage capacitors C_4, C_5 are installed outside the case. The pulse transformer Tp simultaneously performs two functions: it increases the voltage, and at saturation it ensures the recharge of the capacitor C_4 through the secondary winding. The cores of the MPG saturation chokes were made of a tape of permalloy 50 NP with a thickness of 0.02 mm, for commutator cores and induction system–permalloy 50 NP by thickness 0.01 mm.

The induction system is demagnetized through the single-layer inductance L_{p1}, connected to the primary winding of the pulse transformer, and additionally by the charge current of the capacitors of the first stage of the MPG compression. The core of the pulse

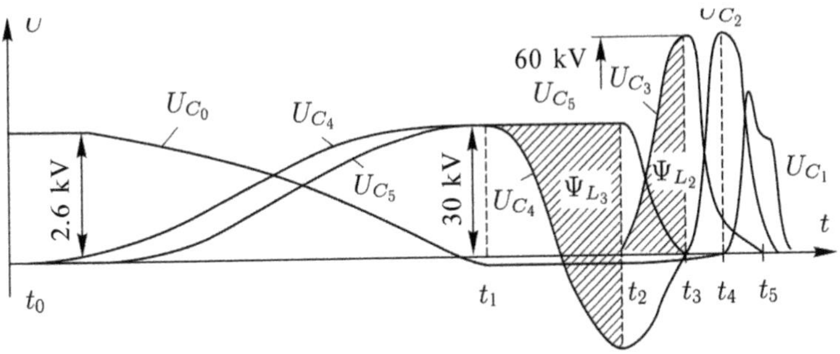

Fig. 3.11. Variation of volatges on elements of LIA,

transformer is demagnetized from the external source through the inductance L_{p2}. The inductance leads are shunted by the capacitors.

The principle of LIA operation on magnetic elements is as follows. Initially, the required current in the circuit Tp determining the magnetic state of the core is established and the rectifiers are switched on to charge the capacitor C_0. With the arrival of the control pulse to the thyristor block T (time instant t_0), the capacitor C_0 is connected to the primary winding of the pulse transformer. The charging of C_5, C_4 begins and the demagnetization current of the induction system is formed. This process proceeds for approximately 28–30 μs depending on the value of the storage voltage C_0. The voltage variation on the circuit elements is shown in Fig. 3.11. The capacitor C_4 is charged directly from the secondary winding Tp, and in the charge circuit C_5 the windings of the saturation chokes L_3, L_2, L_1 and the magnetizing coils of the induction system, whose cores are demagnetized, are turned on.

The discharge time interval C_0 must correspond to the duration of magnetization reversal of the core of the pulse transformer:

$$\pi\sqrt{L_0 C_0 / 2} \approx \Psi_{\mathrm{Tp}} / \langle U_{C_0} \rangle, \tag{3.2.11}$$

where L_0 is the discharge circuit inductance; $\Psi_{\mathrm{Tp}} = W_{\mathrm{Tp}} S_{\mathrm{Tp}} \Delta B$; W_{Tp}, S_{Tp} are the numbers of turns of the primary winding and cross section of the steel of the transformer; ΔB is the induction increment in the core; $\langle U_{C_0} \rangle \approx U_{C_0}/2$ is the average operating voltage on the windings of the transformer; U_{C_0} is the amplitude of the charging voltage of the capacitor C_0.

At time t_1, the capacitor discharging current C_0 goes to zero and the thyristor unit T is turned off. Under the action of the capacitor voltage C_0, the core Tp is saturated and the capacitor C_4 is recharged through the inductance of the secondary winding of the transformer. In this case, the sum of the voltages of the capacitors C_5, C_4 is applied to the turns of the saturation choke L_3. The magnitude of the flux linkage of the saturation choke L_3 is chosen such that by the time t_2, when the capacitor C_4 is completely recharged, saturation of the core occurred, i.e. the following condition is satisfied

$$\pi\sqrt{L_{\mathrm{Tp}}C_4} \approx \psi_{L_3} / \langle U_{C_3} \rangle, \tag{3.2.12}$$

where L_{Tp} is the inductance of the secondary winding of the pulse transformer in the saturated state; C_4 is the capacitance of the capacitor C_4; $\psi_{L_3} = W_3 S_3 \Delta B$; W_3, S_3 are the number of turns and the cross section of the saturation choke steel L_3; ΔB is the range of induction of the choke steel; $\langle U_{C_3} \rangle \approx (U_{C_4} + U_{C_5})/2$ is the average effective voltage at the choke turns L_3; U_{C_4}, U_{C_5} are the amplitudes of the charging voltage of capacitors C_4, C_5.

The capacitors C_5 and C_4 are connected in series with respect to C_3 and will start to discharge into C_3. In the time interval t_2–t_3, the core of the choke L_2 is reversed, and when it is saturated (at time t_3), C_3 will be discharged to the capacitor C_2. During t_3–t_4, the magnetic switch core L_1 is reversed, and C_2 discharges on the capacitance of the forming line C_1. The latter, discharging through the magnetizing turns of the induction system, forms a high-voltage pulse. When operating on an ohmic load, the voltage and current pulses are bell-shaped. In this case, the SFL C_1 serves as the matching line. When working on an electron diode by delaying the appearance of current by the explosive electron emission SFL C_1 is discharged in a state close to the idle forming voltage and current pulses with a flat top.

As follows from the operating principle of the accelerator, the time intervals for the transfer of energy from one stage of compression to the other are determined by the values of the flux linkages of the pulse transformer and the saturation chokes. Choosing the number of turns and the cross-section of the steel of these elements, it is possible to compress the energy of the primary storage ring with a high efficiency and charge the SFL for a time of the order of hundreds of nanoseconds.

Scheme of supply of LIA on magnetic elements and the principle of operation. The basic electrical diagram of the power supply system of LIA 04/4000 is shown in Fig. 3.12 and is divided

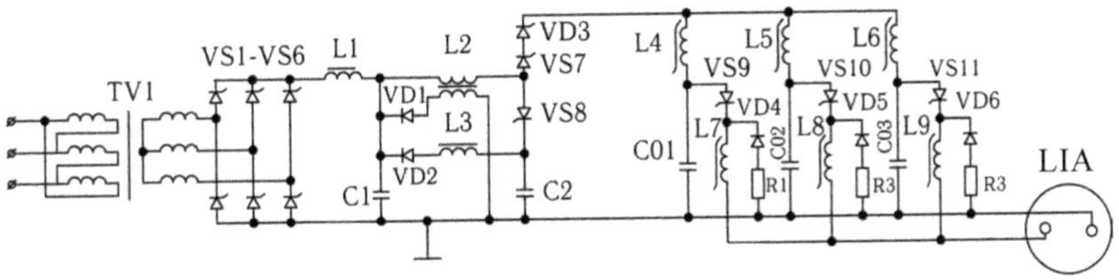

Fig. 3.12. Basic electrical circuit for power supply of LIA 04/4000.

into two functional parts – the input of energy into the LIA and the charging of the capacitive storage.

The energy enters the accelerator by discharging the primary capacitive storages C01, C02, C03, consisting of low-inductance high-frequency capacitors through saturation chokes when three parallel channels are switched on from three serially connected thyristors (VS9–VS11).

The principle of work is the following. The pre-demagnetized throttles L7 (L8, L9) delays the discharge current by 3–4 μs. The current amplitude in the output bus is 13.4 kA with the charging voltage of the storage of 2.6 kV. The duration of the current pulse is ~30 μs. After functioning of the accelerator, some of the energy comes back and after 20–30 μs it emits an emf of opposite polarity at the input terminals of the LIA. This is due to the use of the effect of overlapping phases in the MPG compression links and the incomplete matching of the forming line and the load.

To increase the efficiency of the power circuit, the duration of the control pulses on thyristors VS9–VS11 is increased so that the latter remain conductive at the time the energy returns. The main part of it returns to C01–C03, creating a second pulse of the charging current of the storage. The remaining energy after switching off the thyristors is dissipated by the resistors R1–R3 and the diodes VD4–VD6.

The charging of the capacitive storages C01, C02, C03 to a maximum voltage of 2.6 kV with a cycle frequency of up to 320 Hz is performed from a device powered by a three-phase network with a capacity of 90 kW. The device circuit realizes the principle of oscillatory charging of the capacitor from a constant voltage source. The controllable rectifier VS1–VS6 is fully operational. The rectified voltage is $U_{in} \approx 1.5$ kV. The main control function is to disable control pulses to the thyristors in case of short circuit or overload. Phase control is used to smoothly charge the filter capacitor C1, as well as to manually adjust the rectified voltage.

The capacitors C01, C02, C03 are charged through the choke L2 when VS7 is switched on. The duration of the sinusoidal half-wave current is ≈1.5 ms and the amplitude of the charging current is ~300–500 A, depending on the presence of reverse voltage on C01–C03. The charging voltage is regulated by the interruption of the charging process with a control depth of ~50% of the maximum. The thyristor VS8 turns on when the charging voltage of the primary storages reaches a predetermined level, while the current through L2 switches to C2, and VS7 turns off. The fast-recovery diode VD3 reduces the amplitude of the reverse voltage to VS8. The voltage rise on C2 is limited by the circuit VD1 – secondary winding L2 at level $2U_{in} \approx 3$ kV, since the winding transformation ratio is one. The energy remaining in L2 is returned through this circuit to the filter capacitor C1, the thyristor VS8 is de-energized and switched off. The energy stored in C2 is returned through L3 and VD2 to C1. The activation of VS7 is delayed by 1 ms from the beginning of the charge, when the discharge circuit is already completely de-energized, and the VS9–VS11 thyristors have restored the locking capacity.

The thyristors VS9–VS11 with a large area of structure have a leakage current of up to 10 mA. This current, as well as the current through the equalizing resistors, discharges the storage capacitors, creating an error that increases sharply with decreasing cycle frequency. To compensate for the charge loss, an additional power source of 150 W with a maximum voltage of 3 kV is used. It starts working only after the charger has charged the storage to the specified voltage.

The operational experience of the LIA 04/4000 showed the need to limit the voltage of the C01-C03 storages to improve the reliability of the thyristor unit. It was decided to reduce the charging voltage to 1000 V, i.e. up to the operating voltage level of one thyristor. In this case, the network transformer is excluded from the circuit and the voltage of the three-phase network is fed to the rectifier input. The stored energy in the primary storage is increased due to the use of a capacitance of 1000 μF. This causes an elongation in time of the discharge process and facilitates the working conditions of the thyristors in terms of the rate of current rise. The amplitude of the current commutated by one thyristor is reduced by dividing the discharge circuit into 6 parallel channels. In order to charge the MPG capacitors to their operating voltage, the transformer ratio of the pulse transformer is made equal to 30. To maintain the value of the compression ratio of one link within 3–4 and maintain a high efficiency of energy transfer, the MPG uses an additional compression link. The layout and principal electrical diagram of such an accelerator (LIA 04/6) are shown in Fig. 3.13.

The increase in the time intervals of the charge–discharge processes in the first links of the MPG compression enabled the use of industrial low-inductance high-frequency capacitors 0.1 μF, 40 kV installed in parallel for six pieces for C_4 and C_5 and in series-parallel for 12 pieces for C_3 and C_2. Thus, in a common case, it was possible to place all the elements and greatly simplify the process of assembling the accelerator. Changes in the design of the accelerator associated with the use of uniform magnetizing turns of the ferromagnetic cores of the induction system (previously coils in the form of copper strips) were introduced.

This made it possible to reduce the inductance of the discharge circuit by simultaneously increasing the 'parasitic' capacitance of the induction system. Such a decision was made based on an analysis of the results of [20], in which the influence of switching and explosion emission processes was investigated, as well as the influence of the inductance and capacitance of the discharge circuit on the parameters of the generated HCEA pulses. In particular, a decrease in the inductance of the discharge circuit while simultaneously increasing the 'parasitic' capacitance of the load makes it possible to reduce the voltage surge caused by the delay in the explosive emission of

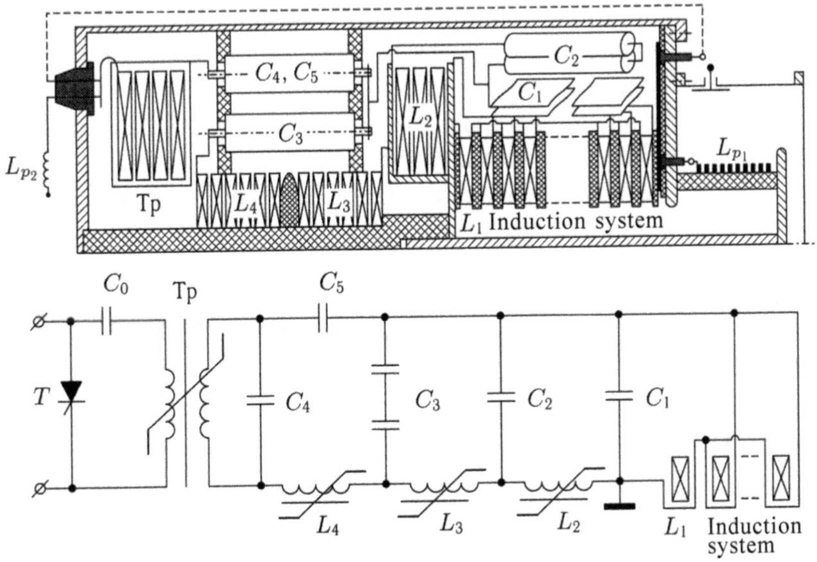

Fig.3.13. Composition and principal electrical diagram of the accelerator on magnetic elements LIA 04/6.

electrons on the cathode surface. This facilitates, on the one hand, the operating mode of the LIA insulator, and on the other hand, improves the process of excitation of the π-mode oscillations of the RM. Since the π-form of the oscillations is the lowest voltage, the appearance at the initial instant of time on the magnetron diode of a voltage exceeding the required synchronous value may lead to the excitation of high-voltage modes.

Engineering calculation of LIA elements on magnetic elements. Let us consider an engineering technique of calculation of elements of the accelerator on an example of the LIA 04/6 and we will justify the choice of their parameters. First of all, it is necessary to determine the total energy compression factor, which consists of the product of the compression coefficients of the individual MPG links:

$$n_{\text{comp}}^{k} = \frac{\Delta t_k}{\Delta t_1}, \qquad (3.2.13)$$

where n_{comp} is the coefficient of compression of one link; k is the number of links; Δt_{k} is the time interval of energy transfer from the primary storage to the capacitors of the first MPG compression link; Δt_1 is the energy transmission time interval of the capacitor of the last compression link of the MPG in the SFL. These data make it possible to determine the number of MPG links, respectively, the weight and dimensions of the LIA. To estimate the first time interval, it is necessary to know the capacity of the primary storage device, the inductance of the discharge circuit, which includes the inductance of the turns of the transformer and the supply lines.

As noted earlier, the primary storage voltage is selected no more than $U_{\text{in}} \sim 1000$ V. The maximum repetition rate ($F = 200$ Hz) is limited by the amount of power drawn from the network, which, say, can not exceed $P_1 = 100$ kW. Consequently,

$$P_1 \leqslant C_0 \frac{U_{\text{in}}^2}{2} F$$

and primary storage capacity will be: $C_0 = 10^{-3}$ F. We select the pulse transformer ratio $K_{\text{Tp}} \approx 30$–32, then the capacitors C_4, C_5 can be charged to 30 kV – their nominal voltage in the pulse–periodic mode. Thus, the capacitor capacitance is $C_4 = C_5 = C_0/2K_{\text{Tp}}^2 \sim 0.55$–$0.6 \cdot 10^{-6}$ F which can be obtained using six parallel-connected capacitors of 40 kV, 0.1 μF.

The inductance of the discharge circuit L_0 is formed by the inductances of leakage of the pulse transformer, the capacitors, the chokes, and the supply of lines and according to the calculations is $L_0 \sim 0.8 \cdot 10^{-6}$ H. Consequently, the discharge time interval C_0 is

$$\Delta t_0 = \pi\sqrt{L_0 C_0/2} \sim 68 \cdot 10^{-6}\ \mu\text{s}.$$

At the last MPG compression link, the capacitance C_1 must be charged in the time not longer than that which can ensure the flux linkage of the magnetic commutator ψ_s. To reduce the size of the magnetic commutator, and hence the inductance of the magnetizing coil, it is advisable to perform it using a single core of the same dimensions as the cores of the induction system. In this way,

$$\Delta t_1 \leq \frac{\psi_s}{\langle U_{C_2} \rangle},$$

where $\langle U_{C_2} \rangle \approx U_{C_2}/2$ is the average voltage acting on the coils of the magnetic commutator, U_{C_2} is the amplitude of the charging voltage of the last capacitor of MPG.

The value of n_{comp} to obtain high values of the energy transfer efficiency must be selected within 3–4 [21]. On the other hand,

$$n_{\text{comp}} = \frac{\psi_{n+1}}{\psi_n}, \tag{3.2.14}$$

where ψ_{n+1}, ψ_n are the flux linkage values of the saturation chokes $n+1$ and n of the MPG links.

The weight and size parameters of the LIA on magnetic elements can be reduce using the effect of overlapping of the discharge phases of the capacitor of the previous MPG compression link and the charge of the next link capacitor (Fig. 3.14).

If the value of the flux linkage of the choke Ψ'_n is less than necessary: ψ_n = 1.1−1,3 Ψ'_n, the efficiency of energy transfer is reduced, but in this case the cross section of the steel core of the saturation choke is noticeably reduced, and, consequently, its weight and energy loss for magnetization reversal decrease. At the same time, the inductance of the saturation coil of the next compression link decreases and the charging and discharging time of the capacitors decreases. When using this effect there appears a difference between

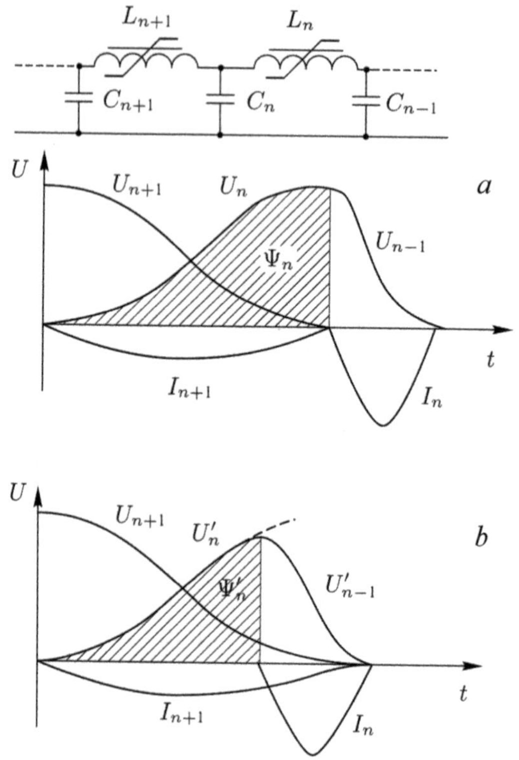

Fig. 3.14. The electrical circuit of the MPG compression links. Diagrams of voltage (U) and current (I) in MPG: a) without overlapping phases; b) with overlapping of the discharge phases of the capacitor C_{n+1} and charge C_n.

the energy transmitting time in the (n + 1)-th link and the saturation choke magnetization reversal time in the n-th link:

$$\Delta t_{n+1} = \pi\sqrt{\frac{L_{n+1}C_{n+1}C_{n+2}}{C_{n+1}+C_{n+2}}} \leqslant \frac{\psi_n}{\langle U_{C_{n+1}}\rangle} = \Delta t'_n. \tag{3.2.15}$$

We choose $\Delta t_{n+1}/\Delta t'_n$ ~ 1.1–1.3, realizing the overlap of the charge C_n and discharge phases C_{n+1}. In this case, the transfer factor of the voltage amplitude from the capacitance C_{n+1} to the capacitance C_n is

$$K_{tr} = \frac{U_{C_n}}{U_{C_{n+1}}} = \frac{\left|1-\cos\dfrac{\pi}{1.1-1.3}\right|}{2} = 0.98-0.87 \tag{3.2.16}$$

in the case $C_n = C_{n+1}$, where $U_{C_n} < U_{C_{n+1}}$ is the voltage amplitude of the voltage capacitors. For high values of $\Delta t_{n+1}/\Delta t'_n$ and the voltage and energy loss becomes unacceptable.

With a decrease in the charging voltage of the primary storage (the operation of the LIA is carried out with reduced output parameters), there is no overlap of the phases, and the transfer of energy from one link of compression to the other occurs without any loss.

So, there are two conditions for selecting the parameters of the choke of the n-th link:

$$\psi_{n+1} = (3.5 \pm 0.5)\psi_n; \tag{3.2.17}$$

$$\pi\sqrt{L_{n+1}\frac{C_{n+1}C_{n+2}}{C_{n+1}+C_{n+2}}} = (1-1.3)\frac{W_n S_n \Delta B}{\langle U_{n+1}\rangle}. \tag{3.2.18}$$

Using the relations (3.2.17) and (3.2.18), it is possible to make estimates for the preliminary selection of the elements of the LIA.

To eliminate voltage losses during the transfer of energy from the capacitor of the last stage of compression of MPG to the SFL, the condition (3.2.17) should have the form

$$\pi\sqrt{L_2\frac{C_2 C_1}{C_2+C_1}} = \frac{W_s S_s \Delta B}{\langle U_{\tilde{N}_2}\rangle}, \tag{3.2.19}$$

where L_2 is the total inductance of the saturation choke turns of the of the last stage of compression and the current leads to it plus the intrinsic inductance of the capacitance C_2; W_s, S_s is the number of turns, the cross section of the steel of the magnetic commutator.

To increase the power released on the load, it is suggested in [22] to use $C_1 < C_2$. In this case, the forming line is charged to a higher voltage, and since its capacitance has become less, it discharges in a shorter time, as illustrated in Fig. 3.15. If a simple equivalent circuit is used for numerical estimation, which is a series connection of a capacitor discharged to inductance and load resistance, then, depending on the ratio of the values of the listed elements, the power growth can reach 40%.

Computer modelling (see section 3.3) with allowance for 'parasitic' inductances, capacitances, losses in steel cores, etc. shows an increase in power by ~30% when the accelerator operates at an ohmic load of 100 Ohm. The dependence of the output power of the LIA on the ratio C_2/C_1 based on the results of calculations on the computer model is shown in Fig. 3.16. For the LIA 04/6,

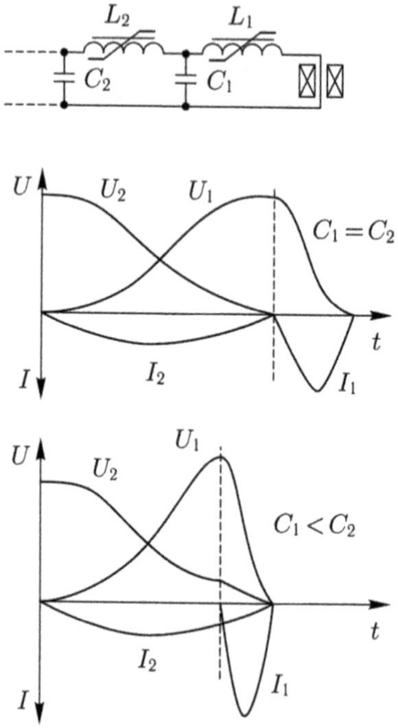

Fig. 3.15. The electrical scheme for connecting the MPG to the SFL. Voltage diagrams (U) and current (I) during the discharge of the capacitor of the last MPG compression link on the SFL and SFL on the load: with $C_1 = C_2$; with $C_1 < C_2$.

$C_2/C_1 = 1.5$ is chosen, since a decrease in C_1 to 0.15 μF causes a rise in the charging voltage of the forming line and the need for additional insulation. In addition, the value of the residual voltage C_2 increases.

An important point in the implementation of this technical proposal $C_2/C_1 = 1.5$ – the residual voltage of the capacitor C_2 after the transfer of energy to the SFL. Therefore, it is useful to introduce the energy transfer coefficient ζ, which is defined as the ratio of the energy stored in C_2 to the moment of saturation of the choke L_2 to the energy stored in the forming line at the moment of saturation of the magnetic commutator. In general, the transmission coefficient is dependent on the energy loss in the coil L_2, in the core L_2, C_2 and C_1, and also on the ratio of the capacitances $\xi = C_2/C_1 = 1.5$. Assuming the absence of losses in the winding and capacitors for the transmission factor we can write

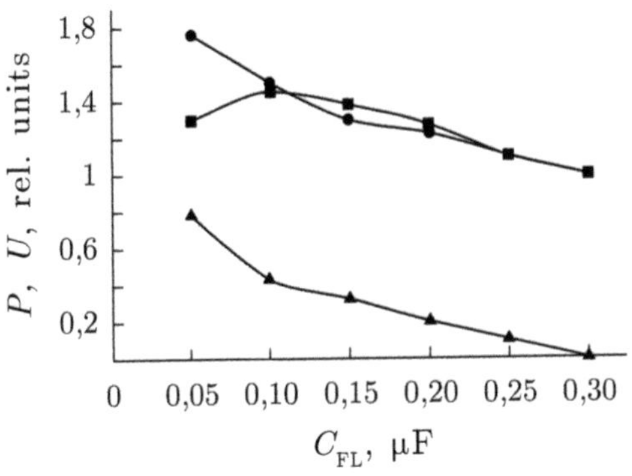

Fig. 3.16. The dependence of the output power (P) of the LIA (■), the charging voltage of the SFL (U_{charge}) (●), the residual voltage of the capacitor of the last MPG compression link (U_{res}) (▲) on the value of the FL capacitance, C_2 = 0.3 μF.

$$\varsigma = \frac{4\xi}{(1+\xi)^2} \approx 96\%. \tag{3.2.20}$$

Nevertheless, an increase in the most important parameter of the LIA, such as impulse power, by 30% with insignificant losses of energy (~4%) makes it expedient to use the unbalance of capacities. However, the process of increasing the pulse power is accompanied by a significant increase in the charging voltage of the forming line, which may exceed the breakdown values. If the condition [23] is imposed on the magnitude of the flux linkage of the magnetic commutator,

$$W_s, S_s \Delta B = \langle U_s \rangle \sqrt{(C_1 + C_2)/L_2 C_2 C_1} \arccos\left(C_1 / C_2\right), \tag{3.2.21}$$

then this leads to the following. As soon as in the discharge of the capacitor of the last link of the magnetic pulse generator the magnitude of the charging voltage of the forming line reaches the charging voltage of the capacitor (~$2\langle U_s \rangle$), the core of the magnetic commutator is saturated and the common charge of the capacitor and the forming line for the load starts to take place. In other words, the overlapping phases of the discharge of the capacitor and the charge of the forming line are simultaneously applied and the unbalance of the capacitances is used.

The pulse power of the accelerator, released on the load when this condition was fulfilled, was calculated using the computer model of the LIA. The value of C_1 relative to C_2 decreased by

2–4 times. The inductance of the discharge circuit comprising the inductance of the coil of the saturation choke of the last circuit of MPG compression, was assumed equal to $L_2 = 23.5 \cdot 10^{-9}$ H, which corresponds to the actual accelerator parameters of the LIA 04/6. The highest calculated power was observed at $3C_1 = C_2$ and the condition (3.2.21) is satisfied. Compared to the case without the use of the phase-overlap effect, with the same capacitance ratio, the power increase was 11%.

Note that since the value of the capacitance of the forming line has significantly decreased, the duration of the discharge–charging processes in the capacitor C2–the forming line circuit has decreased.. This makes it possible to reduce the magnitude of the flux linkage of the magnetic commutator, which automatically leads to a reduction in its size, and hence to a decrease in the inductance of the magnetic-commutator turn. In the calculations, this parameter was changed from $18.7 \cdot 10^{-9}$ to $16 \cdot 10^{-9}$ H, which caused the increase in power to the load by another 3%.

Thus, in comparison with the traditional designs of LIAs on magnetic elements having a set of capacitors of the same capacitance in all links of MPG compression equal to the capacity of the forming line, the implementation of the accelerator, according to the recommendations of [23], increases the power released on the load by ~40%. This option was implemented when creating the LIA 04/6.

The results of engineering calculations were used to select the parameters of the saturation chokes L_4–L_2 of the MPG and the magnetic commutator of the linear induction accelerator LIA 04/6 (Table 3.2).

The values of the inductance of current leads to the elements of the accelerator indicated in Table 3.2 depend on the geometry and mutual arrangement of current-carrying busbars and were calculated using [24]. In the process of computer calculations, the parameters of the accelerator elements and their constructive fulfillment can be varied to adjust the LIA to maxima for the pulse power, efficiency, the quality of the electron beam, etc. (see secton 3.3).

The pulse repetition rate of the LIA on magnetic elements. The most important question is the limiting frequency of the pulse repetition of LIAs on magnetic elements. In general, restrictions are associated with reverse magnetization of the ferromagnetic cores of the MPG saturation chokes, a pulsed transformer, and the damping of the impulse fluctuations [25]. The process of reverse magnetization must be completed by the time the switch of the

Table 3.2.

Element of LIA	Cross-section of steel, m^2	Number of cores	The size of cores	Number of turns	Inductance of windings, H	Inductance of current leads, H	Flux linkage, V·s
Transformer	0.112	4	K500 × × 220 × × 25	33	$29.6 \cdot 10^{-6}$	$0,15 \cdot 10^{-6}$	0.924
Choke L_4	0,0084	6	K250 × × 110 × × 25	14	10^{-5}	$2 \cdot 10^{-7}$	0.315
Choke L_3	0,0084	6	K250 × × 110 × × 25	4	$7.8 \cdot 10^{-7}$	$2.25 \cdot 10^{-7}$	0.084
Choke L_2	0,0084	3	K500 × × 220 × × 25	1	$16 \cdot 10^{-9}$	$7.5 \cdot 10^{-9}$	0.021
Magnetic commutator	0,0021	1	K360 × × 150 × × 25	1	$5.6 \cdot 10^{-9}$	10^{-8}	0.00525

primary energy storage is switched on. Otherwise, the stability of the amplitude–time parameters of the output pulses is violated. Therefore, the task of analyzing the process of reverse magnetization is to estimate its duration. From Table 3.2 we can note the following. First, the saturation of the pulse transformer of the first compression link and the saturation choke L_4 of the second compression link have the maximum flux linkage. It is these two elements that are used to spend the bulk of the energy and time for magnetization reversal. Secondly, these elements have the largest number of turns in comparison with others and are magnetized by smaller values of the currents. With a coercive force $H_c \sim 25–30$ A/m of the 50 NP alloy with a thickness of 0.02 mm to convert the transformer cores and the saturation choke L_4 to a saturated state, the currents 1 A and 1.5 A are demanded, respectively. Thus, the minimum demagnetizing current must be at least 1.5 A. It should be recalled that the saturation chokes L_4–L_2 are demagnetized by the capacitor C_5 charge current. After the transfer of energy from the capacitors C_4 and C_5 to C_3 at the time t_3 under the action of the voltage on the capacitor C_3, the core L_4 starts to magnetize in the opposite direction, and the core L_3 in direct direction. As a result, the demagnetization current of the pulse transformer is closed through the capacitance C_5, forming on it a negative magnetization reversal pulse.

By analogy with the investigations carried out in [25], one can write the expression for the limiting pulse repetition rate in the form

$$F \leqslant \sqrt{k}\sqrt{F_1} n_{\text{comp}}^2 / \sqrt{2} t'_p (t_1 / 2 + t_p), \tag{3.2.22}$$

where

$$F_1 = \pi^2 \mu_0 \mu_{\text{sat}} H_c / 16 \Delta \text{B};$$

k is the number of compression links; μ_{sat} is the relative magnetic permeability of the core in the saturated state; t_p = 1−1.5 is the parameter of period filling with intermittent oscillations; t_1 is the time of charge of the capacitances C_5 and C_4 (Fig. 3.11); t_{p} is the duration of control pulses on thyristors VS9–VS11 (Fig. 3.12) after the end of charging of capacitors C_4 and C_5 to restore the valve properties of the primary storage switch. After substituting the numerical values of the parameters for the LIA 04/6, we obtain the limiting pulse repetition rate $F \leq 3230$ Hz. Naturally, such modes of operation of the accelerator due to the huge power consumption are possible only with the use of several pre-charged capacitor storages each with its own switch discharged in series to a pulse transformer (as implemented for the LIA 4/2) [9].

A possible factor limiting the pulse repetition rate or the number of pulses in a continuous series can be the heating of the elements of a LIA. The thermal conditions of the LIA elements determine the losses: 1) ohmic when the current flows through the conductors, 2) losses in the dielectric of the capacitors, 3) losses due to eddy currents and magnetizing current of the cores of the MPG saturation chokes, the magnetic commutator and the induction system. The estimate of the loss is presented in section 3.3. The computer program allows one to calculate the loss for each individual element of the accelerator. The increase in the temperature of the cores of magnetic elements for a certain time interval, depending on the repetition rate of the pulses, is calculated using the technique for calculating conventional transformers. In general, depending on the matching of the accelerator with the load and the magnitude of the charging voltage, the efficiency of the energy transfer from the primary storage device is ~40–50%. Naturally, all the remaining energy is released in the form of heat in a large mass of the accelerator of the order of 1800–2000 kg. To heat such a mass to a temperature of 60°C, a

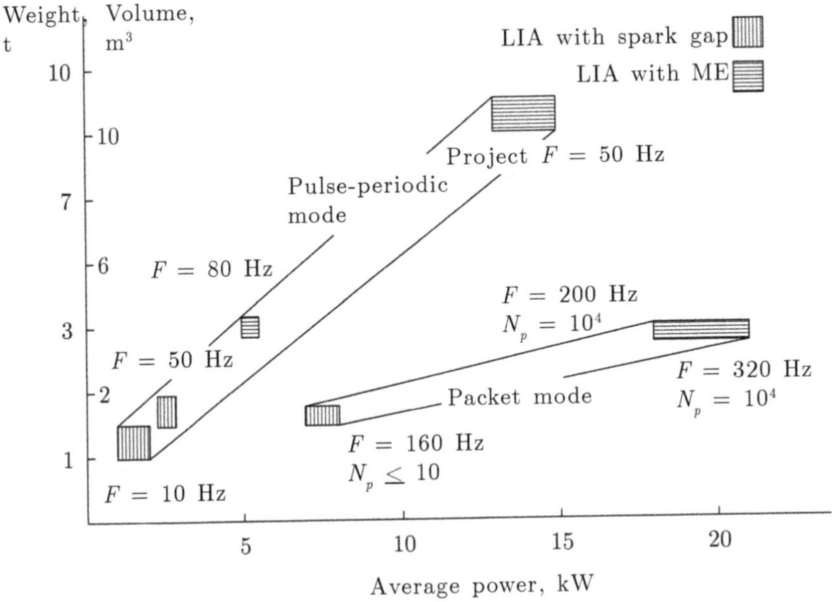

Fig. 3.17. Comparison of the weight and size parameters of various types of LIA (including the power supply system), depending on the average power of the electron beam generated by them (N_p is the number of pulses in the series).

which is a threshold for the industrial capacitors used, 80 Hz without heat sink exceeds $4 \cdot 10^3$ s, which is more than one hour of operation of the accelerator.

Figure 3.17 shows a comparison diagram for the mass-size parameters of the two types. As a result of the picture, the LIA on the magnetic elements can generate electron beams of a much higher average power due to the high repetition rate, although they are inferior to LIAs with a multichannel gas spark. We also note that LIA on magnetic elements have significant weight and high cost due to the need to use magnetic pulse generators to charge the forming lines.

3.3. Simulation of the operation of HCEA and LIA on relativistic magnetrons and other loads

3.3.1. Statement of the problem and the purpose of the work

One of the main tasks of relativistic microwave electronics is to increase the power and energy of microwave radiation pulses when powered by high-current accelerators, when the efficiency of the application and the efficiency of the devices depend significantly on the characteristics of the accelerating voltage pulse. Thus, for example, to reduce the beam losses in the transport of a slowing-down system of a relativistic backward wave tube, a rectangular pulse of accelerating voltage is required, which ensures a high energy homogeneity of the electron beam. In a number of cases it is necessary that the shape of the accelerating voltage pulse be varied according to a predetermined law. A feature of the formation of a voltage pulse in RMs is due to the fact that the regions of formation of the electron beam and the interaction of the beam with the slowing-down waves of the anode block coincide spatially. For such devices, the magnitude of the operating current is determined not so much by the geometry of the electrodes and by the magnitude of the electric field strength as by the intensity of the high-frequency fields, which depend on the properties of the resonator system. In addition, for the effective work of an RM, the ratio of the magnitude of the voltage-induced voltage $U(t)$ in the cathode–anode gap of the magnetron to the value of the constant magnetic field H must satisfy the synchronism condition. At the same time, the value of the anode current essentially determines the value of $U(t)$, since the power source has limited power, that is, the operating modes of the magnetron and the power supply are interdependent. Therefore, such a device should be considered as a common system with a strong feedback. Such a regime imposes stringent requirements on power supplies and microwave generators on the energy conversion efficiency. Since the power source of the RM is laborious in manufacturing and expensive installation, the design stage is decisive.

The purpose of modelling is the study of physical processes in the power source–nonlinear load system the resistance of which depends both on its own characteristics and on the parameters of the supply pulse. We consider a unified model of the power supply, suitable for modelling its operation with any relativistic microwave generator, allowing the optimal choice of the parameters of the elements entering into it. Optimization is carried out by means of numerical modelling, since the whole system has a strong feedback and the calculation of transient processes on the elements is a rather difficult problem related to solving non-linear differential equations. We also note that the model is constructed using the physics of the processes of pulsed magnetization reversal of the steel of ferromagnetic cores of LIA (pulse transformer saturation chokes, magnetic commutator, induction system).

The practical goal is to determine the optimal parameters and the choice of the design of the installation elements. In this regard, it is necessary to create a joint model for calculating the processes in the LIA and in a load with non-linear characteristics, which is the RM. Further, the possibility of solving this problem by modelling based on the representation of sections by LIA-equivalent schemes is shown, and the non-linear elements of the equivalent relativistic magnetron circuit are calculated in accordance with the theory of averaged motion (see Chapter 2).

The modelling problem was solved by a two-stage approach. The first-order calculations are made at the first stage of the development of the outline design of the installation using a simplified equivalent scheme. At the same time, transient processes, energy characteristics and installation parameters are analyzed in general form without unnecessary detail. The operating mode of the power supply and the microwave generator is optimized by a sequential search of the parameters. After clarifying the main regularities of processes and the parameters of the elements of the installation, the second stage of modelling is carried out. To do this, more accurate non-linear and parametric representations of individual nodes are used, and methods for planning a numerical experiment for finding optimal conditions are used, and a final calculation is made.

It should also be noted that various criteria for estimating the optimum mode for a power source and a microwave generator can be selected: full or electronic efficiency, output power, electron beam spectrum on the load, stability of operation.

3.3.2. Construction of the model and calculation of the parameters of the equivalent circuit

To solve the problem, the real electrical circuit of the power source and the load is represented by an equivalent circuit (Fig. 3.18), the parameters of the elements are determined, and the differential equations for voltages and currents are recorded [26]. The processes in the computer model are considered from the moment when the capacitances of the magnetic pulse generator C_{m4} and C_{m5} are charged, the switch K_1 is closed and C_{m4} begins to be recharged through the inductance of the secondary winding of the saturated pulse transformer L_{m5}. Switching on the keys K_1–K_5 in the equivalent scheme of the accelerator simulates the transition of the cores of the saturation chokes from the unsaturated state to the saturated one.

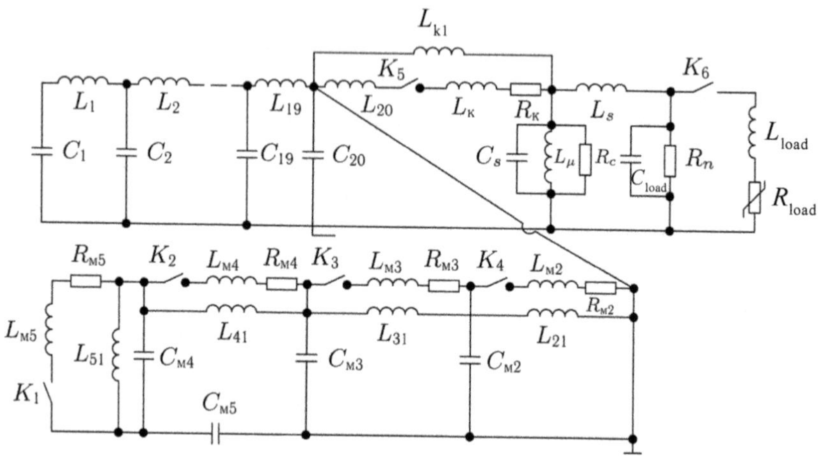

Fig. 3.18. Equivalent circuit of LIA 04/6.

The upper half of Fig. 3.18 shows the equivalent circuits of the forming line, the induction system and the load, on the lower half – the equivalent MPG scheme, formed by four compression links.

1. *The first MPG compression link* consists of a capacitor C_{m4}, C_{m5}, the secondary winding of the pulse transformer (transformer fulfills two functions: increase the voltage up to 30 kV at the discharge of the primary storage device, at saturation (key K_1 is activated) provides recharging of the capacitor C_{m4}). In this scheme:

1.1. L_{m5} is the inductance of the secondary winding of a saturated pulse transformer, defined as [24]

$$L_{m5} = \frac{1}{2\pi} a_5 W_5^2 \ln \frac{D_{ext5}}{D_{int5}}, \tag{3.3.1}$$

where a_5 is the linear size of the secondary winding; D_{ext5}, D_{int5} are the external and internal diameters of the winding; W_5 is the number of turns.

1.2. R_{m5} is the ohmic equivalent of losses, including resistance: winding, connecting current leads, contact connections and loss in the dielectric capacitors.

1.3. L_{51} is the equivalent inductance of the magnetization of transformer. The circuit with inductance L_{51} describes the process of the magnetizing current flowing in the transformer core at charging of of C_{m4}, C_{m5}. The presence of a magnetizing current leads to a partial discharge of the capacitors and incomplete energy transfer from the previous MPG compression link to the subsequent one.

Since this computer model does not cover the capacitor charging process, and the voltage inverting occurs when the transformer core is saturated, the losses in transformer steel are not taken into account, and we can assume $L_{51} \to \infty$.

2. *The second–fourth links of MPG compression* (saturation chokes L_{m4}–L_{m2}, capacitors C_{m5}–C_{m2}) have almost identical equivalent circuits (Fig. 3.18) and consist of:

2.1. L_{m4}–L_{m2} – the inductance of the windings of the chokes in the saturated state. They are determined by formulas analogous to (3.3.1).

2.2. R_{m4}–R_{m2} – ohmic equivalents of total losses: in the steel of the chokes upon their magnetization reversal; the losses in the copper of the coils, in the supply circuits and contact connections; the losses in discharged and recharged capacitors.

2.3. L_{41}–L_{21} – inductance of the magnetization of the cores of the saturation chokes. The core of the saturation choke of the second compression link is magnetized during the recharge time C_{m4} under the action of the total voltage on the capacitors $U_{Cm4} + U_{Cm5}$. The cores of the chokes of the third and fourth compression links are magnetized by a charge of C_{m3} and C_{m2}, respectively. The magnitudes of the magnetization inductances of the cores are determined by the following approximate formula:

$$L_{i1} = \frac{2W_i^2 B_s l_c K t_i (D_{iext} - D_{iint})}{\pi(D_{iext} + D_{iint})[H_{0s}t_i + 2S_1 + S_2]}, \qquad (3.3.2)$$

where W_i is the number of magnetizing turns; l_c is the width of the core steel; B_s is the saturation induction of the ferromagnetic material; K is the coefficient of filling the steel volume of the core; t_i is the duration of the magnetization reversal process; D_{iext}, D_{iint} – external and internal diameters of cores; H_{0s} is the magnetic field of the start; S_1 is the part of the switching coefficient, due to the action of eddy currents; S_2 is the part of the switching coefficient due to magnetic viscosity. Knowing the characteristics of the ferromagnet used (permalloy 50 NP) [6], we can determine L_{i1}.

Enabling the switches K_4–K_2 occurs after obtaining the *necessary* voltseconds by the i-th choke:

$$\psi_i = \langle U_i \rangle t_i = 2W_i S_i B_s K, \qquad (3.3.3)$$

where $\langle U_i \rangle \approx U_{mi}/2$ is the voltage applied to the turns of the chokes; U_{mi} is the amplitude of the applied voltage; S_i is the cross section of steel of the saturation chokes.

3. *Magnetic commutator*. The equivalent circuit of the magnetic commutator is similar to that considered above and consists of the following elements.

3.1. L_{k1} is the equivalent inductance of magnetization of the magnetic commutator:

$$L_{k1} = \frac{2B_s l_k K(D_{ext.k} - D_{int.k})t_2}{\pi(D_{ext.k} + D_{int.k})[H_{0s}t_2 + 2S_1 + S_2]}, \tag{3.3.4}$$

where $D_{ext.k}$, $D_{int.k}$ is the external and internal diameters of the turn of the magnetic commutator; $l_к$ is the linear dimension of the magnetizing coil (Fig. 3.7); t_2 is the charge time of the forming line. In contrast to (3.3.2), here W = 1 and in the formula is not present.

3.2. The inductance L_k is the sum of the inductances of the magnetizing coil of the magnetic commutator, the current leads to it, the inductance of the magnetizing turns of the induction system and its outputs. The inductance of the magnetizing coil of the magnetic commutator is defined similarly to (3.3.1). The remaining components of L_k depend on a specific design and are calculated using [24].

3.3. R_k is the ohmic equivalent of the losses in the magnetic commutator. Switching K_5 simulates the discharge of the forming line through the induction system to the load when a condition similar to (3.3.3) is met.

4. *The induction system* consists of the following elements (Fig. 3.18).

4.1. C_s is the dynamic capacitance of the induction system, which depends on the type of magnetizing coil of the ferromagnetic inductor: whether the coil consists of individual strips ('inductive' inductor), is made in the form of a disk ('capacitive' inductor), an intermediate variant is also possible.

In general, the capacity of the inductor consists of four components: the turn–steel of its 'own' inductor, the turn–steel of the adjacent inductor; the turn–the turn of the adjacent inductor; inductor–inductor, which are calculated using [27].

4.2. The leakage inductance of the inductor L_s is determined by the type of inductor: 'capacitive' or 'inductive'. For a 'capacitive' inductor

$$L_s^{(1)} = \frac{\mu_0 l_c}{2\pi} \ln\frac{D_{ext.c}}{D_{int.c}}, \tag{3.3.5}$$

where l_c is the linear dimension of the inductor core; $D_{ext.c}$, $D_{int.c}$ are the external and internal diameters of the magnetizing coil of the inductor.

For an 'inductive' inductor, setting $\mu\to\infty$, we can assume that the leakage field is concentrated in the volume between the turn and the core of the neighbouring inductor. Outside this volume, the magnetic fields of the adjacent turns partly cancel each other, and the leakage inductance is

$$L_s^{(2)} = \frac{\mu_0(D_{ext.c} - D_{int.c})}{n} \cdot \frac{(l_1 + 2l_2)}{h}, \tag{3.3.6}$$

where l_1 is the thickness of insulation between the core steel and the magnetizing coil; l_2 is the distance between the magnetizing coils of the neighbouring cores; n is the number and h is the width of the strips of the magnetizing coil.

To the inductance $L_s^{(1)}$ or $L_s^{(2)}$ we must add the inductance of the leads of the induction system. Assuming that the leakage field is enclosed between the ferromagnetic material and the copper nearest to the cores of the electrodes of the forming line, we use the formula for calculating the inductance of rectangular turns [24]:

$$L_{lead} = \frac{\mu_0}{nl_{lead}}\left[\frac{\pi(D^2_{ext.lead} - D^2_{int.lead})}{4n} - \frac{(D_{ext.lead} - D_{int.lead})h}{2}\right], \tag{3.3.7}$$

where $D_{ext.lead}$, $D_{int.lead}$ are the external and internal diameters; l_{lead} is the lead length.

In the general view, the leakage inductance of the inductor is: $L_s = L_s^{(1)} + L_{lead}$ ('capacitive'), $L_s = L_s^{(2)} + L_{lead}$ ('inductive').

4.3. The magnetization inductance of the inductor L_μ is defined similarly to L_{41}–L_{21}, taking into account that $W = 1$.

4.4. Ohmic losses in the cores of the induction system R_c are calculated using the equation for magnetization reversal of the steel under the action of a rectangular voltage pulse [6]:

$$R_c = \frac{2(D_{ext.c} - D_{int.c})l_c KB_s}{\pi(D_{ext.c} + D_{int.c})(S_1 + S_2)}. \tag{3.3.8}$$

5. *The load, including the cathode holder*, is represented by an equivalent circuit (Fig. 3.18) containing: the load capacitance C_{load}, which consists of the sum of the capacitances of the cathode holder

within the induction system, the capacitance of the cathode holder in the region of the high-voltage insulator and the load capacitance; L_{load} is the load inductance, including the inductance of the cathode holder; R_{load} is the load resistance (linear or non-linear).

5.1. The capacitance of the cathode holder within the induction system, consisting of N_c ferromagnetic cores, is determined by the formula

$$C_{\text{cat}} = \frac{2\pi\varepsilon_m\varepsilon_0 l_{i.s.}}{3\ln\dfrac{D_{\text{int.lead}}}{D_{\text{cat}}}}, \tag{3.3.9}$$

where D_{cat} is the diameter of the cathode holder; $l_{i.s.}$ is the length of the induction system; ε_m is the dielectric permittivity of the insulation (transformer oil).

5.2. The capacitance of the cathode holder in the region of the high-voltage insulator consists of two components:

$$C_{\text{cat}}^{(1)} = C_{\text{cat}}^{(2)} + C_{\text{cat}}^{(3)}, \tag{3.3.10}$$

where $C_{\text{cat}}^{(2)}$ is the capacitance of the cathode holder between its outer surface and the spiral of the insulator; $C_{\text{cat}}^{(3)}$ is the capacitance between the outer surface of the spiral and the case of the vacuum chamber. $C_{\text{cat}}^{(2)}$ and $C_{\text{cat}}^{(3)}$ are calculated using the formulas analogous to (3.3.9).

5.3. The load capacitance is calculated by the formula for the coaxial conductor, broken into i sections of length l_i with various external $D_{\text{ext}i}$ and internal $D_{\text{in}i}$ diameters:

$$C_{\text{load}} = \frac{2\pi\varepsilon_0 l_i}{\ln\dfrac{D_{\text{ext}i}}{D_{\text{int}i}}}. \tag{3.3.11}$$

5.4. The inductance of the load, including the cathode holder, is equal to the sum of the inductances in individual sections and is calculated from the formulas for a coaxial conductor:

$$L_{\text{load}} = \sum_i L_{ni} + L_{\text{cat}} = \sum_i \frac{\mu_0}{2\pi} l_i \ln\frac{D_{\text{ext}i}}{D_{\text{int}i}} + L_{\text{cat}}, \tag{3.3.12}$$

where L_{cat} is the inductance of the cathode holder within the induction system, caused by counter-flow currents along the cathode holder and along the inner surfaces of the magnetizing coils. Neglecting the gaps between the magnetizing coils, one can write

$$L_{\text{cat}} = \frac{\mu_0 l_{i.s.}}{2\pi} \ln \frac{D_{\text{int.lead}}}{D_{\text{cat}}}. \tag{3.3.13}$$

6. *The strip forming line*. The LIAs on magnetic elements use strip single forming lines – homogeneous two-wire lines with distributed parameters.

Transition process equations for the commutation of the arms of the lines can be obtained from its substitutional circuit which is a series of similar RLC links [28].

When $\Delta t \to 0$ these equations are of the form

$$L_l \frac{\partial i(x,t)}{\partial t} + R_l i(x,t) = -\frac{\partial U(x,t)}{\partial x}; \tag{3.3.14}$$

$$C_l \frac{\partial U(x,t)}{\partial t} + G_l i(x,t) = -\frac{\partial i(x,t)}{\partial x} \tag{3.3.15}$$

and are known as the telegraph equations (C_l, L_l, R_l, G_l are the linear capacitance, inductance, resistance and conductivity of the forming line). The number of elementary cells of the line was assumed to be 20, proceeding from the condition that the discharge time of the elementary capacitance of one cell per neighbouring elementary capacitance be much less than the duration of the front of the propagating wave.

To take into account parasitic leakage currents from the cathode holder, breakdowns along the insulator, etc., the loss resistance R_l shunting the load is introduced in the equivalent circuit. The switch K_6 is also introduced into the equivalent circuit, simulating the operation of the accelerator on a controlled cathode [29], the sharpening discharger or a shock line [30].

The load resistance R_{load} in the computer model can be represented as:

1) linear resistance $R_{\text{load}}(t) = \text{const}$;

2) an electron diode – a non-linear resistance, varying in accordance with the Child–Langmuir law:

$$R_{\text{load}}(t) = P^{2/3} / I^{1/3},$$

where P is the perveance, I is the diode current;

3) non-linear parametric resistance $R_{\text{load}}(t)$, which at each step of integration is defined as the result of solution of the relativistic magnetron model.

For the second and third circuits, the moment when the switch K_6 is turned on corresponds to reaching the threshold voltage for exciting explosive electron emission.

In the magnetron there are two currents – the end current [31], which is determined by the formula

$$I_e = \frac{mc^3}{e}\frac{(\gamma^{2/3}-1)^{3/2}}{2\ln(R_{\text{tube}}/r_c)}, \tag{3.3.16}$$

as the limiting current from the end of the magnetized cathode, and the anode current of the magnetron I, which, depending on the dissynchronism parameter $\alpha = 1 - \beta_p/\beta_e$, is formed from the electrons entering the region of correct phases, R_{tube} is the inner radius of the drift tube. In this case, the electrons can get to the anode only in the case when the quantity α satisfies the condition

$$\ln\left(A+\sqrt{A^2-1}\right)-\sqrt{A^2-1}-ApR'-\text{sh}(pR')\leqslant 0. \tag{3.3.17}$$

Here $p = 2\pi/\lambda\beta_p\gamma$ is the transverse wave number; λ is the wavelength of the radiation; $A = \alpha\gamma\text{sh}\,(pd)E_0/E_{\text{HF}}$, E_0 is the static electric field strength; E_{HF} is the effective amplitude of the basic spatial harmonic of the microwave field on the surface of the anode block; $R' = E_N/(H_N^2 - E_N^2)$ is the Larmor radius of rotation of the electron in the crossed electrical ($E_N = eE_0/mc^2$) and magnetic ($H_N = eH_0/mc$) fields. Calculating the drift velocity of electrons from the region of incorrect phases to the region of correct phases is similar to [32], in the range of limitation of the cathode current by a space charge, one can obtain the following estimate for the anode current of the magnetron:

$$I = \frac{mc^3}{e}NLE_{\text{HF}}\frac{\beta_ð\left[A_1+\text{ch}(pR')\right]}{8\pi\gamma_p\,\text{sh}(pd)(1-\alpha\gamma_p^2)}, \tag{3.3.18}$$

where L is the length of the device; N is the number of resonators of the anode block of the magnetron,

$$A_1 = \begin{cases} A \text{ at } A > \text{ch}(pR'), \\ \text{ch}(pR') \text{ at } A \leqslant \text{ch}(pR'). \end{cases} \tag{3.3.19}$$

The derivation of the relations (3.3.17)–(3.3.19) is given in Chapter 2 of this monograph.

From the expression (3.3.18) it is possible to determine the effective amplitude of the fundamental harmonic of the microwave field E_{HF} and then the power of the generated radiation.

For the operation of the LIA in the optimum mode, it is necessary that the magnetron is a matched load on the peak of the generated voltage pulse. This is achieved by selecting the appropriate Hartree voltage threshold values U_{th} and the settling time (rise) of the oscillations τ_{set}, which in turn depend on the geometric dimensions of the magnetron and the constant magnetic field.

The rise time of the oscillations can be estimated from consideration of perturbations of the near-cathode flux of electrons as a result of their interaction with the synchronous harmonic and using the theory of resonator excitation [32]:

$$\tau_{est} \approx \frac{N/2+7}{30} \frac{\mathrm{sh}(pd)}{\mathrm{sh}(pR')} \sqrt{\frac{2\gamma_p V}{E_0 S_c}}, \tag{3.3.20}$$

where S_c is the area of the cathode; V is the volume of the resonator.

Starting from the time t_0, when the voltage reaches the value U_{th}, a radial (anode operating current) will flow through the magnetron diode and the resistance of the magnetron will vary exponentially:

$$R_{load}(t) = R_{t0} \exp\left[-\frac{(t-t_0)}{\tau_{est}\delta}\right], \tag{3.3.21}$$

where R_{t0} is the resistance of the magnetron, determined by the end current I_t at time t_0 of reaching the level of the threshold voltage U_{th}; t is the current time; δ is a constant determined from *a priori* data. At this time point, until the resistance of the magnetron R_{load} becomes equal to R_c – the self-consistent resistance of the LIA, the condition (3.3.17) is not satisfied and there is no radiation. Then, with

$$t_m = t_0 + \tau_{est}\delta \ln(R_{t0} / R_c)$$

the conditions for the excitation of the magnetron begin to be fulfilled, and subsequently the dynamic resistance of the magnetron is determined by the sum of the anode and end currents.

Thus, the general equivalent circuit of a LIA and a RM, for which the equations of currents and voltages in the differential form are compiled according to the method of contour currents, is shown in Fig. 3.18.

3.3.3. Results of simulation of a relativistic magnetron with a power source – LIA on magnetic elements

The model problem was solved in stages: the output of the LIA was connected with an ohmic resistance, an electron diode, an relativistic magnetron. When using the ohmic resistance and the electron diode, the results of numerical simulation (amplitude and time parameters of the output pulses) corresponded with almost 100% accuracy to the results of experimental studies of the simulated accelerator. This circumstance made it possible to draw a conclusion about the adequacy of the description of the computer model of the LIA of the processes that are taking place and to proceed to the investigation of the LIA–RM system.

Calculation by the model showed that due to the strong feedback in the power source–magnetron load system an oscillatory regime is generated, formed in oscillations of the microwave power, voltage and current of the accelerator. This result is a consequence of the application of the analytical formulas of the stationary theory of the relativistic magnetron for the calculation of non-stationary processes. Disruptions of microwave generation are associated with a fast output of the device from the synchronism regime, which is observed experimentally only in the suboptimal settings. Such behaviour of the device is explained as follows: as soon as the voltage on the magnetron begins to exceed the Hartree threshold voltage, the oscillations begin to rapidly increase. This causes the appearance of a high anode current, which reduces the dynamic resistance of the RM. Since the LIA is a power source with limited power, reducing the load resistance leads to a reduction in the induced voltage below the threshold level and the exit of the device from the synchronism mode. Introduction of the a priori constant for the generating device (δ) made it possible to smooth out these oscillations in the computer model. Physically, the value of δ determines the inertia of the processes of accumulation and scattering of the space charge in the interelectrode gap of the magnetron. Thus, the use of the model allows us to impose limitations on the Q-factor of the resonator system.

Practical interest in the computer model of the LIA–RM system is due to the possibilities:

1) optimization by efficiency and output power of the parameters of the LIA and the RM;
2) calculation of the amplitude and envelope of the microwave pulses of a RM when fed from the LIA;
3) estimating the effect of a dynamic load of the relativistic type magnetron on the discharge processes in the forming line of the power source.

Simulation allows to determine the optimal geometric characteristics of a relativistic magnetron (the diameters of the anode block and cathode, the length and type of the resonator system), the design parameters of the linear induction accelerator (type and quantity cores of the induction system, characteristics of the forming line, energy of the primary storage, etc.), the necessary value of the magnetic field induction. Also, the model experiment made it possible to rethink the concept of creating LIA from the point of view of the balance of capacitive (C_s) and inductive (L_s) characteristics of the induction system and the forming line.

Below are the results of the application of the model and their comparison with the experimental data for an installation based on a 6-cavity relativistic magnetron powered by a linear induction accelerator on magnetic elements. The results of calculations are presented in tabular and graphical form. Figure 3.19 shows the processes in the magnetic pulse generator LIA – current and voltage diagrams, amplitude values of currents and voltages, characteristic time intervals for the 'ideal' configuration of the circuit elements. In particular, it can be seen from the figure that the magnitude of the voltage (upper diagrams) on the capacitors of the magnetic pulse generator circuits remains practically unchanged and the energy is compressed by increasing the current (lower diagrams), the amplitude of which increases. The computer model determines the coefficients transfer of energy from one compression link to another, which allows to calculate the thermal processes of the individual elements of the accelerator. Figure 3.20 shows the results of calculating the output pulses of the voltage and current of the LIA and the microwave power of the 6-resonator relativistic magnetron.

Testing the results of calculations is carried out by comparison with the experimental data. Moreover, such extreme cases as the work of the accelerator in modes close to short-circuit and idle (magnetic field is absent or much higher than the synchronous value) were also checked, as well as the formation of an electron beam with

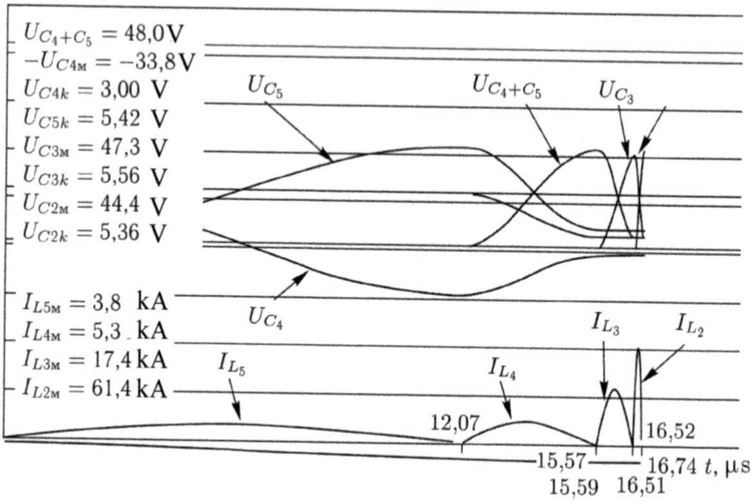

Fig. 3.19. The results of the process calculations in a magnetic pulse generator of LIA 04/6. The voltage diagrams on the capacitors (top) and the currents through the windings of the saturation chokes (bottom), the amplitude value of the voltage (U_{C2m}–U_{C3n}) and the residual voltage (U_{C2k}–U_{C5k}) on the capacitors of the compression stages and amplitudes currents (I_{L2m} – I_{L5m}) in the windings of the saturation chokes, characteristic time intervals.

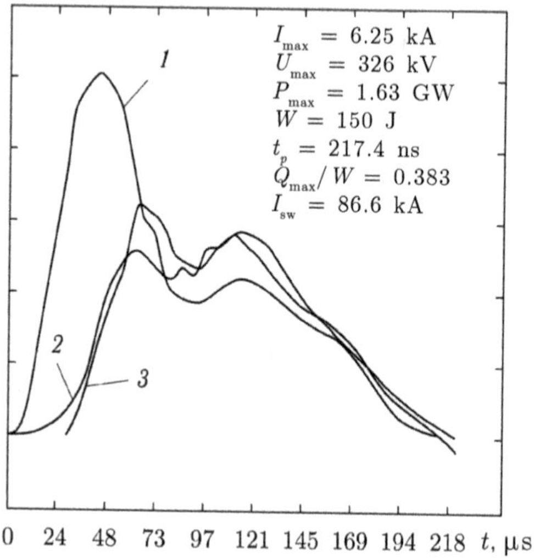

Fig. 3.20. Results of computer simulation of a 6-cavity RM [4]: voltage (*1*), current (*2*), microwave power (*3*). I_{max} and U_{max} are the amplitudes of the current and voltage of an RM at the time of the maximum generated microwave power; P_{max} – the maximum power of the electron beam on the load; W – the energy stored in the primary storage; t_p – the duration of the current pulse on the load (on the base); Q_{max}/W – efficiency of energy transfer from the primary storage device to the load; I_{sw} – current amplitude through a magnetic commutator.

a magnetically insulated diode (there is only the end current of the magnetron). In these cases, the discrepancy between the measured and calculated values of voltage and current (amplitude and duration of pulses) did not exceed 10%. Calculations of the level of generated power and the electronic efficiency of the RM in the region of synchronous magnetic fields differed from the measured values by not more than 20 and 10%, respectively, which can be attributed to not taking into account the influence of the output device of microwave radiation from the anode block.

Thus, the processes in the power supply–microwave oscillator system with a strong feedback are investigated by computer simulation. Modelling such a complex system allows us to examine in more detail the physical processes taking place in different elements of the system and analyze the effects of the mutual influence of the load and the power source. The results are obtained using a model based on equivalent power source circuits and formulas of the analytical theory of the averaged motion of a relativistic magnetron. The load (RM) is represented in the form of a non-linear–parametric resistance, which at each step of integration of differential equations is determined as the result of solving the relativistic magnetron model. Note that an analytical description of processes in such a system is hardly possible. Therefore, the implemented method of computer modelling is very fruitful.

The method allows to quickly adjust the microwave source to extremes in terms of output power, electronic and full efficiency. In particular, the application of the model made it possible to justify the use of 'capacitive' type inductors in linear induction accelerators intended for feeding relativistic magnetrons, and also to use the unbalance of capacitances of the last link of MPG compression and the forming line [22].

Note that the power supply in the model can be represented by a more complete equivalent circuit, but its complexity should be justified, since in the real design of the installation there are difficult to take into account parameters such as the leakage current from high-voltage electrodes mentioned above, the wide variation in the performance of ferromagnetic inductors, etc.

3.3.4. Results of modelling of LIA with multichannel dischargers and HCEA

The element base of such accelerators includes a multichannel spark gap, the commutating forming line with a small (units or fractions of ohm) wave impedance. Usually several parallel spark gaps are used, which are not structurally connected, but are united by a common start circuit. In this case, the time constant of the commutator-line circuit is much longer than the switching characteristic of the discharger. Thus, we can assume that at the initial time (units of nanoseconds) the discharge processes are determined by the properties of the discharger itself and its spark channel is formed due to the discharge of interelectrode capacitances. The law of change in the current of the discharge circuit is then determined by the values of its resistance R_k, inductance L_k and wave resistance of the line ρ_l. Therefore, the transient model is chosen for calculations at which the switching of the discharger from the non-conductive state occurs after a finite time $t_{fr} = 2.2 \cdot \tau_{com}$, where τ_{com} is the switching time during which the voltage on the discharger falls from 0.9 to about 0.1. The change in voltage is approximated by formula

$$U_k(t) = U_0 \exp(-t/\tau_{com}), \qquad (3.3.22)$$

where U_0 is the charge voltage of the forming line.

The current I_k commutated by the spark gap is described by the parametric equation

$$L_k \frac{dI_k}{dt} = U_c(t) - U_k(t) - R_k I_k, \qquad (3.3.23)$$

where $U_c(t)$ is the voltage change at the end of the electrode of the forming line to which the commutator is connected.

Simulation of the magnetron is carried out in the same way as described above. Figure 3.21 shows the results of calculations for the described model for a 6-resonator RM with a power supply – a LIA with a multichannel discharger.

3.4. Formation of pulses of increased voltage for the supply of RMs

It is of interest to investigate RMs at increased electric field strengths, which was achieved using systems with a reflective breaker (switch) [33] (Fig. 3.22 *a*). The voltage pulse for RM was formed by a system accumulating magnetic energy. The reflex breaker was located axially in relation to the magnetron on the side opposite to the power source, i.e. the cathode of the magnetron K_1 was extended and served as the cathode of a reflex triode – switch. In the closed

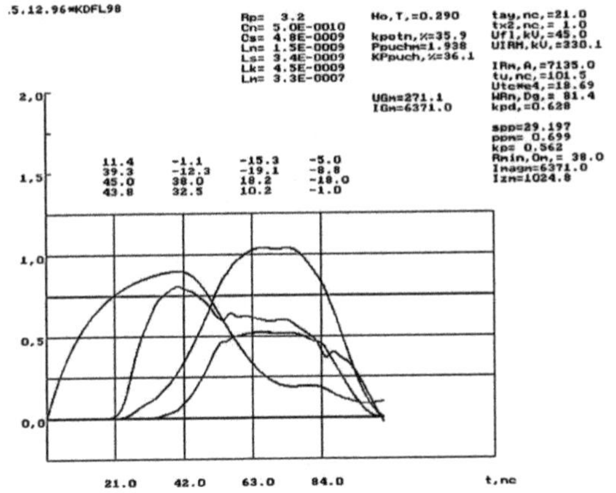

Fig. 3.21. The results of computer simulation of the |RM with a power supply of a LIA with a multichannel discharger.

phase, the power source (pulse voltage generator and 5-ohm forming line) charges the coaxial inductance formed by the cathode and the anode. A low-impedance (4 ohm) reflex triode forms counterflows of electrons and ions. At the same time, the voltage on the RM is below the threshold for excitation of microwave oscillations with an applied magnetic field. The open phase begins when the cathode–anode gap of the triode is short-circuited by the plasma (an additional cathode K_2 is electrically connected to the anode). The breaker opens, and the voltage on the magnetron cathode K_1 sharply (for ~10 ns) rises to a synchronous value of ~1.24 MV. The electric power pulse in the device is increased 2 times, reaching 100 GW. The main requirement for the implementation of such systems is the need to obtain high currents for the operation of the breaker, which leads to the occurrence of significant axial currents along the cathode of the magnetron, the formation of an azimuthal magnetic field, and the removal of electron spokes from the interaction space. As a consequence, a low efficiency of the device (~600 MW with an efficiency of less than 1%). The authors of [33] note that the microwave power is sharply reduced when the axial currents begin to exceed the level of 10–15 kA.

Therefore, the continuation of the work [33] was aimed at excluding the current of the circuit breaker through the cathode of the magnetron, which was later placed before the RM. In the first scheme (Fig. 3.22 *b*), the beam was emitted by a secondary cathode and propagated through the drift tube into the magnetron. This scheme is based on the results of experiments [34] with an RM with an external injection of an electron beam, when it is out of the interaction space. The authors of [33] repeated this experiment at voltages of 1.2 MV, a current of 4–8 kA, and obtained a microwave power of 0.1–1 GW. High efficiency was not achieved, since an additional anode was used at the end of the anode block, causing scattering of electrons and a change in the distribution of the longitudinal and transverse electron velocities in the injected beam. The second circuit (Fig. 3.22 *c*), using the circuit breaker and the so-called inverse diode, was more interesting because of the elimination of the azimuthal magnetic field formed by the flowing current. In this scheme, the current from the reflex triode was received at a collector electrically connected to the cathode of the magnetron. The microwave power level was low (~200 MW), which was explained by the insufficient level of voltage developed on the magnetron for efficient generation.

It is characteristic that in practically all experiments there was a difference in the duration of the voltage pulses and microwave

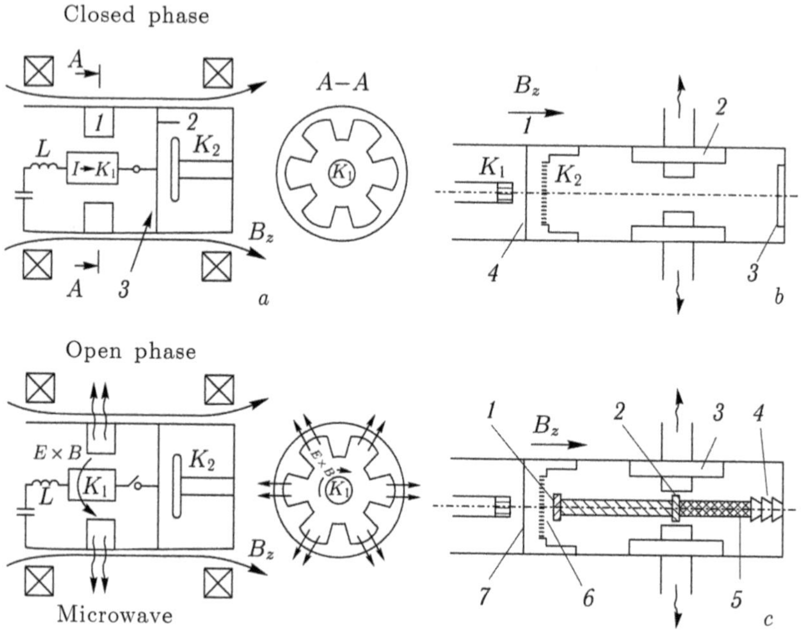

Fig. 3.22. The scheme of experiments using the reflex switches: *a*) power magnetron with a reflex switch: *1* – magnetron, *2* – anode, *3* – reflex switch; *b*) a magnetron with external beam injection from a diode with a reflex switch: *1* – anode, *2* – magnetron, *3* – collector, *4* – reflex switch ; *c*) power supply to the magnetron from the inverse diode with a reflex switch: *1* – graphite collector, *2* – cathode magnetron, *3* – magnetron, *4* – insulator, *5* – nylon rod, *6* – inverse diode, *7* – reflex switch.

radiation. Possible causes of this phenomenon were: the bridging of the resonators by the anode plasma which occurs when the lamellae are heated by the anode current, the high-frequency breakdown of the resonator of the output device; criticality of the device to the shape of the anode voltage. However, only detailed studies with an increased (microsecond) pulse duration allowed to determine the true reason for the limitation of the duration of the microwave pulse (see section 5.1).

In conclusion, we note that the LIAs on magnetic elements presented in this chapter [35] provide a pulse repetition rate of hundreds of hertz with a high repeatability of the amplitude and shape of the output pulses. The use of the original layout scheme, application of the effect of phase overlap, unbalance of capacitance of the last link of the compression of the magnetic pulse generator and the capacity of a single forming line makes it possible to significantly reduce the weight parameters of accelerators in comparison with existing analogues. The chapter presents a method for computer simulation of processes in the power supply system – a microwave generator with strong feedback. The model is based on equivalent power source circuits and formulas of the analytical theory of the averaged motion of the RM. The load (RM) is represented in the form of a non-linear–parametric resistance, which at each step of integration of differential equations is determined as the result of solving the RM model. The method allows to quickly adjust the microwave source to extremes in terms of output power, electronic and full efficiency

The pulse–periodic relativistic magnetrons based on LIAs developed by the TPU can reliably and efficiently operate with a high average microwave power, have a long service life, and have high stability of the generated microwave oscillations (for more details, see Chapters 4 and 6).

References

1. Markov M.A., Usp. Fiz. Nauk, 1973. Vol. 111. No. 4. P. 719.
2. Vorob'ev G.A., Mesyats G.A., Technique for the formation of high-voltage nanosecond pulses. Moscow, Gosatomizdat, 1963.
3. Mesyats G.A., Generation of powerful nanosecond pulses. Moscow, Sov. radio, 1974.
4. Didenko A.N., et al., Powerful electron beams and their application. Moscow, Atomizdat, 1977.

5. Didenko A.N., et al., in: Proc. doc. 6 th All-Union. Symp. on High-current Electronics. Tomsk, 1986. Part 2. P. 197–198.
6. Vakhrushin Yu.P., Anatsky A.I., Linear induction accelerators. Moscow, Atomizdat, 1978.
7. Furman E.G., Pri. Tekh. Eksper. 1987. No. 5. P. 26-31.
8. Furman E.G., Vasilyev V.V., *ibid,* 1988. No. 1. P. 111–116.
9. Furman E.G., et al., *ibid,* 1993. No. 6. P. 45-55.
10. Kiselev Yu.V., Cherepanov V.P., Spark dischargers. Moscow, Sov. radio, 1976.
11. Fourman E.G., Vintizenko I.I., in: Proc. 12 Int. Conf. on High Power Particle Beams. 1998, Israel, Tel-Aviv. P. 107.
12. Meerovich A.A., et al., Magnetic pulse generators. Moscow, Sov. radio, 1968. 476 p.
13. Dolbilov G.V., et al., Prib. Tekh. Eksper. 1984. No. 4. P. 26-31.
14. Birx D.I., IEEETransactions on NuclearScience. 1985. V. NS-32. P. 2743–2747.
15. Vintizenko I.I.,et al., in: Proc. 12 Int. Symposium on High Current Electronics. Russia, Tomsk, 2000. V. 2. P. 255–258.
16. Butakov L.D., et al., Prib. Tekh. Eksper. 2000. No. 3. P. 159–160.
17. Butakov L.D., et al., *ibid,* 2001. No. 5. P. 104–110.
18. Harjes H.C., et al., Proc. 9 Int. Conf. on High-Power Particle Beams. Washington, 1992. V. 1. P. 333–340.
19. Ashby S., et al., Proc. 8 Int Conf. on High-Power Particle Beams. Washington, 1992. V. 2. P. 1855–1860.
20. Mesyats G.A., Movshevich B.Z., Zh. Teor. Fiz. 1989. V. 59. No. 5. P. 39–50.
21. Dolbilov G.V., et al., Prib. Tekh. Eksper. 1984. No. 4. P. 26–31.
22. Vintizenko I.I., Linear induction accelerator, Patent for invention No. 2178244 of the Russian Federation. Publ. BI. 2002. No. 1.
23. Vintizenko I.I., Linear induction accelerator, Patent for invention No. 2185041 of the Russian Federation. Publ. BI. 2002. No. 19.
24. Kalantarov P.L., Tseitlin L.A., Calculation of inductances. Leningrad, Energoatomizdat, 1986..
25. Gordeev V.G., et al., Prib. Tekh. Eksper. 1980. No. 5. P. 117-119.
26. Vasil'ev V.V., et al., Izv. VUZ, Fizika, 2003. No. 10. P. 14-23.
27. Yossel' Yu.A., et al., Calculation of electrical capacity. Leningrad, Energoizdat, 1981.
28. Ginzburg S.G., Methods for solving problems on transient processes in electrical circuits. Moscow, Vysshaya shkola, 1967.
29. Tomskikh O.N., Furman E.G., Prib. Tekh. Eksper., 1991. No. 5. P. 136–138.
30. Dubiev A.I., Kataev I.G., Prib. Tekh. Eksper. 1979. No. 4. P. 172–173.
31. Sulakshin A.S., Zh. Teor. Fiz. 1983. P. 53. No. 11. P. 2266–2268.
32. Nechaev V.E., et al., Relativistic magnetron, Relativistic high-frequency electronics. Gorky: IPF Academy of Sciences of the USSR, 1979. P. 114–130.
33. Benford J., et al., IEEE Trans. 1985. V. 13. No. 6. P. 538–544.
34. Craig G., et al., Abstr. IEEE Int. Conf. on Plasma Science, Montreal, 1979. P. 44.
35. Vintizenko I.I., Izv. VUZ, Fizika, 2007. No. 10/2. P. 136–141.

4

Pulse–Periodic Relativistic Magnetrons

To increase the average power of electromagnetic radiation sources, it is possible to use relativistic magnetrons with moderate values of the pulse power and duration, but operating with a high repetition rate. However, there arises a set of scientific and technical problems, connected with the choice of RM power sources, determination of the parameters of electron beams that do not lead to destruction of the anode blocks, calculation and design of RMs that allow operation of the device with a high pulse repetition rate. In addition, further requirements are imposed on the cathode unit of the RM, magnetic and vacuum systems. This chapter is devoted to the discussion of these questions.

4.1. Studies of RMs in pulse–periodic mode with the use of LIAs with multichannel spark dischargers

First studies of the relativistic magnetron in the pulse–periodic mode of operation were carried out in Russia (Tomsk, the Tomsk Polytechnic University) in 1986 [1, 2]. Later, in the early 90s, linear induction accelerators (LIA) for the supply of RMs began to be used in the USA, their parameters were in the range of 0.3–0.75 MV, current 3–10 kA, pulse repetition rate up to 100 Hz, average power of microwave radiation to 6.3 kW [3–5]. The appearance of the first installation of the TPU is shown in Fig. 4.1 [1]. The scheme of the experiment is presented in Fig. 4.2. The power source was an LIA with 12 ferromagnetic cores and with a 24-channel spark discharger switching 8 parallel DFLs. In the nominal mode, the charging voltage

Fig. 4.1. Appearance of a pulse–periodic relativistic magnetron with a power source – LIA with a multichannel discharger, 1986.

of the lines was 40–42 kV, which provided a voltage at the cathode of ~400 kV at a load of 130 Ohm. The accelerator operated with a pulse repetition rate in a continuous mode up to 50 Hz, in a packet mode of 5–10 pulses – up to 160 Hz. The pulse frequency in the continuous mode was determined by the rate of gas pumping of the operating volume of the discharger. In the packet mode, the frequency limits were determined by the time characteristics of the charging scheme of the DFL. As noted above, LIA has an internal resistance, which is in good agreement with the impedance of the RM. In the formulation of research, this circumstance was of principal importance. In the experiments, a 6-cavity magnetron of the 10 cm wavelength band was used. The magnetic field was formed by two coils as a Helmholtz pair. The power supply circuit of the magnetic system operated in the single-pulse mode, however the long duration of the current pulse ~ 100 ms allowed to realize a packet of three 80 nanosecond pulses of the voltage with a repetition frequency of 160 Hz at practically constant magnetic field value.

In the experiments [1, 2], a protective screen was mounted on the cathode holder *9* to eliminate the 'parasitic' reverse current, which leads to the breakdown of the radial high-voltage insulator *8* of the electron gun. The outer diameter of the screen was chosen so that the lines of force of the magnetic field emerging from the interaction space of the magnetron *11* crossed its surface. This design allowed to intercept the reverse electron flux from the interaction space of

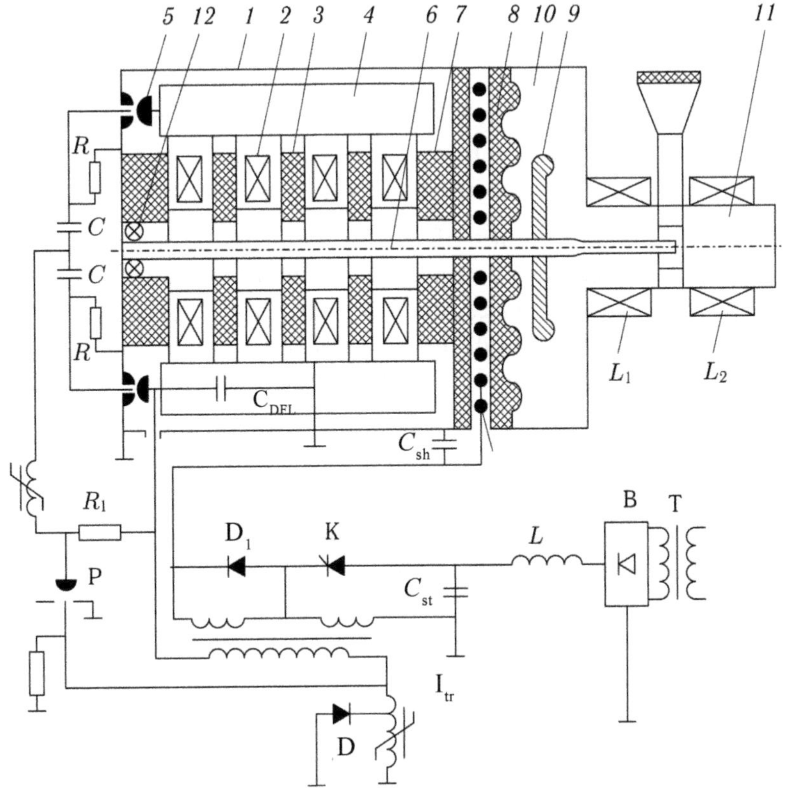

Fig. 4.2. The layout of the LIA and the principal electrical circuit for feeding the LIA for a pulse–periodic relativistic magnetron: *1* – the case of the LIA; *2* – ferromagnetic core; *3* – magnetizing turn; *4* – DFL; *5* – anode of a multichannel discharger; *6* – cathode holder; *7* – insulator between the cores; *8* – high-voltage insulator; *9* – protective screen; *10* – the vacuum chamber; *11* – relativistic magnetron; *12* – Rogowski coil; L_1, L_2 – coils of the magnetic system; C_{DFL} – forming line; C_{sh} – shunting capacitor of the output of a demagnetization spiral; *P* – start-up discharger; *R*, R_1, *C*, Dr_n – elements of a circuit of start of the multichannel discharger; *T* – mains transformer; *B* – rectifier; *L* – charge inductance; C_{st} – storage capacitance; *K* – thyristor commutator; I_{tr} – the impulse transformer; D_1, L_d – diode and inductance of the demagnetizing circuit; *D* – diode of the starting circuit of the starting discharger.

the RM and solve the problem of breakdown of the insulator, which made it possible to conduct a long series of experiments.

The maximum power of the microwave radiation of an RM reached 360 MW which corresponds to the electron efficiency ~30%. Since there was no voltage sensor in the experimental scheme, estimates of the magnitude of the accelerator voltage were made on the basis of measurements of the synchronous magnetic field under the assumption $\beta_p \approx \beta_e \approx 0.34$ and amounted to ~300 kV.

A characteristic feature of the operation of an RM during feeding from an LIA in this experiment was the growth of the operating current in the region of synchronous magnetic fields. This phenomenon, typical for high-efficiency low-voltage magnetrons, was noted earlier only for inverted relativistic magnetrons (see Chapter 5). Such a mode of current resonance is possible:

1) at comparable values of the strength of the static electricity field and electrical components of the high-frequency fields;
2) at low current losses due to 'parasitic' emission from the surface of the cathode holder and the small value of the end current owing to the relatively low cathode–anode voltage.

It is due to these factors that the high efficiency of the relativistic magnetron was achieved.

In the experiments, a good repeatability of the amplitude and shape of the recorded pulses was noted. Analogous pulses were obtained also with a single operating mode of the accelerator. One should note a close match of the current pulse duration and the microwave signal, which indicates a high accelerator efficiency and a short time of the rise of the oscillations not exceeding a few nanoseconds.

Thus, the first experiment demonstrated the possibility of a relativistic magnetron operating under power from an LIA in a pulse–periodic mode with a repetition frequency of at least 160 Hz. Subsequently several RMs were constructed using LIAs with multichannel dischargers for conducting experiments with the summation of microwave power from the two outputs of the magnetron, and the power of summation of the two magnetrons powered by two LIAs (see Chapter 1).

The result of the first studies was the development in 1988–1989 of a pulse–periodic relativistic magnetron placed on a mobile platform (Fig. 4.3) [6]. The layout the power supply circuit based on LIA 04/7 differed in that there was no solid insulator separating the cathode holder from the case (see Fig. 4.4). As shown by the first experiments with the RM, this element is the least reliable, since it is subjected to intensive action from the cathode–anode gap (ultraviolet irradiation, bombardment by charged particles). The insulator is made in the form of separate plexiglass rings between the ferromagnetic cores. The thickness of the rings was 16 mm and they had a developed surface, the electrical strength exceeded 80 kV/cm. The magnetizing coils of the inductors were made solid (a

Fig. 4.3. Appearance of a pulse–periodic relativistic magnetron with a power source – the LIA located on a mobile platform.

version of the 'capacitive' inductor was realized – see section 3.3.2), which together with the insulators formed the vacuum volume of the accelerator. Since the turns have an inner diameter smaller than the diameter of the insulators, they shield the surface of the insulators from the action of the above factors.

Special conditions were imposed on the design of the power supply system of the installation from the on-board network with a power consumption of not more than 40 kW. The power supply of the LIA should provide:

1) charging of the forming lines;
2) formation of a demagnetizing field of inductors;
3) formation of a pulsed magnetic field;
4) synchronous activation of the channels of a multichannel discharger.

When selecting and developing the supply, it was taken into account that its most energy-intensive part is the power supply system of the magnetic field coils. The basis was an economical unipolar circuit with energy recovery. The schematic circuit diagram is shown in Fig. 4.4 [6]. The main parts of the scheme are as follows:

1) a power source consisting of a charging inductance L, mains transformer T, rectifier B;
2) capacitive storages of the magnetic field former C_1, C_2 and charge of DFL C_{st};
3) switching block of thyristors T_1 and ignitron I;
4) pulse charge transformer for DFL I_p;

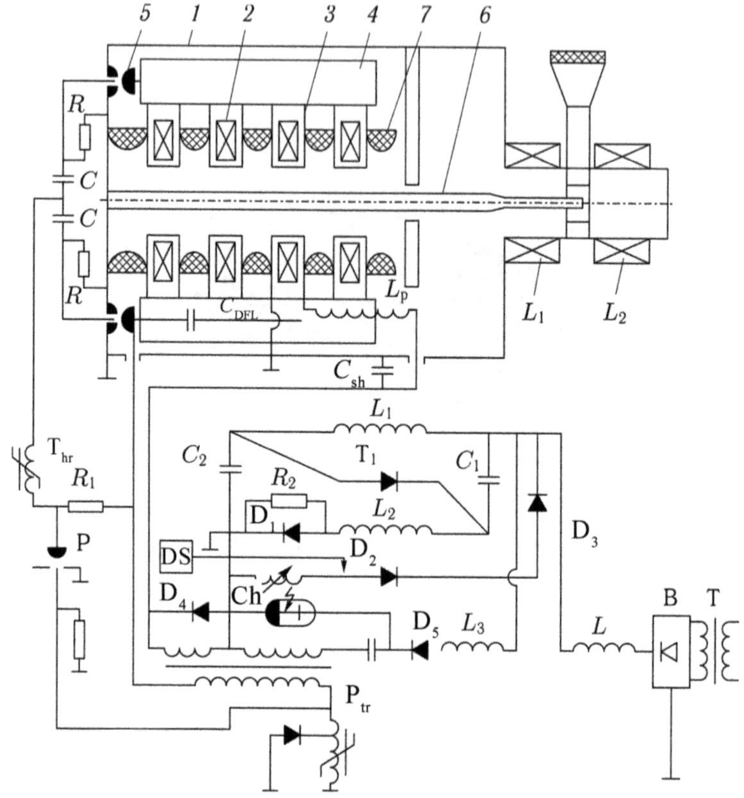

Fig. 4.4. The layout of the LIA and the principal electrical circuit for feeding the LIA and the magnetic system for the pulse–periodic relativistic magnetron on the mobile platform: *1* – the case of the LIA; *2* – ferromagnetic core; *3* – magnetizing turn, *4* – DFL; *5* – anode of a multichannel discharger; *6* – cathode holder; *7* – insulator; L_1, L_2 – coils of the magnetic system; C_{DFL} – forming line; C_{sh} – shunting capacitor of a conclusion of a spiral of demagnetization; P – starter; R, R_1, C, T_{hr} – elements of a circuit of start of the multichannel discharger; T – mains transformer; B – rectifier; L – charging inductance; C_n – storage capacity; I – an ignitronic switch; P_{tr} – the impulse transformer; D_1, L_p – diode and inductance of demagnetization circuit; D – diode of the starting circuit of the starting gap.

5) diode D_1 and resistor R_2, performing protective functions and preventing oscillations in the circuit – capacitive storage C_1 and C_2 – inductance of coils of the Helmholtz pair L_1 and L_2;
6) the diode D_4 connected to demagnetizing inductance L_p of the induction system;
7) starting device of the multi-channel discharger that includes the bias circuits of the multichannel discharger R, R_1, C, the start discharger P (the operation of the starting device is described in section 3.2.3).

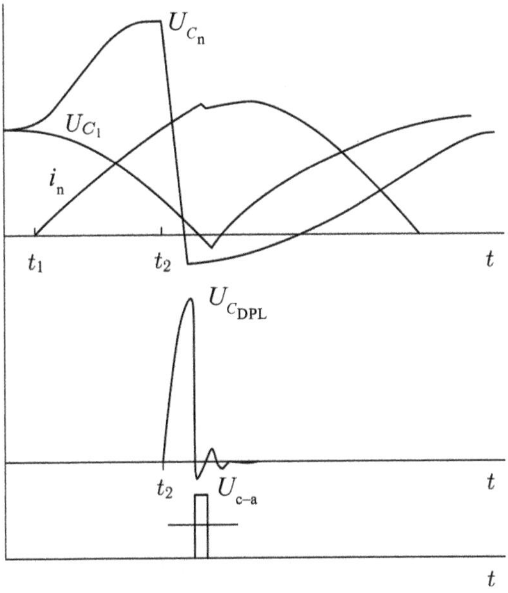

Fig. 4.5. Diagram of voltage on elements scheme of the power supply of the LIA: U_{C_1}, U_{C_n}, $U_{C_{DPL}}$ – diagrams of voltage on capacitive storage C_1, C_n and of the double forming line; i_n – current through the coils of the magnetic systems; U_{c-a} – voltage on cathode–anode gap of the RM with duration τ.

The power scheme works as follows. From AC mains through transformer T, rectifier B and inductance L energy is supplied to the storage capacitors C_1 and C_2 of the magnetic field former and storages C_n of the charge of the DFL. When the thyristor unit T_1 is turned on – time instant t_1 (Fig. 4.5) – the current begins to flow through the turns of the magnetic field coils L_1, L_2 and through the circuit L_3, D_5 and the storage C_n is recharged to a voltage close to $UC_1 + UC_2$. When the voltage reaches the desired value (set by the control system of the installation), the ignitron is activated. The storage C_n begins to discharge through the primary winding of the pulse transformer, charging the DFL and simultaneously forming the demagnetizing current of the cores of the induction system by means of an additional winding of the transformer, inductance L_p. To reduce the amplitude of the induced voltage on the diode D_4, the demagnetizing spiral is shunted by a capacitor C_{sh}. At time t_2, the DFL is charged to the specified voltage. The charging current of the line goes through zero, the ignitron turns off, the starter generates the trigger pulse of the multichannel discharger. Switching off the thyristor block occurs under the action of reverse voltage with a partial overcharge of the storages C_1 and C_2. The recharge amount (the duration of reverse voltage) is determined by the magnitude of the flux linkage of the choke Ch, previously demagnetized from the external source DS. When the thyristor unit is turned off, the current of the magnetic field coils passes into the capacitive storage devices C_1 and C_2 along the circuit D_1, D_2, D_3. The demagnetizing current is closed via diode D_4. The magnitude of the induction of the magnetic field is regulated by the voltage and capacitance of the storage devices C_1 and C_2.

The ignitron is turned on at angles $60° \leq \omega_{cir} t \leq 80°$, where $t = t_2 - t_1$, ω_{cir} is the eigenfrequency of the discharge circuit: C_1, $C_2 - L_1$, L_2. The magnetic field varies according to the law

$$B \approx U_{C_1}^{M} \rho_{cir}^{-1} K_1 \sin(\omega_{cir} t), \tag{4.1.1}$$

where ρ_{cir} is the wave impedance of the circuit; K_1 is the coefficient determined by the geometry of the coils; $U_{C_1}^{M}$ is the amplitude of the charging voltage of the storage C_1. At time t_2, the DFL is charged to the specified voltage:

$$U_{CDFL} \approx 2U_{C_n}^{M} K_{tr} \cos(\omega_{cir} t), \tag{4.1.2}$$

where K_{tr} is the transformer ratio of the pulse transformer.

The impulse of the output voltage U_i is proportional to U_{CDFL} with the coefficient N_c (the number of cores of the induction system). Thus, in the described scheme with the selected parameters: K_1, K_{tr}, N_c, ρ_{cir}, the ratio

$$U_i / B = \rho_{cir} K_{tr} N_c \, \mathrm{ctg}(\omega_{cir} t) / K_1 \tag{4.1.3}$$

does not depend on the voltage of the storages and is determined only by the moment of activation of the ignitron. In the range $60° \leq \omega_{cir} t \leq 80°$ the output voltage of the LIA changes by 2–3 times with insignificant changes in the magnitude of the magnetic field (5–10%). Using such a power scheme makes it easy to adjust the voltage and magnetic field to fulfill the synchronism condition in the RM.

The frequency of the pulses is limited by the duration of the current pulse of the magnetic system and increased to 20 Hz in comparison with the traditional scheme without energy recuperation (10 Hz). The decrease in the duration of the current pulse in systems with energy recuperation is

$$\Delta t = \frac{\pi(\Omega - 2\omega_{\text{cir}}) - 2\Omega\omega_{\text{cir}}t}{2\Omega\omega_{\text{cir}}} + \frac{2\pi R}{\rho_{\text{cir}}\omega_{\text{cir}}}, \qquad (4.1.4)$$

where Ω is the circular circuit frequency: storage C_n – DFL; R is the equivalent loss resistance in the LIA supply circuit. The thermal losses in the coils of the magnetic system are also reduced in proportion to the reduction in duration.

The power supply circuit of the LIA made it possible to generate output voltage pulses with an amplitude of up to 300 kV at a current of ~ 3.3 kA, duration ~70 ns, with a repetition frequency of 20 Hz. The magnetic system created a magnetic field with an induction up to 0.4 T. In general, the power scheme was located in a support with the dimensions 1.2 × 0.9 × 1 m. The output parameters of the relativistic RM reached 200 MW, which corresponded to an electron efficiency of ~20%.

In addition, the pulse–periodic RM was fitted with a harmonic filter for non-linear radar electronic devices and a pyramidal antenna of high directivity of microwave radiation.

The systems described above were used to study the effect of high-power electromagnetic radiation on various semiconductor elements, radioelectronic devices, and biological objects.

4.2. The basic elements of the pulse–periodic RM

A new stage in the development of pulse–periodic RMs occurred when LIAs on magnetic elements were constructed in the early 90s [7,8]. Examples of such sources are two options of the microwave-generators based on LIA 04/4000 [9,10] and LIA 04/6 [11]. The design, the principle of operation, and the elemental basis of such accelerators are described in section 3. Since the LIAs on magnetic elements allow to significantly increase both the repetition rate of pulses and the number of pulses in a continuous series, this introduces additional requirements for the elements of the relativistic magnetron, the magnetic and vacuum systems. Long research studies have been devoted to their solution, which will be discussed below.

Conventionally, the following elements of the installations can be distinguished (block diagram of the installation is shown in Fig. 4.6):

- The linear induction accelerator, consisting of a high-voltage generator and a power source, is used in the installation to operate with the pulse repetition frequency up to 320 Hz at a

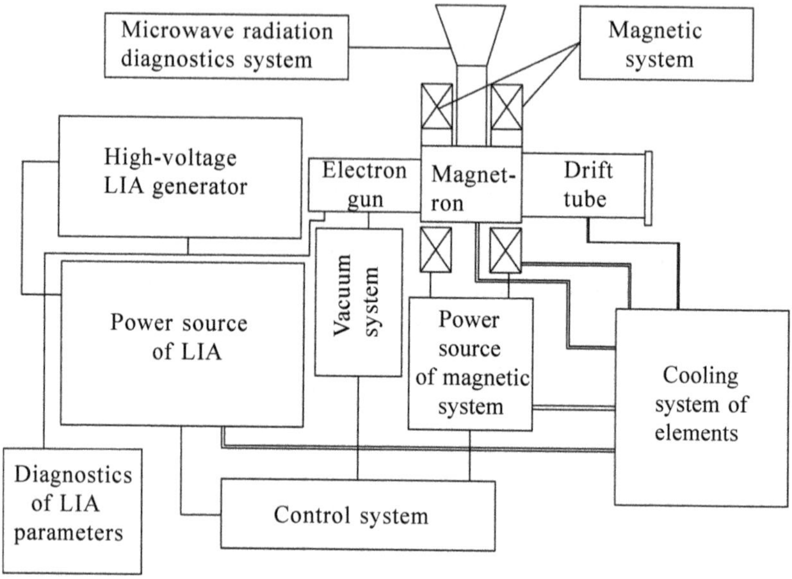

Fig. 4.6. The block-diagram of the elements pulse–periodic relativistic magnetron.

high repetability of the amplitude–time pulse characteristics (instability of less than 5%).

- The high-voltage generator of the LIA is designed to generate high-voltage pulses at the central electrode (the cathode of a RM), consists of a magnetic pulse generator, forming lines, and an induction system (for details see Chapter 3).
- The vacuum system is designed to produce a vacuum no higher than $2 \cdot 10^{-5}$ Torr in the vacuum chamber, has additional channels for evacuating the interaction space on the drift tube side and waveguide power output.
- The power supply of the LIA with stabilization of the voltage level of the charging pulses of the high-voltage generator of the LIA and energy recovery.
- The control system of the installation determines the pulse repetition rate, the number of pulses in the packet, the duration of the pause between packets, the automatic or manual activation of the packet, sets the amplitude of the charging pulse voltage, the magnetic field, and is provided with locks that exclude the misuse of the installation.
- The system for diagnostics of parameters consists of capacitive voltage dividers and Rogowski coils for measuring the output parameters of the LIA.

- Diagnostics of the microwave radiation used for measuring the amplitude, power and frequency parameters of the generated microwave pulses is composed of detectors, directional couplers, attenuators, and tunable attenuators, bandpass filters so on.
- The relativistic magnetron contains a graphite cathode of special geometry for reducing the loss of end current, a water-cooled anode block with a device for outputing microwave radiation.
- The magnetic system is a Helmholtz pair, powered from a constant current source, creates a uniform magnetic field in the interaction space of the magnetron with induction to 0.59 T, ensures a continuous operation of the installation. The magnetic system is made of a hollow copper tube with a central hole for the flow of cooling water.
- The water cooling system establishes the necessary thermal regime of the anode block, drift tube, magnetic field coils, semiconductor elements of the LIA power system and the magnetic system.

4.3. The construction of the anode block with the device of output of microwave radiation

The slowing-down systems of the magnetron are based on 6- and 8-cavity anode blocks of the sector type. Differing in simplicity of design, the slowing structure of this type nevertheless satisfactorily adapts to the operating conditions with the power source – LIA. A characteristic feature of LIAs on magnetic elements is the change in the amplitude of the anode voltage during the pulse. At the initial time, the accelerator operates at the idle mode, since the anode current of the relativistic magnetron is small. Then, as the oscillations develop, the current increases, the impedance of the load falls, which leads to a decrease in the amplitude of the voltage. Under these conditions, for the stable operation of the magnetron it is necessary to maximize the separation of the operating mode of oscillations from 'parasitic' in frequency and phase velocity. It was shown above that the corresponding choice of the size of the anode block can be achieved. In addition, a very important condition for relativistic magnetrons is the possibility of operating at low values of the magnetic field, such that electromagnetic systems can be used with direct current. This condition is necessary when working with a high frequency of repetition of the pulses. In all this there are significant limitations on the choice of the geometry of the resonator system.

One of the most stringent requirements for a pulse–periodic magnetron is the durability of the cathode and the anode block, both in the total number of pulses and in the number of pulses in a separate series, with high stability of the generated frequency. These parameters of the magnetron are connected with the processes of destruction of the anode surface (see section 1.3) and changing emissivity of the cathode material.

As a result of the calculations and recommendations, the following geometric dimensions of the resonator system of the magnetron were chosen: R = 2.15 cm, D_r = 4.30 cm, N = 6, 2Θ = 40°, r_c = 1.0–1.2 cm, L = 7.2 cm (N = 8, 2Θ = 22.5°, other dimensions are the same). In this case, the magnitude of the slowing-down of the electromagnetic wave for the working π-mode $\beta_p^{N=6} \sim 0.45$, $\beta_p^{N=8} = 0.366$ the required magnetic field strength is 3–5 kG for the output voltage level of the LIA of 300–400 kV.

The relativistic magnetron with an 8-cavity block was designed to increase the efficiency of the installation by increasing the electron efficiency. The excitation of the operating π-mode of oscillations in this magnetron occurs at higher synchronous magnetic fields, with the radius of the cyclotron rotation of the electrons decreasing, which should lead to an increase in the electronic efficiency of the device. For a correct comparison of the experimental results, both anode blocks have the same geometric dimensions (cathode and anode radii, internal resonator radius), which ensures the same load of the power source at the pulse front.

Table 4.1 shows the results of calculating the wavelengths of the modes of oscillations and the magnitude of the slowing-down of the electromagnetic wave of the operating, the nearest (N|2 – 1), its (−1) and (+1) harmonics of the modes of oscillations, as well as the results of ‘cold’ measurements at a low power level using a microwave panoramic meter (in italics).

Some difference in the calculated and measured wavelengths of oscillations is caused by the presence of a waveguide output of power, which was not taken into account in the calculations.

As seen from the table, the separation of oscillatory modes in magnitude to the phase velocity of the 6-resonator RM is $\beta_\pi/\beta_{2\pi/3} \approx$ 27% of its (−1) harmonic $\beta_\pi/\beta_{2\pi/3}^{-1} \approx$ 31%. Separation of the oscillation modes for the 8-resonator anode block is less than for the 6-cavity anode by about a third and is $\beta_\pi/\beta_{3\pi/4} \approx$ 21% and $\beta_\pi/\beta_{3\pi/4}^{-1} \approx$ 24%. Out. 4.1 it is clear that the slowing-down of the mode of oscillation of

Table 4.1.

$N = 8$		$N = 6$	
$3\pi/4$-type of oscillations		$2\pi/3$ type of oscillations	
$\lambda_{3\pi/4}$, cm	9.305	$\lambda_{2\pi/3}$, cm	10.093
$\beta_{3\pi/4}$	0.448	$\beta_{2\pi/3}$	0.669
β_{-1}	0.29	$\beta_{2\pi/3}^{-1}$	0.335
β_{+1}	0.132	$\beta_{2\pi/3}^{+1}$	0.167
$\lambda_{3\pi/4}$, cm	*9.68*	$\lambda_{2\pi/3}$, cm	*10.4*
π-type of oscillations		π-type of oscillations	
λ_{π}, cm	9,223	λ_{π}, cm	9.77
β_{π}	0.366	β_{π}	0.461
β_{-1}	0.366	β_{π}^{-1}	0.461
β_{+1}	0.122	β_{π}^{+1}	0.154
λ_{π}, cm	*9.43*	λ_{π}, cm	*10.6*

the 8-cavity anode block in comparison with the 6-cavity anode block is approximately 26% higher and higher efficiency can be expected from this block. On the other hand, the large amount of slowing-down necessitates the operation of the magnetron at higher values of the synchronous magnetic field, which required the modernization of the magnetic system.

The anode block without a water cooling jacket and the working drawing of the 6-cavity anode block are shown in Fig. 4.7. Following the results of thermal calculations, the block is entirely made of non-magnetic stainless steel. All permanent joints in the construction are made by argon-arc welding. The anode block is a cylinder 86 mm high with an external diameter of 116 mm. A technological boss is welded to the side surface of the anode block intended, first, for precision docking to the anode block of the microwave energy output device, and secondly, to ensure, if possible, welding of the anode block with the output device without deforming the block.

The internal cavity of the block in cross section consists of a central anode hole with a diameter of 43 mm and six sector-shaped resonators with an external diameter of 86 mm and angular dimensions of 20°. The height of the working part of the anode block is 72 mm. The inner surface of the anode block is manufactured

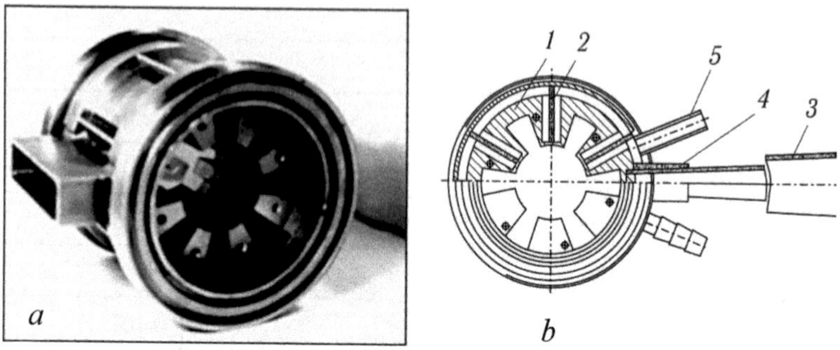

Fig. 4.7. Appearance of the anode block without a water cooling jacket and power output (*a*); anode block of the 6-cavity pulse–periodic relativistic magnetron (*b*): *1* – anode block; *2* – diaphragm; *3* – power waveguide output; *4* – boss; *5* – nozzle.

by electroerosion technology, which allows obtaining the necessary accuracy and quality of the product. In a series of 6 anode blocks, the resonant frequencies of the modes and the values of the loaded *Q*-factor were compared in the range of 0.3%. The highest requirements are imposed on the geometric dimensions of the central anode aperture – diameter $43^{+0.03}$ mm and slits of the resonators – $7.4^{\pm 0.08}$ mm, and on the symmetry of the structure: the angle between the axes of the slits is 60°±10′; slit asymmetry 0.05 mm; the flatness of the side walls of the lamellae is 0.02 mm; on the arithmetical mean value of the profiles of the anode aperture and the walls of the resonators it was 0.63 μm. To reduce the edge strength of the electric field in order to increase the durability of the anode block, the radius of the lateral edges of the lamellae is set to 0.5 mm. For the output of microwave energy from the anode structure the rear wall of one of the resonators contains a rectangular window 10 mm wide, height 72 mm and thickness 2 mm. In the boss there is a profiled hole, the size and shape of which provide an accurate connection between the output device and the anode block communication slit.

The cooling system of the anode block consists of internal and external circuits and a regulating device. The internal circuit is formed by rectangular cavities milled in the anode block lamellae from its outer side. The external circuit is formed by the outer wall of the anode block and the inner surface of the jacket. The regulating device has six special shape diaphragms, partitioning the inner and outer circuits along the radius and ensuring the sequential flow of cooling water through each of the six cavities in the lamellae and regulating the water flow along the radius and height of the cavities. Cooling water is supplied through the nozzle *5*.

The microwave energy output device *3* is 345 mm in length and consists of a rectangular waveguide of variable cross section and a choke flange welded thereto. The internal cross-section of the waveguide smoothly varies from 10 × 72 mm^2 at the point of contact with the lead-out window of the anode block up to 34 × 72 mm^2 at a length of 312 mm. The microwave radiation is outputted into free space using pyramidal horns. Since the radiation from the magnetron is carried out at the lowest wave of a rectangular waveguide, it is not difficult to calculate similar antennas with a given directionality. The antennas are made evacuated – the aperture is covered by a dielectric window made of plexiglass. To prevent surface discharge, one must select the appropriate window sizes. Under the assumption of an operating power level of ~300 MW, the window size was 320 × 300 mm. If the electric strength of the dielectric on the external (air side) surface is ~30 kV/cm and a pulse duration of microwave-radiation ~100–150 ns, the antenna output window of a predetermined size is able to transmit tentatively without breakdown the pulse power of ~600 MW. The length of the horn was chosen to be 730 mm, which formed the opening angle in the *H*-plane equal to 25°. When excited by a wave of H_{10} (λ ~ 10 cm), the calculated beamwidth of the directional pattern between the zero values of the main lobe was 60°. During the 'cold' measurements of the antennas, the effective area of the aperture, the directivity coefficient, the width of the radiation patterns in the *E*- and *H*-planes at different wavelengths corresponding to the modes of the oscillating system were determined. Microwave probes (loops) were placed on the lateral surface of the antenna for recording the output power. There was also a branch pipe for additional vacuum pumping of the interaction space of the magnetron through the communication gap, power output and an antenna. To determine the power, in some experiments the vacuumed directional couplers were installed between the output waveguide and the antenna.

4.4. Cathode unit

Structurally, the magnetron diode is attached to the LIA module by means of a cantilever connection. The mounting point of the console is the output flange of the LIA. The distance from the module flange to the centre of the anode block is 740 mm, and to the end flange of the drift tube is 1100 mm. Taking into account the fact that the individual elements are rigidly interconnected through the gaskets from the vacuum and oil resistant rubber (high-voltage insulator of the electron gun of the LIA), it is very difficult to ensure that the cathode and the multi-resonator anode block are aligned at such a large length during assembly. This is illustrated by photographs of the interelectrode gap made during the experiments through the vacuum window of the drift tube. Figure 4.8 *a* shows that the volume discharge in the space between the anode and the cathode is inhomogeneous in azimuth. At the same time, one can judge the radial displacement of the cathode from the axis and the decrease in the interelectrode gap at this point. This circumstance substantially worsens the condition for establishing electromagnetic oscillations in the resonators of the relativistic magnetron and affects the formation of electron spokes in the interaction space. Figure 4.8 *b* shows the oscillograms of the total current pulse, voltage and microwave radiation during the operation of the magnetron with the radial displacement of the cathode. A distinctive feature is the strong modulation of pulses. We also note that an asymmetric discharge in the interaction space leads to local overheating and destruction of the surface of the anode block and cathode.

Figure 4.9 shows a photograph of the interelectrode gap of the RM when the cathode is installed using an alignment device and the corresponding waveforms of the pulses. It can be seen that the discharge has a more homogeneous azimuth character and the oscillograms do not have modulations. This device showed great reliability, providing high reproducibility of the output parameters of the RM when replacing the cathodes and the durability of the anode block.

The thermal regime of the cathode is determined by the electrons which are not trapped in the generation mode and catch into the accelerating phase of the microwave wave and then fall on the cathode. According to the data known from the literature, 5–10% of the power supplied to the magnetron is spent on heating the cathode by 'inverse' (incorrect phased) electrons. However, there is no reliable data on the energy of the incident electrons. If we assume that the energy of the 'inverse' electrons is greater than 20 keV, then the impulse temperature of the stainless steel cathode surface will reach 500°C and the permissible number of pulses, even at an average cathode temperature of T_0 = 200°C, will exceed 10^5. In this case, a great danger for the stable operation of the magnetron is the loss of the emission capacity of the cathode at a large number of pulses. Experimental data published recently [12,13] indicate that in

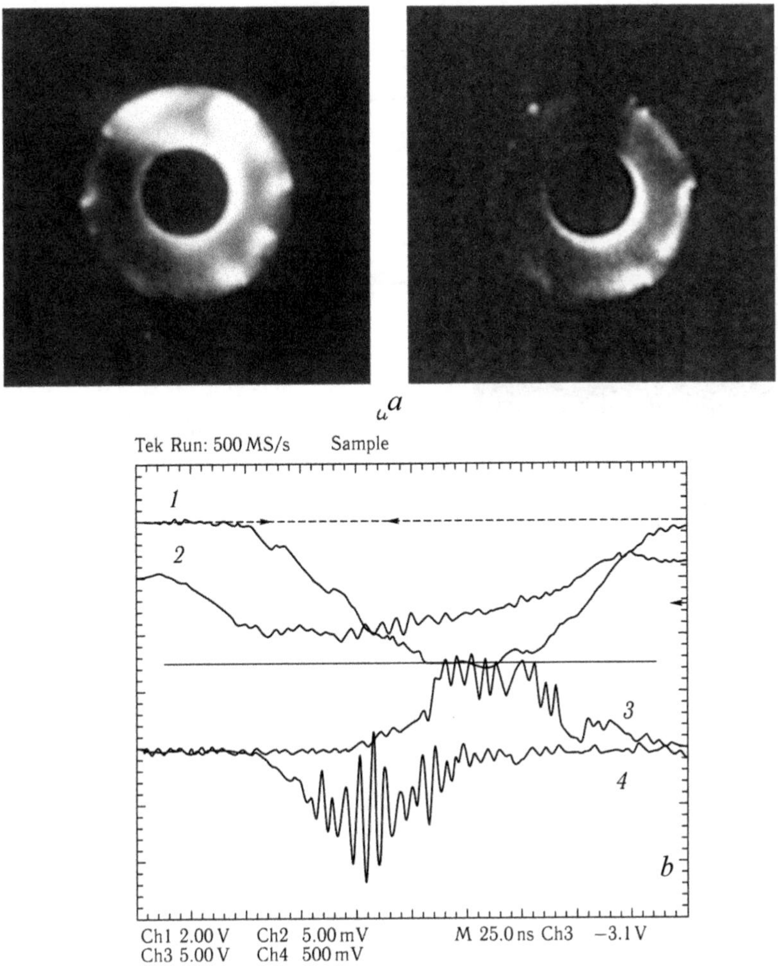

Fig. 4.8. Glow of plasma in the interelectrode gap of a relativistic magnetron at a radial displacement of the cathode relative to the axis of the anode block. Number of pulses 20 and 10 (*a*). Oscillograms of the pulses during radial displacement of the cathode: *1*) current, *2*) voltage, *3*, *4*) microwave detector signals (*b*).

planar and coaxial diodes at voltages close to the conditions of these studies, but with a pulse duration of ~20 ns, the permissible number of pulses for cathodes made of different metals, does not exceed 10^3–10^5. Therefore, considerable attention was paid to the choice of the cathode material. The results of these tests are described below.

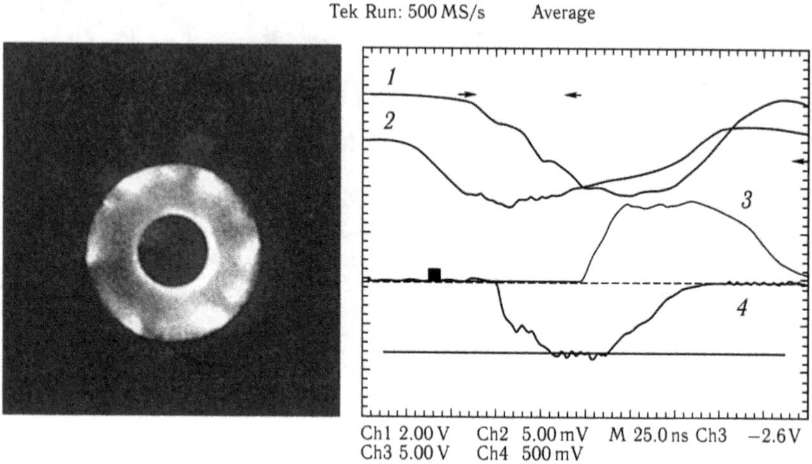

Fig. 4.9. The glow of plasma in the interelectrode gap of the RM using the alignment device and the corresponding pulse waveforms: *1*) current, *2*) voltage, *3*, *4*) microwave detector signals.

4.4.1. The cathode material for a pulse–periodic relativistic magnetron

Cathodes made from stainless steel, graphite, and also metal-dielectric were studied. Cathodes made of stainless steel are traditionally used for relativistic magnetrons, beginning with the very first experiments. Their advantages are high resistance to electron bombardment by incorrect phased electrons, high heat capacity, small material loss. The main disadvantage of such a cathode is a small longevity in terms of emissivity. As the operating time increases, the amplitude of the voltage increases and the current decreases with a simultaneous fall in the rate of rise of the front of the current. Also noticeable is the shortening of the duration, the appearance of instability of the amplitude of the signals from the shot to the shot.

A cathode was proposed in [14] for planar vacuum diodes, in which the metal–insulator contact serves as the emission surface. For such a cathode, the threshold for the formation of explosive electron emission is lower, as a result of this, a large number of emission centres are formed in comparison with cathodes made of homogeneous materials. Metal–dielectric cathodes for pulse-periodic RMs were produced as follows. Copper was interlaced with one or two mica gaskets. The outer diameter of the cathode was machined on a lathe, and the surface was etched with nitric acid to remove copper burrs. From the point of view of obtaining high output power and generation stability at large (~10^5) numbers of pulses these cathodes exceed the stainless steel cathodes. However, they have limitations on the pulse repetition rate. Continuous operation of an RM with a metal–dielectric cathode does not exceed units of hertz because of the deteriotation of the vacuum due to the large amount of the evaporated cathode material both by explosive electron emission and by bombardment with incorrect phased electrons. The work of a magnetron with such cathodes at repetition frequencies is possible up to 80 Hz or in the regime of short bursts of pulses (no more than 20 pulses).

The best parameters for the criteria of durability, stability, low threshold of formation of explosive electron emission were recorded to the cathodes from pyrolytic graphite. Judging by the oscillograms (Fig. 4.9), it can be noted that the duration of microwave pulses is close to the current pulse duration. However, in long-term continuous operation of the RM (more than 10^6 pulses) there was an increase in the voltage pulse amplitude, decrease of the accelerator current with a corresponding drop of the microwave generation power. These facts indicated a reduction in the emission capacity of the cathode and require abrasive treatment of the cathode surface.

4.4.2. Effect of cathode dimensions on the output characteristics of RM

The tests of each RM begin with the determination of the outer diameter of the cathode and its length, providing the maximum output parameters, with synchronous values of the magnetic field. Figure 4.10 shows the dependence of the microwave power of the RM on the cathode diameter. The maximum of the curve $P(d_k)$ is explained by better impedance matching of the linear induction accelerator and the relativistic magnetron and also by the physical processes of interaction of electrons with the microwave waves of the anode block discussed in Chapter 2. Further experiments were carried out only with cathodes with a diameter of 19 mm.

The length of the cathode and the profile of its ends make it possible to reduce the current leaving the interaction space (the end current of the drift tube). For pulse–periodic magnetrons operating at relatively low currents of 3–6 kA, the end current is determined by the electric field strength at the end of the cathode. In this connection, it has been proposed to use a cathode longer than the length of the anode block [15]. The end of the cathode is rounded

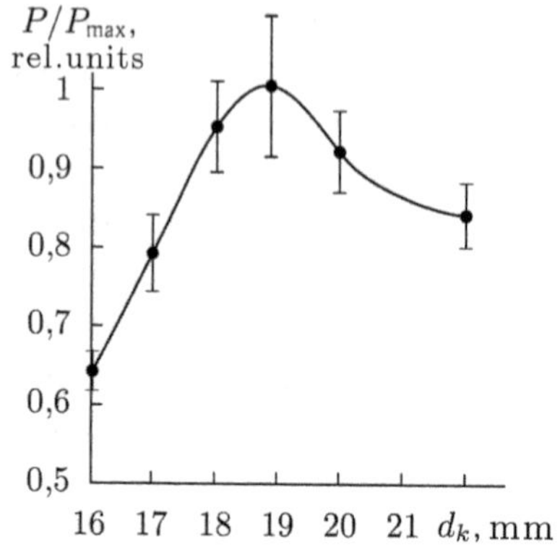

Fig. 4.10. The dependence of the microwave radiation power of the RM on the cathode diameter, P_{max} = 170 MW.

with a radius, equal to the radius of the cathode. Due to this, an uniform electron beam of the end current is formed [16]. It is known that the limiting transport current of an uniform beam in the drift tube is less than the limiting transport current of a cylindrical beam.

Figure 4.11 shows the dependence of the output power of the RM on the cathode length, with the the maximum of the microwave power corresponding to a cathode length of 110 mm. Increase in length above 110 mm led to the fact that the end of the cathode fell into the edge magnetic field systems and current losses increased.

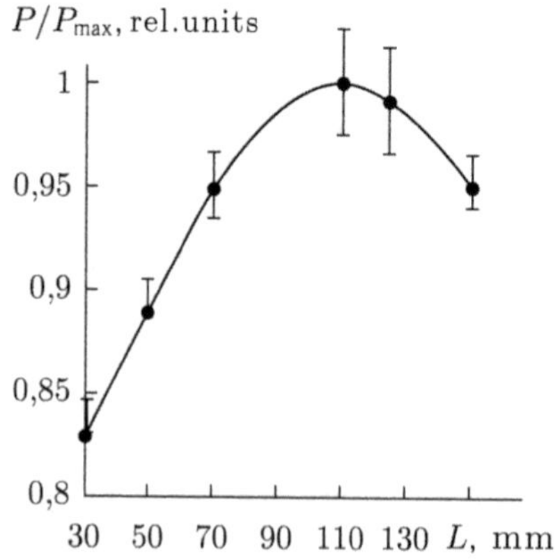

Fig. 4.11. The dependence of the microwave radiation power of the RM on the cathode length, P_{max} = 200 MW.

4.5. Magnetic system

For effective operation of magnetrons, including relativistic ones, a magnetic field with a high circular symmetry and maximum uniformity along the length of the interaction space is necessary, localized in a volume whose diameter is larger or commensurate with the length. When designing, manufacturing and operating it is necessary to minimize fluctuations in the magnitude of the field along the anode block [17]. The production of an uniform magnetic field is based on the principle of superposition: two or more sources of the magnetic field are located relative to a given volume in such a way that the vector sum of the source field strengths is constant in this volume. The resulting magnetic field is uniform with a certain degree of accuracy, depending on the design of the system.

To create an uniform magnetic field in a certain range (in this case in the operating region of the magnetron), it is necessary to fulfill the conditions for locating and selecting the geometric dimensions of the magnetic coils that create this field (the Helmholtz condition). Helmholtz coils are a system of two coils forming a uniform field in the axial region when the following conditions are fulfilled:

a) the distance between the geometric centres of the coils is equal to their average radius

$$A = R_{\text{av}} = \left(R_1 + R_2\right)/2,$$

where R_1 is the inner radius, R_2 is the outer radius of the coils;

b) the thickness of the windings of the coils $K = R_2 - R_1$ and its length L_{coil} into small compared to A. If the cross section of the winding is chosen so that

$$31K^2 = 36L_{\text{coil}}^2,$$

then the magnetic field represented as an expansion in the Legendre polynomials will be uniform up to the expansion term containing r^4/A^4, where r is the radius of the internal volume ($0 < r < R_1$). If this condition is satisfied, we have:

$$K / L_{\text{coil}} = 1.077$$

(the ratio of the thickness of the winding to its length). The area in which it is required to create a uniform magnetic field by induction to 0.5T is determined by the dimensions of the RM: the diameter of the cathode is 18–22 mm; internal diameter of the anode is 43 mm; the length of the anode is 72 mm.

In our case, to the geometric parameters of the magnetic system under study there are several additional requirements due to its design features. This is, first of all, a limitation on the outer radius of magnetic coils (R_2), which is associated with the overall dimensions of the installation and is limited by tightening studs. The inner radius (R_1) is chosen as the result of a compromise solution of several requirements that are opposite in their functions. On the one hand, a decrease in R_1 leads to the possibility of creating the necessary magnetic field due to additional turns without a sharp increase in the current density in the magnet winding. On the other hand, this violates (more precisely, worsens) the first Helmholtz condition, since with the existing design parameters, the value of A (the distance between the geometric centres of the coils) already exceeds the value of the average radius R_{av}. In addition, this causes a decrease in the transverse dimensions of the drift tubes, hence, the loss is due to the end current of the RM. Further reduction in the distance between the coils is limited by the dimensions of the waveguide for the microwave energy output whose width is equal to 80 mm.

Analogous arguments can also be put forward regarding the length of each magnetic coil. Thus, an increase in L_{coil} in order to add additional ampere-turns and, therefore, to obtain the required field of 0.5 T without significant growth of current in the coils leads to an increase in the distance between the geometric centres of the coils (A) and, naturally, to a disruption of the uniformity of the magnetic field.

Within the framework of these limitations, the following dimensions of the magnetic system of the relativistic magnetron were chosen with the maximum conservation of the Helmholtz coil configuration:

- coil radius: inner – 100 mm;
- outer – 236 mm
- width of each coil – 100 mm;
- the distance between the coils – 90 mm.

The calculated axial inhomogeneity of the magnetic field in the working area of the RM did not exceed 3%.

4.5.1. Calculation of the thermal conditions of the coils of the magnetic system

For the RM to work at an output voltage of the LIA of ~400 kV the magnetic field of 0.4–0.5 T is required, which corresponds to a current density in the conductor j_{req} of ~8.4–10.5 A/mm^2 (preliminary calculations were made for a copper tube with a cross section of 3.05 × 3.28 mm^2). To remove the heat generated in the magnetic system it is necessary to ensure efficient cooling [18, 19], therefore, calculations of the thermal state of coil windings were performed at the design stage using the numerical simulation method based on the solution of finite-difference equations of heat conduction with internal heat sources [20]. The task was to determine the maximum temperatures in the windings in order to select the optimal design of the cooling system for the implementation of the continuous operating mode of the RM.

The temperature field in the magnet coils is described by the heat conduction equation of the form

$$\frac{\partial T}{\partial \tau} = a\left(\frac{\partial^2 T}{\partial r^2} + \frac{v}{r}\frac{\partial T}{\partial r} + \frac{\partial^2 T}{\partial x^2}\right) + \frac{q_v}{\rho_m C_p}, \tag{4.5.1}$$

which uses the effective thermophysical properties of the coil, considered as an uniform body: a is the coefficient of thermal diffusivity; ρ_m is the density of the material; C_p is the specific heat of the coil material; v is the coefficient of the shape taken for the cylinder to be unity; q_v is the power of internal heat sources. The range of the arguments is of the form

$$0 < \tau \le \tau_{pr}, \quad R_1 < r < R_2, \quad 0 < z < L_{\text{coil}},$$

where τ_{pr} is the predetermined calculation time.

The initial conditions are given by the following: for $\tau = 0$, $T = T_0$. On the cylindrical surfaces of the coil, the boundary conditions have the following form:

$$-\lambda_{\text{mat}}\left(\frac{\partial T}{\partial r}\right)_i = \alpha_i\left(T_{ni} - T_{\text{c}}\right), \quad i = 1, \dots, 4, \tag{4.5.2}$$

where α_i are the heat transfer coefficients on the corresponding surfaces; λ_{mat} is the coefficient of thermal conductivity of the material of the coils; T_n, T_{c} are the temperatures of the coil surfaces and the environment.

We replace the region of continuous variation of the arguments by the grid region the finite nodes of which go beyond the boundaries of the coil:

$$\{k\Delta\tau; \quad r_i = R_1 + (i-1.5)h_r; \quad z = (j-1.5)h_z\},$$
$$k\Delta\tau < \tau; i=1,...,N; \qquad j=1,...,M;$$

k, i, j are respectively the time and space coordinates; $\Delta\tau$ is the time step; h_r, h_z are the steps along the radial and longitudinal coordinates:

$$h_r = \frac{R_2 - R_1}{N-2}, \quad h_z = \frac{L}{M-2}. \tag{4.5.3}$$

The approximation of equation (4.5.2) on a rectangular grid according to the explicit scheme will be as follows:

$$\begin{aligned} T_{i,j}^{k+1} = T_{i,j}^{k} + p_r\left[T_{i+1,j}^{k} - 2T_{i,j}^{k} + T_{i-1,j}^{k} + \frac{\nu h_r}{2r_i}(T_{i+1,j}^{k} - T_{i-1}^{r})\right] + \\ + p_z(T_{i,j+1}^{k} - 2T_{i,j}^{k} + T_{i,j-1}^{k}) + W, \end{aligned} \tag{4.5.4}$$

where

$$p_r = a\Delta\tau / h_r^2; \quad p_z = a\Delta\tau / h_z^2; \quad W = q_v\Delta\tau / (c_p\rho_m).$$

The resulting difference equation were solved by the sweep method [21], which ensures the stability of numerical calculations when the parameters of the problem vary widely.

A numerical realization of the solution of equation (4.5.1) with respect to the implicit difference scheme was also carried out. Comparison of the calculation results by two methods guaranteed the absence of errors.

Figure 4.12 shows the results of calculations for the following versions of the design of the coils of the magnetic system:

– water cooling of the outer surface (curve *1*; j_{req} = 8.4 A/mm^2);

– outer water cooling and a single disc-shaped cooling channel inside each winding (curve *2*; j_{req} = 9.46 A/mm^2);

– an outer water cooling and two circular cooling channels inside each coil (curve *3*; j_{req} = 11,5 A/mm^2).

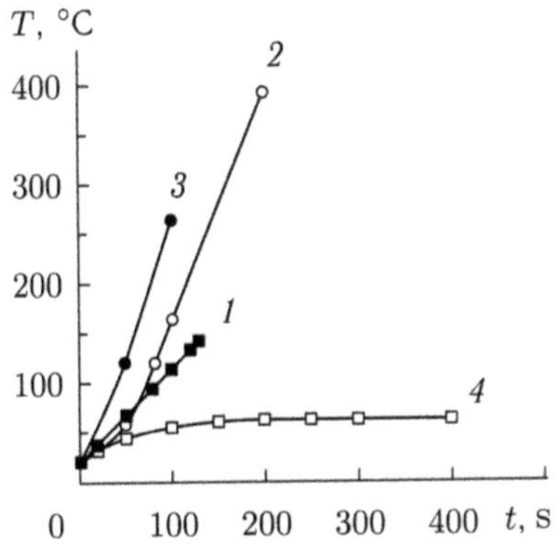

Fig. 4.12. The results of calculating the temperature regime of the winding of a magnetic system made of a copper tube with water cooling of the outer surface: *1* – without the use of additional disc cooling channels inside each winding; *2* – one channel; *3* – two channels; *4* – two channels and reduction of current density to 0.7 j_{req}.

The use of internal cooling channels reduces the copper filling factor of the coil volume, which explains the increase in the required current density.

Calculations showed that none of the above options yields positive results and the stationary thermal mode of operation of the magnetic system is possible only with a significant reduction in the current density in the conductor. For example, for the coils with two cooling channels (Fig. 4.12, curve *4*) the stationary regime is feasible if the current density is reduced to 0.7 j_{req}. For other variants, an even lower current density is required [22].

4.5.2. The design of the magnetic system of pulse–periodic RMs

In the final version, a coil design was selected using a copper tube with a square cross-section of 8.5 × 8.5 mm^2 with a central hole of 5 mm in diameter for the flow of cooling water. In order to create a magnetic field with the induction of 0.4 T in each coil, it is necessary to lay 196 turns and pass a current of 490 A (current density 10.1 A/mm^2). The width of the coils increased to 140 mm, which led to a certain increase in the non-uniformity of the magnetic field in the working zone of the magnetron to 5%. As a result, each winding consists of seven planar helix sections (Fig. 4.13), connected in series and according to the direction of currents and in parallel according to the direction of the cooling liquid. The length of the tube in each

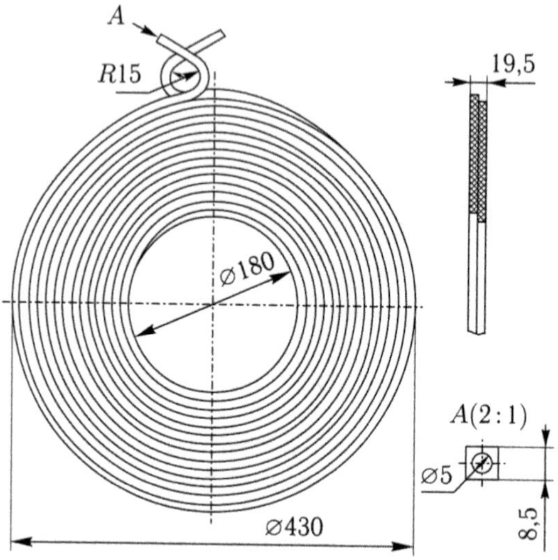

Fig. 4.13. Drawing of the section of the winding of the magnetic system, made of a square copper tube with a central hole for cooling.

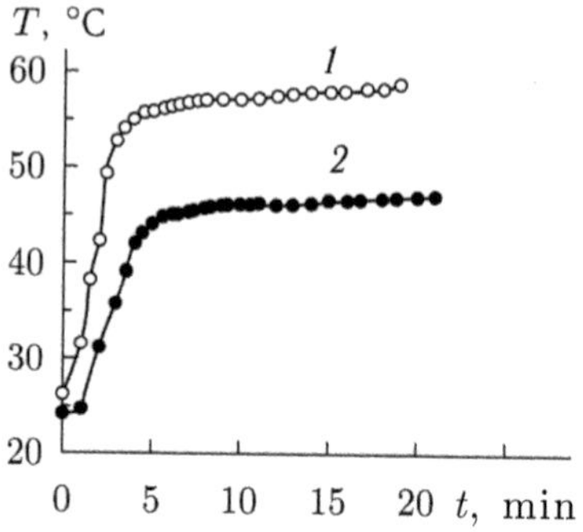

Fig. 4.14. The temperature regime of the windings of the magnetic system at a current, *A*: *1* – 500, *2* – 400.

section is 28 m. The cooling circuit of the magnetic system is made as a closed system, and desalted water is used as a coolant. By adjusting the water flow it is possible to achieve the necessary heat removal [23] and ensure stationary mode of operation also at high values of current density.

Figure 4.14 shows the change of the cooling water temperature at the output of the winding central section in the continuous mode at a current intensity of 500 A (j_{req} = 10.3 A/mm^2, curve *1*) and 400 A (j_{req} = 8.24 A/mm^2, curve *2*). The water flow was 0.32 l/s at a water inlet pressure of 2.5 atm. As follows from the presented curves, the

Fig. 4.15. Appearance of the magnetic system.

magnetic system ensures the continuous operation of the RM during an arbitrarily long time interval. The slow rise in water temperature is explained by its heating in the closed cooling system.

The power supply of the magnetic system is made according to the scheme of a transformer bridge rectifier and has the following parameters:

- maximum direct voltage is 90 V;
- maximum current is 500 A
- limits of current regulation is 50–500 A;
- maximum power consumed from the 380/220V network is 50 Hz–52 kW;
- pulsation of the magnetic system – is less than 0.1%.

Thus, the calculated and manufactured magnetic system provides continuous operation of RMs fed from the LIA 04/4000 and LIA 04/6.

The appearance of the coils of the magnetic system of a pulse–periodic RM is shown in Fig. 4.15.

4.6. Vacuum system

The vacuum system was developed on the basis of the following requirements:

- the working vacuum in the vacuum chamber should be better than $5 \cdot 10^{-5}$ Torr, and in the cathode–anode gap of the RM – better than $1 \cdot 10^{-4}$ Torr;
- the partial vapour pressure from organic materials should not exceed 10% of the vacuum pressure in the chamber;
- the vacuum system must provide a high pumping speed;
- materials with a low operating temperature (less than 100–150°C) must be excluded from heated areas;
- high-vacuum evacuation means should allow the production of an oil-free vacuum, or the installation should be equipped with a freezing nitrogen trap.

The developed vacuum system is schematically shown in Fig.4.16.

It consists of the pumped volume of the vacuum chamber *1* with a nozzle system, the anode block with the microwave output device *3*, an antenna *4*, a drift tube (collector) of end current *5* and an output device *6*. An additional path *9* is used to drain the antenna, and also the cathode–anode gap of the magnetron through the microwave output device. The figure shows the location of the measuring sensors *2*. The pumped volume is mainly made of stainless steel. Permanent joints are made by argon-arc welding, and temporary by sealed gaskets from oil-resistant vacuum rubber, simple vacuum rubber and teflon. The high-voltage insulator *7* of the electron gun and the exit window of the antenna *8* are made of plexiglass.

The intrinsic gas load, consisting of gassing flows of structural materials and equipment, as well as the flow of leakage gas from the outside, according to estimates [24, 25] should be in the untrained vacuum system equal to $Q_{intr} = 1\text{–}1.1 \cdot 10^{-2}$ l · Torr/s.

When the magnetron is operating, especially in the repetition rate regime, a significant additional gas load is possible in the vacuum system due to thermal desorption of the heated cathode and desorption of the surfaces of magnetrons that are bombarded by x-ray radiation and electrons. It is rather difficult to accurately estimate the amount of additional gas evaporation. Assuming that the cathode of the magnetron, which is a stainless steel cylinder 80 mm in length and 20 mm in diameter, is heated to 500°C, and the contribution of both components of the additional gas evolution is approximately the same, it is possible to obtain a numerical value of Q_{add}, and also the value of the total gas evolution in the system $Q_\Sigma = Q_{intr} + Q_{add}$ as a function of time after the establishment of the quasi-stationary operating mode of the RM.

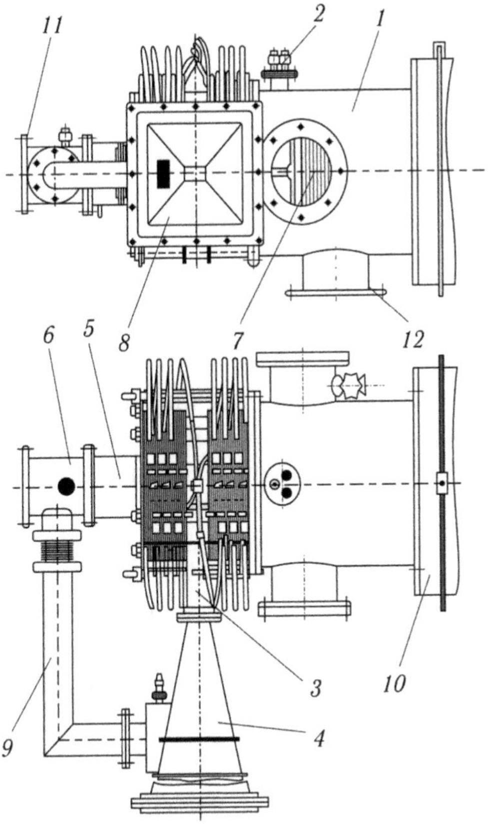

Fig. 4.16. Vacuum system of pulse–periodic RM: *1* – vacuum chamber with a system of nozzles; *2* – measuring sensors; *3* – the anode block of the magnetron with the microwave output device; *4* – antenna; *5* – end current collector; *6* – the output device; *7* – high-voltage insulator of the electron gun of the LIA; *8* – vacuum window of the antenna; *9* – additional pumping path; *10* – module case of the LIA; *11* – connecting flange ODP-500 l/s; *12* – connecting flange ODP-2000 l/s.

These data were used to determine the required rates of pumping out the chamber and the magnetron and a decision was made to pump it from two sides. The throughput of the magnetron can be increased by means of holes in the end caps of the anode block. To do this, each cap contained six holes located in the centre of the resonators. The maximum area of a hole that can be tolerated without reducing the output power level and save the frequency parameters of microwave radiation is ~90 mm^2. The second option for increasing the throughput of the magnetron is the use of end caps made of metal mesh with high geometric transparency (70–80%). In this case, a good contact of the mesh with the ends of the resonators is ensured by soldering or by using profiled end caps.

High-vacuum aggregates include oil–vapour diffusion (ODP) or turbomolecular pumps (TMP). The only advantage of TMP is that they pollute the total volume of organic substances with a lower rate than ODP. This is important in the case where the pumped volume does not contain organic materials or other sources of contamination. Since its own gas load is determined by 80% by the desorption of gases and vapours from the surfaces of organic materials facing the inside of the pumped volume, therefore, there is reason to believe that the 'purity' of the vacuum should not depend significantly on the type of high-vacuum pumps. At the same time, the oil–vapour diffusion pumps should better cope with variable (impulse) gas loads and evacuation of hydrogen, which is an essential component of additional gas evolution.

The high-vacuum evacuation system consists of two parallel branches: the main pumping of the system through a vacuum chamber and additional evacuation through the output device. The main branch is based on a high-vacuum oil–vapour diffusion pump with a capacity of 2000 l/s, a nitrogen trap and a mechanical two-rotor pump. At the inlet of the unit the pumping speed is 710 l/s. The additional branch consists of a high-vacuum ODP with a productivity of 500 l/s, a nitrogen trap and a mechanical rotor pump. The pumping speed at the inlet of this unit is ~200 l/s. The additional vacuum path was used in the cavity for the exhaust of the interaction of the RM through the waveguide output of power and the coupling gap in the resonator. In experiments, the vacuum in the chamber before the start of the packet was:

1) with a single ODP 2000 l/s – $6 \cdot 10^{-5}$ Torr;
2) with two ODPs 2000 and 500 1/s – $1.8 \cdot 10^{-5}$ Torr;
3) with two ODs 2000 and 500 l/s with nitrogen vacuum traps – $6 \cdot 10^{-6}$ Torr.

In the vacuum chamber there is a system for diagnosing the accelerator parameters: in the end flanges of the vacuum chamber there are two Rogowski coils for measuring the current and a capacitive voltage divider is installed on the inner surface of the chamber case. The Rogowski coils and the voltage divider are calibrated when connecting an ohmic load with a known resistance at the output of the LIA.

The drift tube of the end current also has a water cooling jacket to reduce the temperature of the pulsed heating by the electron beam and, therefore, to preserve the gas evolution at the natural level.

4.7. Experimental investigations of the RM using LIA on magnetic elements

At different stages RM experiments were carried out at two facilities: on the basis of LIA 04/4000 and LIA 04/6. The appearance of one of the installations is shown in Fig. 4.17. The differences between the two plants are as follows.

The power source of the accelerator LIA 04/6 has higher output parameters with smaller weight and size. The magnetic system is

Fig. 4.17. Appearance of a pulse–periodic relativistic magnetron with a power source LIA on magnetic elements, 2000.

made of a hollow copper tube of a larger cross-section 13 × 13 mm^2 with an internal hole 6 mm in diameter so that it was possible to reduce ohmic losses and increase the magnetic field. With the same dimensions of the magnetic system and the same power of the power source, the maximum induction of the magnetic field reached 0.59 T.

4.7.1. Scheme and methodology of experimental research

The scheme of experimental studies is shown in Fig. 4.18. The process of investigating relativistic magnetrons can be conditionally divided into two stages. The first is to set the device to the maximum output parameters in a single mode. The second is to study the

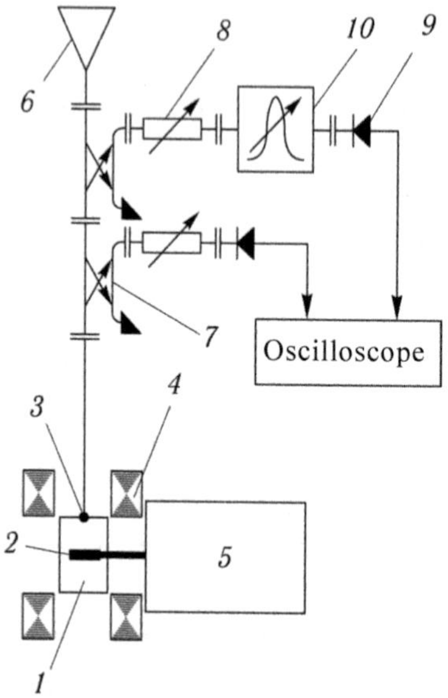

Fig. 4.18. Structural diagram of the experiment with RM. *1* – anode block, *2* – cathode, *3* – power output, *4* – magnetic system, *5* – LIA, *6* – antenna, *7* – directional coupler, *8* – variable attenuator, *9* – detector, *10* – bandpass filter.

parameters of a relativistic magnetron at a high pulse repetition rate.

The microwave power of the RM was measured with two vacuum waveguide directional couplers *7* with the coupling coefficient −40 dB connected in series with each other. The couplers were located between the output of the magnetron and the radiating horn antenna *6*. The channels of each of the branch power couplers incorporated waveguide variable attenuators *8*. Further, the output of each of the attenuators was connected to lamp or semiconductor microwave detectors *9*, the signals from which were sent via coaxial cables to the oscilloscope input. Thus, the power of the microwave radiation of the RM was determined as the average of the power measured by the detectors, taking into account the attenuation of the directional couplers and the attenuators and calculated in accordance with the expression

$$P_{\mathrm{RM}} = \frac{\left(\alpha_{C_1} + \alpha_{A_1}\right)P_{d_1} + \left(\alpha_{C_2} + \alpha_{A_2}\right)P_{d_2}}{2}, \tag{4.7.1}$$

where P_{RM} is the power of the relativistic magnetron; α_{C1}, α_{C2} – attenuation of the vacuum directional coupler; α_{A1}, α_{A2} – attenuation of the waveguide attenuator; P_{d1}, P_{d2} – the powers registered by the first and second detectors, respectively.

The scheme of the measurement of the spectral composition of microwave radiation differs slightly from the circuit used for power measurement. In this case between one of the attenuators *8* and the corresponding microwave-detector *9* there was a tunable bandpass waveguide filter *10*. The second detector was used to record the reference signal, which made it possible to significantly reduce the influence on the spectral characteristics of the radiation of the amplitude spread between the pulses. The procedure for measuring the radiation spectrum was as follows. From the shot to the shot, the resonance frequency of the filter *10* changed discretely. At each frequency the signal powers from each of the detectors were measured. The current spectrum was calculated from the ratio of the power of the signal passed through the filter at a given resonant frequency to the power of the signal from the output of another detector (reference signal) at a fixed time:

$$S(t_i, \Delta t) = \frac{P_f(t_i, \Delta t)}{P_0(t_i, \Delta t)}. \tag{4.7.2}$$

In spectral measurements of the short pulse microwave radiation using a narrowband bandpass filter it must be considered that, besides the forced oscillation process, therein are also excited natural oscillations. The duration of the transient process is comparable with the filter time constant τ_k, which proportional to the Q-factor of the filter Q: $\tau_k = \frac{2Q}{\omega}$. This makes it difficult or even impossible to register spectral parameters during time intervals that are shorter than τ_k. However, a decrease in the quality factor broadens the analysis band and reduces the resolving power of the filter, which does not allow us to evaluate the spectra of narrowband processes. The band of the used filter in the generation frequency region of the magnetron (2750 MHz) is 30 MHz, which corresponds to $Q \approx 90$ and $\tau_k \approx 10$ ns.

The experiments were carried out at different charging voltage of the primary storage of the accelerator U_{C0} from 500 to 900 V, which made it possible to vary the output voltage and current of the power source. Measurements of microwave power, total current and voltage of the accelerator were carried out in the magnetic field induction range 0.24–0.55 T. In the process of measuring the power and spectral composition of microwave radiation 10–50 shots were performed for each value of the magnetic field. Such a volume of experimental data, due to the automated statistical processing of the results, made it possible to significantly reduce the error and improve the accuracy of measurements.

4.7.2. The results of experimental studies

The results of measurements of the output characteristics of pulse-periodic relativistic magnetrons are summarized in Table 4.2.

From Table 4.2 it follows that the higher output power of the RM was achieved both by increasing the number of anode block resonators (increasing the slowing-down of the electromagnetic wave of the working type), and by increasing the output parameters of the power source. This process is accompanied by a slight decrease of the stability of microwave pulses by increasing the spread parameters of the magnetron power supply. It should be noted that the value of the pulsed power of the magnetron was repeatedly checked by various methods, including when using the measuring equipment of customers of similar devices.

Table 4.2.

Parameters of pulse-periodic relativistic magnetrons	LIA 04/4000	LIA 04/6
Radiation power, MW	200 ($N = 6$)	300 ($N = 6$) 350 ($N = 8$)
Frequency of radiation, MHz	2840	2840 ($N = 6$) 3030 ($N = 8$)
Radiation band on level –3 dB, MHz	40–50	50 ($N = 6$) 60 ($N = 8$)
Voltage of LIA, kV	300	400
Current of LIA, kA	2.6	3.6
Current pulse duration, ns	160	170
Microwave radiation pulse duration, ns	120	110
Instability of microwave power amplitude	12 %	15 %
Pulse repetition frequency, Hz	0.4–320	0.4– 200
Average power of microwave radiation at the maximum pulse repetition frequency, kW	3	4.1

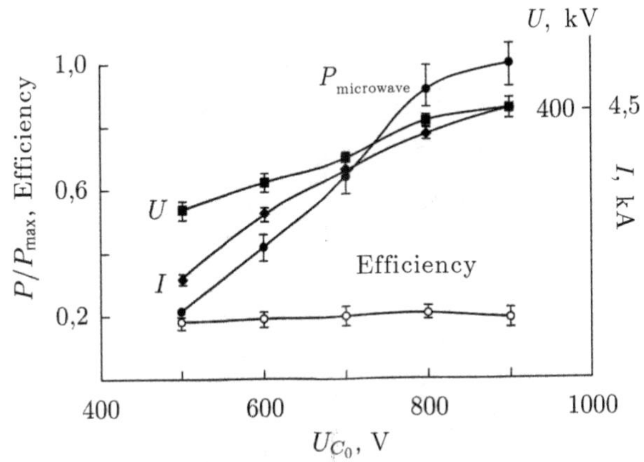

Fig. 4.19. Dependence of microwave power ($P_{microwave}$), the total efficiency of the relativistic magnetron, voltage (U), and total current (I) of the LIA 04/6 on the magnitude of the charging voltage of the primary storage, P_{max} = 350 MW.

Figure 4.19 shows the dependence of the power of the relativistic magnetron, its total efficiency and the voltage of the accelerator on the charge voltage of the primary storage of the LIA. The power growth has a linear character, the efficiency is constant, which agrees with the theoretical estimates (see (2.1.37)). It can be noted that the increase in the generated power is mainly due to the increase in the anode current – the steepness of the current rise above the corresponding increase in the output voltage.

The stability of the operation of RM shows a set of waveforms – signals of microwave detectors – recorded at different pulse repetition frequencies (number of superimposed pulses 100) (Fig.4.20). As can be seen, the amplitude, duration, and pulse shape in the frequency range of 80–320 Hz practically do not change, which allows one to assume the possibility of operation of a relativistic magnetron with an even higher frequency. Figure 4.21 shows a series of 200 current pulses (upper waveforms) and microwave detector signals at 80 Hz repetition frequency (waveforms were recorded at peak detection mode, i.e. each peak – single pulse, the intervals between the pulses are not registered). The instability of the current amplitude generated by the LIA is less than 5%, and the microwave amplitude instability does not exceed 12%. When using two ODPs, stabilization of the vacuum (at the level of 2–3.2 × 10^{-5} Torr) and of the performance of the device occurred approximately after 100–150 pulses after the start of the packet. Exclusion of nitrogen traps of the ODP from the vacuum system resulted in an increase in the initial vacuum level, however, by increasing the pumping rate it was possible to maintain

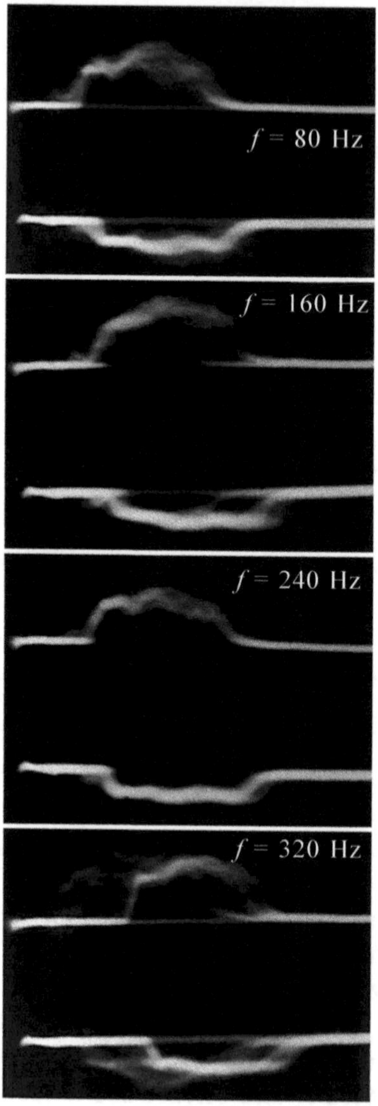

Fig. 4.20. Oscillograms of signals of the microwave detector at different pulse repetition frequency of the relativistic magnetron.

the vacuum level ~$4 \cdot 10^{-5}$ Torr during the pulse burst acceptable for stable operation of the RM. The use of only one ODP with a productivity of 2000 l/s was obviously insufficient: there was a drop in vacuum to $2 \cdot 10^{-4}$ Torr after about 350 pulses of a packet with a frequency 80 Hz and then microwave generation was interrupted.

As demonstrated by long-term operation, the durability of the anode block proved to be extremely high. During the operation at

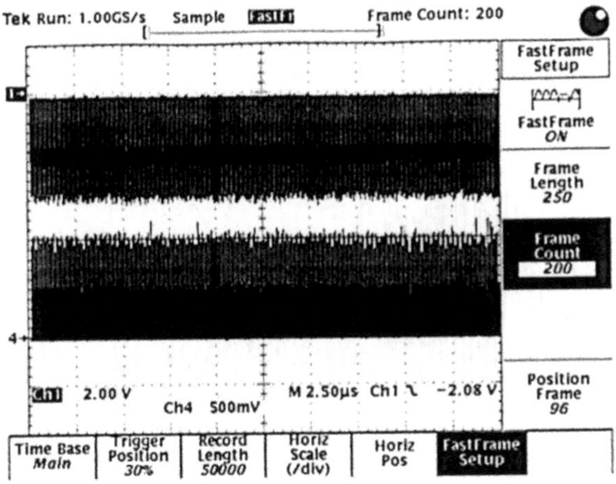

Fig. 4.21. Series of 200 pulses with a repetition frequency of 80 Hz. Upper – current of LIA, lower – the microwave detector signals.

both facilities about 10^7 pulses were applied and not a single block had been destroyed. This is due to the fact that the parameters of the electron beam are chosen in accordance with the results of thermal calculations and do not exceed the limit values. In addition, measures have been taken to increase the durability of the anode blocks by applying cooling and rounding the edges of the lamellae. The durability of the elements of the relativistic magnetron is also influence by specific features of the operation of LIA. It means that the energy of the repeated pulses of the LIA is not scattered in the magnetron diode, but is spent on the magnetization reversal of the cores of the induction system. This feature of the LIA, which was very important for the RM, was emphasized earlier, which predetermined the choice of them as power supplies for pulse–periodic RMs.

The maximum pulse repetition frequency of an RM based on LIA 04/6 was limited by the power of the supply cables to the experimental room with 200 Hz. In this mode, the power consumption of the LIA and the magnetic system from the network reached 150 kW. The number of the pulses in the continuous series was limited to 3200, i.e. ten seconds of continuous operation of the installation. However, the limiting frequency with which the LIA 04/6 accelerator can operate is 400 Hz and is determined by the duration of the charge–discharge processes of the primary storage (this mode of operation of the accelerator was checked with an ohmic load). The

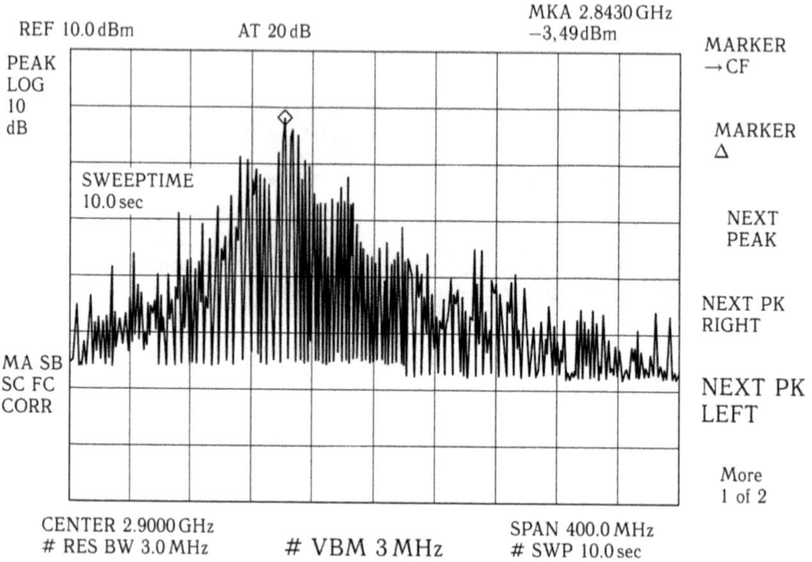

Fig. 4.22. Frequency spectrum of radiation of a pulse–periodic RM, measured by a frequency spectrum analyzer, P_{max} = 200 MW.

number of RM pulses in the series at 200 Hz reached 10^4 at the generated power of the microwave radiation of 300–350 MW. Such a number of pulses can be considered sufficient for the output of the elements of the installation to the stationary thermal regime.

Typically, modes with a high pulse repetition rate are generated in series, between which a pause is required to cool the elements of the relativistic magnetron, semiconductor devices and the winding of the elements of the power supply systems of the accelerator and the magnet, as well as to restore the vacuum conditions in the working volume.

The duration of a pause (in seconds) between series of pulses is estimated by the empirical formula

$$T_p = 0.1 N_{\text{pulse}} \left(\frac{1}{F_{\text{cont}}} - \frac{1}{F} \right), \tag{4.7.3}$$

where N is the number of pulses in a series; F_{cont} = 8 Hz – the frequency of the pulses of a relativistic magnetron in continuous mode (mode not limited by time); F is the repetition rate of the pulses.

The pulse–periodic mode of operation of the RM makes it possible to use frequency analyzers of spectra to record the spectral characteristics of the generated microwave pulses. The results of one of the measurements are shown in Fig. 4.22. The registered bandwidth of the radiation at half-power level was about 50 MHz, which correlates well with the results of spectral measurements with a bandpass filter.

4.8. Relativistic magnetron of the Physics International Company

The studies [3,5] describe experimental studies of a relativistic magnetron of the 30-cm wavelength band, performed on a linear induction accelerator Compact Linear Induction Accelerator (CLIA) with the output characteristics of 750 kV voltage, a current of 10 kA, a pulse duration of 60 ns, a repetition rate of 200 Hz for about one second.

The starting point for these experiments was the results obtained at the TPU by using a relativistic magnetron as a power source for the RM. However, unlike the LIA of the TPU, the CLIA used magnetic elements to switch the forming lines. The research task consisted in obtaining a sequence of 100 pulses with a frequency of 100 Hz with a peak power of each pulse of the order of 1 GW.

Power source – Compact LIA. Compact LIA [4] is a ten-section accelerator (see Fig. 4.23) with a cathode holder, which summarizes the voltage of the inductors applied to the load. Each section consists of an inductor and a single forming line (SFL) with water insulation and magnetic commutation. A two-link magnetic pulse generator (MIPG) charges all the SFLs from the intermediate thyratron-switched energy storage and the common resonant charge units. Elements of the CLIA carry out pulse compression, increasing the voltage from 40 kV to 150 kV, and then using an induction system, the output voltage is up to 750 kV. This scheme allows switching of the primary storage at moderate voltage with the help of hydrogen thyratrons, and forming lines by magnetic commutators.

Figure 4.23 shows the block diagram of the CLIA. DC power supply has the output characteristics: power 300 kW, output voltage 50 kV, and charges capacitors up to 40 kV. The energy is extracted from this block, compressed in time by the circuits of the resonance charging blocks, switched by thyratrons. From the output of these devices a magnetic pulse generator is charged. Further, the two-stage MPG additionally compresses the energy and increases the voltage using magnetic commutation and a transformer with a factor of 2:1.

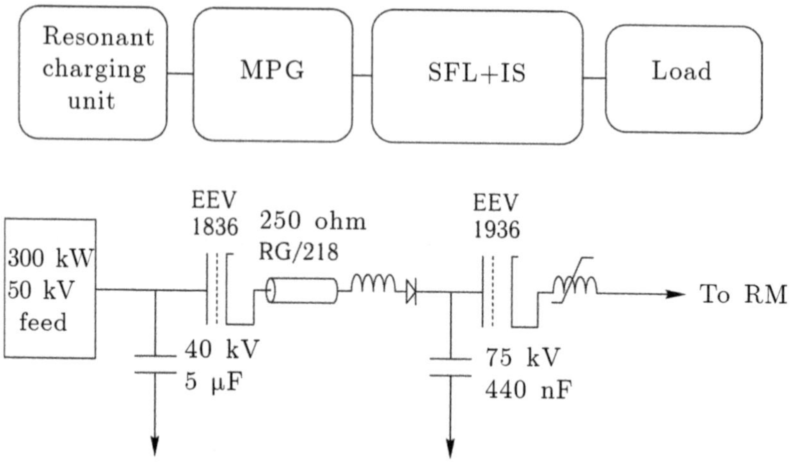

Fig.4.23. The block diagram of the CLIA.

From the output of the MPG ten parallel SFLs with a water dielectric are charged. The forming lines are then discharged through magnetic switches to the induction system of the accelerator. At the output of the accelerator a pulse is formed with a voltage of 750 kV, a current of 10 kA. When working at a matched load at a frequency of 250 Hz voltage pulses with an amplitude of 600 kV were obtained.

The authors of [4] note that the power source–forming lines circuit was designed with a 65% energy reserve necessary to charge the lines. This increase in energy was done in order to compensate for losses in the elements of the installation, if they are greater than expected. Figure 4.23 schematically shows the power supply –resonance charging unit circuit. The first capacitor is a primary energy storage. It closes through the hydrogen thyratron EEV 1836 to the next capacitive storage, which charges up to 75 kV per 100 μs. Then this storage is discharged through the thyratron EE 1936 to the MPG storage. The magnetic commutator shown in the diagram operates as a diode that prevents reverse current through the thyratron.

Figure 4.24 shows a diagram of the following parts of the CLIA: MPG–SFL. The energy from the resonant charging unit charges the first MPG storage to 75 kV for 4 μs. This storage is discharged through a two-turn magnetic switch to a step-up transformer 2:1. The output of this transformer charges a water storage with the capacity

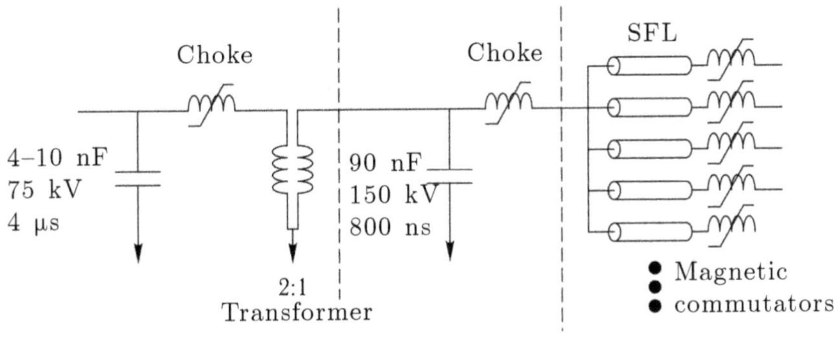

Fig. 4.24. Electrical circuit of CLIA elements – magnetic pulse generator – forming lines.

of 90 pF to 150 kV, which, in turn, is discharged through another two-turn magnetic commutators to 10 parallel-connected SFLs.

SFL is a coaxial structure with water insulation with a wave impedance of 6.8 ohm. The line is charged up to 150 kV each, forming an output pulse with an amplitude of 75 kV, with a duration of 60 ns. Each line has its own magnetic commutator at the output. The windings of the commutators are connected to the magnetizing coils of the inductors of the induction system. The length of the induction system in the assembly was 1 m, which gives an acceleration rate of 0.75 MW/m.

The cores of the magnetic elements of the CLIA, made of the Allied Metglas 2605CO tape, return to the original magnetic state by direct currents from separate demagnetization sources. The first current source is turned on at the point immediately after the second thyratron and demagnetizes the magnetic commutator of the first stage of the MPG and the primary transformer. A second current source is connected at a point between the water storage and the magnetic commutator of the second stage of the MPG for demagnetizing the transformer cores, the commutator of the second stage of the MPG, the output switches of the SFL and the induction system of the accelerator. If an insufficient current was used to restore the magnetic state of the accelerator cores, the effect of a slow decrease in the output pulse was observed from pulse to pulse over a period of about 50 pulses until it disappeared altogether.

Relativistic magnetron with the 30 cm range. The experiments with a high pulse repetition rate were carried out using a modified previously manufactured magnetron of the 30-cm wavelength band (the frequency of the radiation was 1.1 GHz at π-type oscillations).

It has six resonators, a cathode with a radius of 1.27 cm, an anode with an internal radius of 3.18 cm, and resonators with an external radius of 8.26 cm. When working with power from the HCEA, the magnetron had an output power of 3.6 GW. The authors expected significantly lower power values, because, in addition to the fact that the CLIA has lower output characteristics, the magnetron also has a matched resistance of ~25 Ohm, while the CLIA ~75 Ohm.

Changes in the anode block of the RM for the frequency operation mode consisted in organizing the cooling of the anode block lamellae through water channels located 3 mm from the surface. In addition, special attention was paid to the creation of a good electrical contact between the parts of the magnetron, the accelerator and the case in order to avoid possible current losses. A cryogenic pump was used to exhaust the vacuum system to avoid possible contamination with oil (vacuum $4 \cdot 10^{-6}$ Torr). Previous experiments in a single pulse mode show that the peak power of the magnetron increases with decreasing pressure in the system.

As shown in Fig. 4.25, microwave radiation comes from two opposed resonators that are attached to standard WR650 waveguides through quarter wave transformers, and is absorbed by the ohmic load of the vacuum. The couplers designed for high microwave power with coefficient ~80 dB, allow to diagnose the amplitude and frequency of the microwave pulses. Signals were recorded in two ways: a semiconductor detector that detected each pulse and directly on a high-speed oscilloscope. In this case, one pulse (not necessarily the first one) inside the pulse packet was selected. The only diagnostics of the parameters of the accelerator was a Rogowski coil, which measures the total current of the RM. The voltage was determined from the current set by the charging voltage of the CLIA and the previously measured load characteristic of the accelerator.

Results of experiments. At 10 Hz repetition rate a packet of microwave pulses of the RM was produced with a power of ~1 GW, duration 50 ns, energy 44 J each, giving as a result 4.4 kW average power. Figure 4.26 *a* displays the current pulses and the amplitude of the microwave (each peak – single pulse, the system does not register intervals between pulses) for a packet of 50 'shots'; Fig. 4.26 *b* is an enlarged image of one pulse from the middle of the packet. The microwave signal was detected by a detector from one of the two output waveguides. The total capacity was estimated by doubling the readings of the detector since the previous experiments

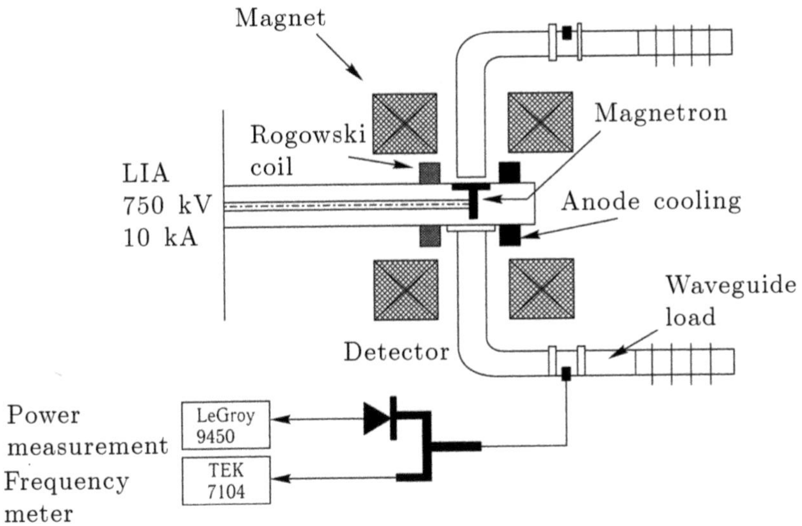

Fig. 4.25. Schematic of experimental investigations of the relativistic magnetron,

showed that the powers in the output waveguides are approximately equal.

In Fig. 4.26 it can be seen that all pulses in the series are approximately equal in amplitude, which allows one to assume the possibility of the magnetron working even for a longer duration of the packet. An essential feature of the magnetron using CLIA is that the duration of microwave pulse (~50 ns) is only slightly less than the current pulse duration of the accelerator (~60 ns). This circumstance was also noted in the experiments conducted at the TPU. The authors note that when using the HCEA in most experiments with RM the typical ratio of the duration of the microwave pulse to current pulse is approximately 1/3.

It can be seen from the oscillograms in Fig. 4.26 that the first few pulses have a larger amplitude, which correlates with the higher amplitudes of the current pulses. This is conditioned by the time which is necessary for establishing a stable energy state in the LIA system. This effect becomes more pronounced when the repetition rate increases, causing a decrease in peak power, although the average power nevertheless increases. As shown in Fig. 4.27, at a frequency of 200 Hz the peak power was reduced to 700 MW, while the average power increased to 6 kW. At 250 Hz, the observed trend was continued – 600 MW peak power and 6.3 kW average power. The authors investigated the work of a relativistic magnetron in

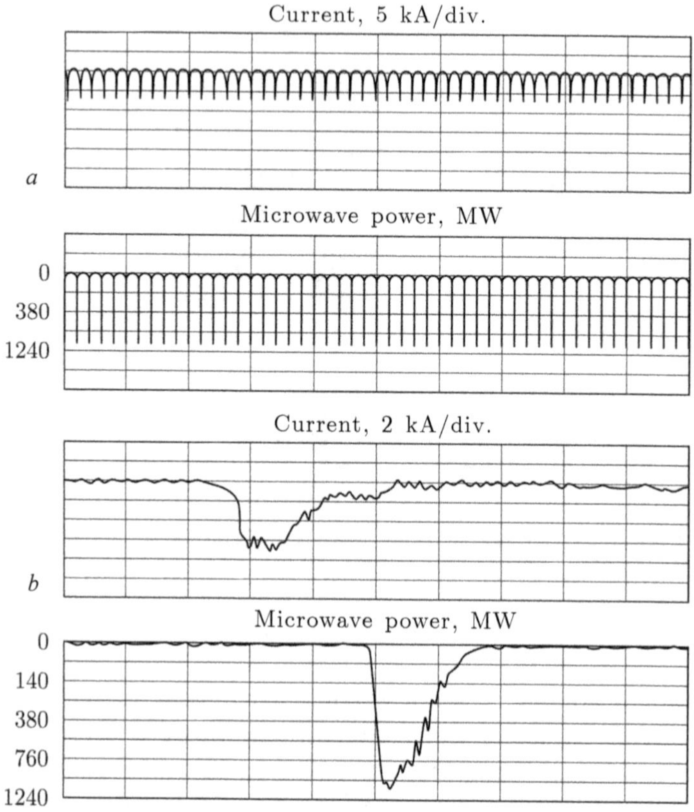

Fig. 4.26. *a*) Oscillograms of current and power of microwave radiation of a series of 50 pulses with a frequency of 50 Hz, *b*) oscillograms of current and power pulses.

the regime of 5 'shots' in a packet at a frequency of 1000 Hz to determine if there is a minimum recovery time between pulses. Figure 4.28 illustrates the operation of the magnetron at 1000 Hz. From the oscillograms it can concluded be that a time interval of one millisecond is sufficient to remove the products of explosive electron emission from the cathode–anode gap of the relativistic magnetron. Using the characteristics of the third pulse in the series, the authors calculated that the average power of the magnetron at a repetition rate of 1000 Hz should be ~25 kW of the peak power of a separate 600 MW pulse.

The results of experimental studies of a relativistic magnetron of a 30-cm wavelength band are presented in Table 4.3.

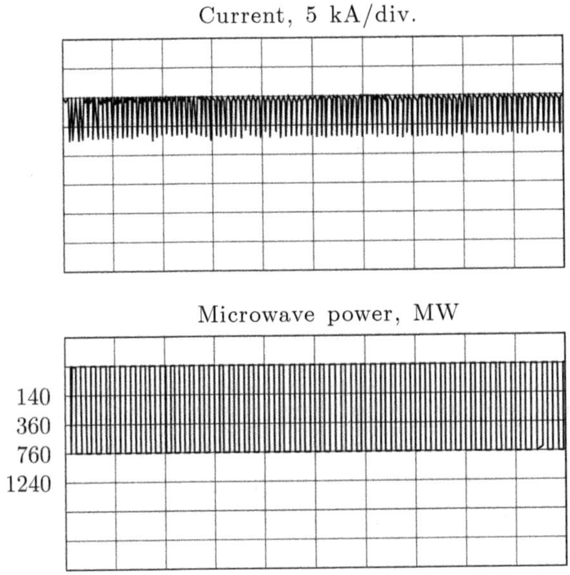

Fig. 4.27. Oscillograms of current and power of microwave radiation of a series of 100 pulses with a frequency of 200 Hz.

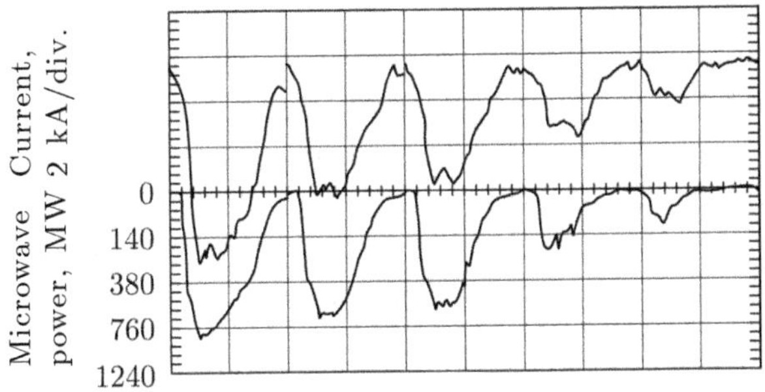

Fig. 4.28. Oscillograms of current pulses and microwave power in the series of 5 pulses with a frequency of 1000 Hz.

The studies completed at the Physics International Company widen the upper limit of the average and peak power for relativistic microwave sources. The authors noted that there is no accumulation of gases in the magnetron, which can lead to a drop in the impedance of the cathode–anode gap and no erosion of the anode block lamellae after even several thousand pulses at the cathode in the CLIA.There was no reduction of the emissivity of the cathode after 1000 pulses. Since the durations of the microwave pulse and current pulse are

Table 4.3.

Pulse repetition frequency, Hz	Microwave radiation power, MW	Average power, kW	Number of pulses in the series
100	1000	4.4	50
200	700	6.0	100
250	700	6.3	100
1000	600 (measured 3 pulses)	25	5

approximately equal, then there is reason to be optimistic that the magnetron can generate longer microwave pulses. It can be assumed that the RM can produce a higher average power of microwave radiation when operating at pulse repetition rate of more than one kilohertz.

This chapter mainly describes the studies performed in the TPU, related to the development, calculation, design, testing of individual components and the entire installation for generating microwave radiation by the RM with high pulse repetition frequency. In the first part, the results of RM studies in the pulse–periodic mode are presented using linear induction accelerators with multichannel dischargers, which experimentally demonstrated the possibility of operating at a high pulse repetition rate – a packet of 3 pulses with a power of 360 MW with a repetition rate of 160 Hz (1986). A pulse–periodic relativistic magnetron with a combined power supply of an accelerator and a magnetic system is then described. The peculiarity of the scheme lies in the fact that the energy stored in the inductance of the coils of the magnetic system is taken to charge the primary storage of the LIA. Such a scheme minimizes the total power consumption and allows the magnetron to operate in a continuous mode with a repetition rate of up to 20 Hz and an output power of ~200 MW (1989). This complex was located on a mobile platform powered by an autonomous source.

Based on the analysis of the experimental data obtained during the first tests of the RM and LIA with multichannel dischargers, the necessity of structural differences between the elements of the installation based on the LIA on magnetic elements is substantiated. First of all, they are connected with the use of magnetic systems on direct current, the use of powerful systems of vacuum pumping, water-cooled anode blocks.

The results of thermal calculations of magnetic systems made using traditional technology from solid copper bars are presented. It is shown that such systems, even with the introduction of water cooling channels, do not allow the magnet to function for long periods of time. A magnetic system, powered from a constant current source, is made for a pulse–periodic relativistic magnetron. A copper tube was used with an internal hole for the flow of cooling water. The magnetic system creates a magnetic field with an induction up to 0.55 T with an axial inhomogeneity in the RM interaction space of about 5% and a temporary current pulsation of less than 0.1%.

The calculation and construction of the vacuum system are described. It is shown that when the RM operates at a repetition frequency above 80 Hz, it is necessary to pump the device from the side of the vacuum chamber and the drift tube, and also to use an additional vacuum path that pushes out the interaction space through the communication gap in the anode block resonator, a smooth waveguide transition and an antenna.

A relativistic magnetron adapted to the pulse–periodic operation mode is described: the anode block has water cooling, rounded corners of the anode block lamellae to reduce the intensity of the electric fields in order to increase the durability. The cathode is made of graphite, which provides a short excitation time for explosive electron emission and the possibility of working at high temperatures.

The results of experimental studies of relativistic magnetrons are presented for feeding from LIA on magnetic elements. To enhance the stability of output power and frequency of microwave generation a charging circuit for the LIA primary storage with voltage stabilization was designed. Using the above-described installation elements, the output pulses were highly stable (the instability of the current pulse amplitude did not exceed 5%, and the instability of the generated magnetron power was no more than 12% when operating at a repetition frequency of 80 Hz).

The LIAs on magnetic elements were used as a basis in producing pulse–periodic RMs and the following modes of operation (1997–2011) were experimentally realized [26–28] (Table 4.4). Note that the pulse–periodic relativistic magnetrons are designed as a common complex comprising a microwave generator, the LIA with the power system, a magnetic system with a power supply, a vacuum system and a cooling system for the components.

The last section of the chapter presents the results of studies of the relativistic magnetron of a 30-cm wavelength band conducted

Table 4.4.

Mode of operation	Parameters of pulse-periodic RMs		
	Power, MW	Pulse repetition frequency, Hz	Number of pulses in series
Continuous	300–350	0.4–8	Not limited
Pulse–periodic	300–350	12–80	10^5
Packet	200 300	120–320 120–200	10^3–10^4

at the Physics International Company, which extend the upper boundary of the average (25 kW) and peak power (600 MW) for microwave sources operating in the pulse-periodic mode. The authors noted that in the relativistic magnetron a higher average power of microwave radiation can be attained provided that power supplies of the relativistic magnetrons are created with a pulse repetition rate of more than one kilohertz.

References

1. Vasiliev V.V., et al., Pis'ma Zh. Teor. Fiz., 1987. V. 13. No. 12. P. 762–766.
2. Didenko A,N,, et al., in: Abstr. 7 Int. Conf. on High-Power Particle Beams. Karlsruhe, 1988. P. 331.
3. Aiello N., et al., in: Proc. 9 Int. Conf. on High Power Particle Beams, Washington, 1992. P. 203–210.
4. Sincerny P., Deeney C., Ashby S., et al. High average power modulator and accelerator. NATO Series, 1993. V. G34. Part A. 12 p.
5. Smith R.R., et al., in: Proc. Int. Workshop on High Power Microwave generator and Pulse Schortering. Edinburgh, 1997. P. 1–9.
6. Vintizenko I.I., et al., In: Proc. 8th All. Symp. on High-current electronics. Sverdlovsk, 1990. Part 3. P. 133–135.
7. Furman E.G., et al., Prib. Tekh. Eksper. 1993. No. 6. P. 45–55.
8. Butakov L.D., et al., *ibid,* 2000. No. 3. P. 159–160.
9. Butakov L.D., et la., Pis'ma Zh. Teor. Fiz., 2000. V. 25. No. 13. P. 66–71.
10. Vintizenko I.I., et al., in: Proc. 12 Int. Symposium on High Current Electronics. Russia, Tomsk, 2000. V. 2. P. 255–258.
11. Butakov L.D., et al., Prib. Tekh. Eksper., 2001. No. 5. P. 104–110.
12. Gunin A.V., et al., Pis'ma Zh. Teor. Fiz., 1999. V. 25. No. 22. P. 84–94.
13. Belomyttsev C.Ya., et al., Zh. Teor. Fiz., 1999. V. 69. No. 6. P. 97–101.
14. Bykov N.M., et al., in: Proc. 10 IEEE Int. Pulsed Power Conf., Albuquerque. 1995. P. 71–74.
15. Vintizenko I.I., Relativistic magnetron, Certificate of utility model No. 13036. Publ. BI, 2000. No. 16.
16. Isakov P.Ya., Prib. Tekh. Eksper. 1988. No. 2. P. 27–29.

17. Melnikov Yu.A., Permanent Magnets of Electrovacuum Microwave Devices. Moscow, Sov. radio, 1967. 183 p.
18. Karasik Yu.R.. Physics and Technology of Strong Magnetic Fields. Moscow, Nauka, 1964.
19. Montgomery D., Obtaining strong magnetic fields by means of solenoids. Moscow, Mir, 1971.
20. Paskonov V.M., et al., Numerical simulation of heat and mass exchange processes, Moscow: Nauka, 1984.
21. Loginov S.S., et al., Izv. VUZ, Elektromekhanika, 1999. No. 4. P. 117–119.
22. Vintizenko I.I., et al., Works of the Research Institute of Computer Technologies 'Mathematical Modeling'. Khabarovsk: Publishing house -KhGTU, 2000. Vol. 10. P. 80–86.
23. Miheev M. A., Mikheeva I.M., Fundamentals of heat transfer. Moscow, Energiya, 1977.
24. Pipko A.I., et al., Design and calculation of vacuum systems. Moscow, Energiya, 1979.
25. Danilin B.S., Mikhaichev V.E., Fundamentals of designing vacuum systems. Moscow: Energiya, 1971.
26. Vintizenko I.I., et al., Izv. TPU, 2003. V. 306. No. 1. P. 101–104.
27. Vintizenko I.I., *ibid*, 2006. V. 309. No. 6. P. 47–50.
28. Vintizenko I.I., Izv. VUZ, Fizika, 2006. No. 9. Appendix. Pp. 271–275.

Application

In conclusion, I present a comparison table of the specific weight and dimensional characteristics of various relativistic microwave generators, developed in different scientific and research centers. We note that the values of the mass and volume of the experimental setups are taken approximately on the basis of the visual estimates of the author.

As will be seen from the table below, the relativistic magnetrons have rather high specific characteristics in comparison with other relativistic microwave generators, which allows us to hope that they can be used efficiently in practice.

The abbreviations used:

P_{imp}, f – the pulse power and frequency of relativistic microwave generators;

F, τ_{MW} – repetition frequency and pulse duration;

P_{av} – calculated average power of the microwave of relativistic microwave generators;

M, V – approximate mass and volume of the installation;

P_{MW}/M, P_{MW}/V – specific characteristics of installations overall weight and dimensions, including relativistic microwave devices, a power source, a magnetic system with a power source;

IHCE – Institute of High Current Electronics;

PIC – Physics International Company, USA;

Varian – Varian Associaties USA, Advanced Technology Group, USA;

RFNC-VNIIEF – Russian Federal Nuclear Center – All-Russian Scientific Research Institute of Experimental Physics;

AFRL – Air Force Research Laboratory, USA;

SNL – Sandia National Laboratory, USA;

IE – Institute of Electrophysics, Russia

References

1. Rostov V.V., et al. Repetitively pulsed operation of the relativistic BWO, Proc. 12 Int. Symposium on High Current Electronics. Tomsk, Russia, 2000. V. 2. P. 408-411.
2. Phelps D., Estrin M., Woodruff J., Sprout R., Observations of a repeatable rep-rate IRES-HPM tube, Proc. 7 Int. Conf. on High-Power Particle Beams. Karlsruhe, 1988. WP112. P. 1347–1352.
3. Spang S.T., Anderson D.E., Busby K.O., et al., IEEE Trans. on Plasma Science.1990. V. 18. No. 3. P. 586–593.
4. Phelps D.A., IEEE Transactions on Plasma Science. 1990. V. 18. No. 3. P. 577–579.
5. Schnitzer I., et al., SPIE. 1995. V. 2843. P. 101–109.
6. Vasil'ev V.V., Pis'ma Zh. Teor. Fiz. 1987. V. 13. No. 12. P. 762–766.
7. Aiello N., et al. in: Proc. 9 Int. Conf. on High Power Particle Beams, Washington, 1992. P. 203–210.
8. Sincerny P., et al. High average power modulator and accelerator. NATO Series, 1993. V. G34. Part A. 12 p.
9. Smith R.R., et al., in: Proc. Int. Workshop on High Power Microwave generator and Pulse Schortering. Edinburgh, 1997. P. 1–9.
10. Butakov L.D., et al., Pis'ma Zh. Teor. Fiz. 2000. Vol. 25. Issue. 13. pp. 66–71; Vintizenko I.I., et al., Pis'ma Zh. Teor. Fiz. 2005. V. 31. Issue. 9. P. 63–68.
11. Treado T.A. in: Proc.2 Int. Vacuum Electronics Conference. Netherlands, 2001. P. 59–60.
12. Kitsanov S.A., in: 12 Int. Symposium on High Current Electronics. Tomsk, Russia, 2000. V. 2. P. 423–428.
13. Korovin C.D., Vircators, in: Vacuum microwave electronics. N. Novgorod, 2002. P. 149–152.
14. Wardrop B.. Marconi GEC Journal of Technology. 1997. V. 14. No. 3. P. 141–150.
15. Dubinov A.E., et al., Inzh. Fiz. Zhurn.1998. V. 71. No. 5. P. 899–901.
16. Korovin S.D., in: Proc. 13 Int. Symposium on High Current Electronics. Tomsk, Russia, 2004. P. 218–223.
17. Rostov V.V., et al. in: Proc. 13 Int. Symposium on High Current Electronics. Tomsk, Russia, 2004. P. 250–253.

Table. Comparison of the specific weight and dimensional parameters of relativistic microwave-generators

Microwave generator	Microwave generator parameters					Parameters of installations				Reference
	P_{imp}, MW	f, GHz	F, Hz	T_{MW}, ns	P_{av}, W	M, kg	P_{MW}/M, W/kg	V, m^3	P_{MW}/V, W/m^3	
O-type back-ward-wave tube IHCE	700	9–10	200	15–30	900	5000	0.18	10	90	1
RM IRT Corp. Cornell Univ.	200	4.4	1	90	6	20000	0.0003	30	0.5	2
RM AAI Corp.	700 325	4.63 4.27	2	50	22 12	20000	0.0011 0.0006	30	0.73 0.4	3
RM General Atomics	300	4.4	1	100	10	20000	0.0005	30	0.33	4
RM Rafael Israel	100	2.5	10	70	17	15,000	0.0001	24	8.5	5
RM TPU	360	2.9	160	80	1550	1500	1	4	388	6
RM TPU	200	2.9	20	50	70	1500	0.047	4	17.5	

Microwave generator	Microwave generator parameters					Parameters of installations				Reference
	P_{imp}, MW	f, GHz	F, Hz	T_{MW}, ns	P_{av}, W	M, kg	P_{MW}/M, W/kg	V, m³	P_{MW}/V, W/m³	
RM PIC	1000 700 600	1.1	100 200 250	50	4400 6000 6300	9000	0.49 0.67 0.7	14	314 430 450	7 8 9
RM TPU	400	2.84	320	160	6000	3500	1.71	8	750	10
RM CPI (Varian)	60	1,873	10	600	350	No of data				11
Vircator IHCE	100	2.65	50 20	20–25	350 140	10000	0.035 0.014	20	17.5 7	12 13
O-type backward-wave tube IHCE	500	10	150	5	200	10000	0.02	10	20	14
Vircator RFNC-VNIIEF	150	10	10	20	30	2000	0.015	6	5	15
Vircator IHCE	200 350	3	10	20–25	350 140	10000	0.035 0.014	20	17.5 7	16
O-tyoe backward-wave tube IHCE	600	10	100	1	60	2000	0.03	3	20	17
O-tyoe backward-wave tube IHCE	2200	10	730	0.8	2500	5000	0.5	8	312	17

5

Relativistic Magnetrons with Increased Duration of Microwave Radiation Pulse

The high-energy microwave pulses can be produced due to an increase in their duration. Such pulses at relatively low levels of generated power and repetition frequency can be used in installations with a high average power of microwave radiation, which have prospects for practical use. Since the amplitude of the impulse temperature of the surface of the anode blocks is proportional to the square root of the duration of the electron beam pulse (see relation (2.3.7)), this makes it possible to remove in part the limitations associated with heating the surface of the anode blocks and their possible destruction.

5.1. Relativistic magnetrons of microsecond duration

The impossibility of using the traditional designs of magnetron devices for obtaining high-energy microwave pulses has already been shown by the first experimental studies of the RM with microsecond voltage pulses [1], in which pulses of 800 MW power and several times shorter than the duration of the applied voltage were obtained. The mechanism for limiting the duration of pulses, both voltage and microwave radiation, was revealed in subsequent experiments [2] in the installation based on the Tonus-2M high-current electron accelerator, shown in Fig. 5.1 *a*. To study the effect of electrostatic and high-frequency electromagnetic fields on the limitation of the duration of the generation process, investigations were carried out to study magnetron diodes with smooth and multiresonator anodes of the same internal diameter.

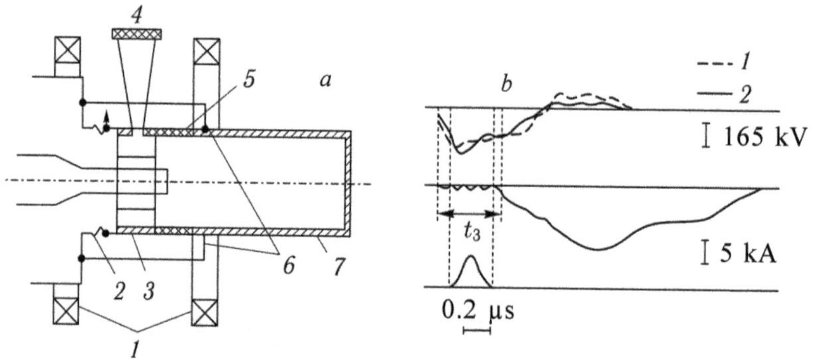

Fig. 5.1. The scheme of the experimental setup: *1* – magnetic field coils; *2* – reverse current shunt; *3* – magnetron; *4* – transmitting antenna; *5* – insulator; *6* – return conductive liners; *7* – drift tube (*a*). Oscillograms of pulses of microwave radiation, voltage, anode current (*b*).

To estimate the duration of the voltage pulse, sometimes it is also called the interelectrode gap time of closing, a shunt is installed in front of the magnetron diode, which records the current of the electrons incident on the anode across the magnetic field. A sharp increase in this current indicates the filling of the interelectrode gap by the plasma, i.e. short-circuiting of the diode. The current drawn from the diode into the drift tube bypasses this shunt through metal liners. The oscillograms of voltage, anode current, and microwave radiation are shown in Fig. 5.1 *b*. During the time t_3, when generation occurs, a small anode current (up to 0.5 kA) flows across the magnetic field to the anode (Fig. 5.1 *b*).

Estimates of the energy release at the anode in a time t_3 before the onset of a sharp increase in the anode current show that in all the experiments the specific load on the anode was less than the threshold (< 150 J/g [2]). This means that in the interval t_3 the anode plasma does not participate in the closing of the gap and the discharge is apparently of a single-electrode nature. The subsequent increase in the anode current is explained by the approach of a highly ionized plasma from the near-cathode zone to the anode at a distance for which the value of the critical magnetic field is higher than the applied one, as a result of which the magnetic insulation is violated. The dependence of the closing velocity v of the gap as the ratio of the interelectrode distance to the time t_3 on the magnitude of the magnetic field induction B is shown in Fig. 5.2.

From curve *1*, for a smooth anode, it follows that with increasing B the velocity decreases monotonically to the value $1 \cdot 10^6$ cm/s.

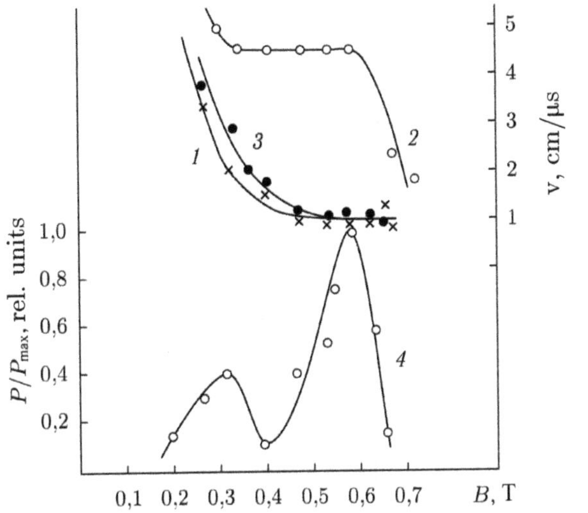

Fig. 5.2. Dependences of the cathode–anode gap speed of closing v and the microwave power on the intensity of the magnetic field.

The dependence of the speed of closing of the interelectrode gap for the 6-cavity magnetron is shown by curve *2*, the power of its microwave radiation by curve *4*. It is clearly seen that the presence of a high-frequency field greatly increases the speed of closing of the interelectrode gap. At the same time, in the radiation region, the average speed of closing is $4.5 \cdot 10^6$ cm/s and practically does not depend on the intensity of the magnetic field.

The maxima of curve *4* correspond to different types of oscillations. They can be determined from the condition that the phase velocity of the electromagnetic wave and the electron drift velocity are equal. Curve *2* makes it possible to determine the anode–boundary of the cathode plasma distance by the time when the microwave pulse appears and to calculate the magnitude of the slowing-down of the wave for these modes. The estimates obtained are in good agreement with the slowing-down values taken from the calculated dispersion characteristic. The first maximum of curve *4* corresponds to π-type oscillations with a wavelength of ~12.7 cm (radiation power $P \sim 40$ MW), the second to $2\pi/3$-type oscillations with a wavelength of 13.2 cm ($P \approx 100$ MW).

In the range of magnetic fields 0.2–0.7 T, the duration of the microwave pulse is 200–300 ns. In this case, the moment of radiation failure with an accuracy of ~20% coincides with the moment of closing of the interelectrode gap.

For the purpose of additional verification whether the increase in the speed of closing is due to the azimuthal heterogeneity of the anode block, some of the magnetron resonators were filled with metal inserts. The experiment showed that in this case the dependence

$$V = f(B)$$

(curve *3* in Fig. 5.2) approaches curve *1* for a diode with a smooth anode. The duration of the microwave pulse in this case starting from B = 0.48 T becomes comparable with the duration of the voltage pulse (maximum microwave power is 100 kW).

Thus, in the first experiment [1], the microwave pulse duration of ~800 MW was about 300 ns at a voltage pulse duration of ~800 ns and was determined by a rapid decrease in voltage below the threshold level. Based on the conducted studies, an important conclusion was drawn that the presence of powerful electromagnetic fields in the interelectrode gap of the magnetron diode leads to a significant (up to 4–5 times) increase in the rate of voltage reduction and the limitation of the duration of the pulses of microwave radiation. It is important that the processes of expansion of the near-cathode plasma cause a reduction in the voltage pulse and a distortion of its shape due to a change in the diode impedance and the exit of the device from the synchronism regime. Moreover, the increase in the duration of the voltage pulse in the relativistic magnetron was observed only when the level of the generated power was reduced to 40 MW.

The results of experiments presented in [3–6] are very revealing for the topic discussed. The RM anode block with an increased number of resonators up to 12 was investigated. A good separation of the modes was achieved by using special transforming devices to shift the microwave radiation 'forward' along the axis of the system. The experiment aimed to achieve a pulse power of ~1 GW for a duration of ~1 μs. To exclude the formation of explosive-emission cathode plasma, it has been proposed to use a tungsten oxide cathode. In reality, the device was excited with a power of ~20 MW for ~35 ns. In the opinion of the authors, short pulses were caused by the formation of an anode plasma with the deposition of space-charge spokes of the electrons. The ions of the anode plasma cross the interelectrode gap and neutralize the space charge of the electron cloud. In formulating the requirements for new designs of relativistic magnetrons, the authors propose to choose the value of slowing-down the electromagnetic wave $\beta_p \sim 0.1$, greatly increasing the synchronous magnetic field to reduce the energy of the electrons bombarding the anode. However, low values of β_p require either a transition to a longer wavelength range, or an increase in the number of resonators 3–4 times, which is associated with the difficulties in separating the oscillation modes.

5.2. Centrifugal instability of the plasma on the central electrode of a coaxial diode with magnetic insulation

The mechanism of motion of the cathode plasma across the magnetic field in a coaxial diode with magnetic insulation was considered in [7, 8] and is based on the development of centrifugal instability in the near-cathode plasma layer. The hydrodynamic expansion of cathode plasma jets formed by individual emission centers, after a time when the voltage pulse is applied, leads to the formation of a uniform plasma layer around the cathode. This layer is unstable with respect to the development of centrifugal instability, since plasma undergoes drift in the azimuthal direction (electrons and ions drift in different directions with different velocities). Any random perturbation of the plasma density distribution uniform in azimuth due to the difference in the particle rotation velocities leads to the appearance of a polarization field. The drift velocity of the plasma in this inhomogeneity (in crossed polarization and external magnetic fields) will be directed to the action of the centrifugal force, i.e., in the radial direction to the anode. Thus, according to the model, the time of closing of the interelectrode gap is the sum of two time intervals: $t_k = \tau_0+\tau_{cf}$, where τ_0 is the time of formation of the plasma layer, τ_{cf} is the time of development of the centrifugal instability.

In the framework of the model of the centrifugal instability of a cathode plasma, the dependence of the time of closing on the magnitude of the induction of the magnetic field has three characteristic regions:

1) For $B < B_{cr}$ there is no magnetic insulation.
2) $B_{cr} < B < B_\delta$ – the magnetic insulation is disturbed during the formation of a plasma layer of thickness δ around the cathode, where

$$B_\delta \approx B_{cr} + \sqrt{B_{cr} + 1/4d^2}\,/\,d + 1/2d^2, \tag{5.2.1}$$

d is the value of the cathode–anode gap; B_{cr} is the critical induction of the magnetic field, below which the magnetic insulation of the gap is violated.

3) $B > B_\delta$ – the breakdown of the diode is determined by the development of a centrifugal instability in the cathode plasma, and the value of t_k is determined by the sum $t_k = \tau_0 + \tau_{cf}$. Comparison of the experimental dependence $t_k(B)$ with the centrifugal instability calculated for the parameters of numerous experiments with coaxial diodes with magnetic insulation indicates their good agreement and the presence of a characteristic maximum at $B = B_{opt}$. Its existence is determined by the competition between the two processes. In the initial section $B_\delta < B < B_{opt}$ the increase of t_k is related to a decrease in the thickness of the plasma layer around the cathode. In higher magnetic fields $B > B_{opt}$ the polarization field becomes stronger, i.e., there is a reduction in the development time of the centrifugal instability τ_{cf}.

A simpler connection of t_k with the experiment parameters such as the interelectrode gap size, cathode radius, electric and magnetic field strengths can be obtained from the qualitative representation of the model [8] if we assume that for higher magnetic fields the breakdown of the diode is mainly determined by the rate of development of centrifugal instability ($\tau_0 < \tau_{cf}$). Using this simplifying assumption, it is shown that

$$E_0 t_k^{1/2} / U^{1/2} \approx \left(4\sqrt{2}\pi n m_i c / B\right)^{1/2} \left(r_c / d\right)^{1/4}, \tag{5.2.2}$$

where E_0 is the electric field strength between the electrodes; U is the voltage across the diode; n, m_i are the density and mass of cathode plasma ions; r is the cathode radius. The left-hand side of expression (5.2.2) completely coincides with the empirical relation derived in [9] on the basis of a considerable number of experimental facts:

$$E_0 t_k^{1/2} / U^{1/2} \approx \alpha, \tag{5.2.3}$$

where $\alpha \sim 0.3$–0.5.

A distinctive feature of the coefficient α is its independence from the geometric parameters of the diode. A similar weak dependence demonstrates the right-hand side of (5.2.2), which serves as an indirect proof of the validity of the model of centrifugal instability. It should be emphasized that this breakdown mechanism is associated with the development of instability on an internal electrode, i.e., on the cathode of a magnetron diode of direct geometry. An anode plasma in such a structure does not play an important role, which is confirmed by the conclusions of [10]. The presence of powerful microwave fields of the RM anode block accelerates the development of centrifugal instability, strengthening the azimuthal non-uniformity of the near-cathode electron layer due to the growth of the polarization field.

The radial motion of the outer boundary of the cathode plasma leads to an increase in the diameter of the electron beam with a characteristic velocity $\sim 4 \cdot 10^5$ cm/s. Since the magnetic field does not affect the expansion of the cathode plasma along the lines of force, the velocity of its motion can reach 10^7 cm/s and is explained by the ambipolar acceleration of ions in the field of the space charge of plasma electrons [11].

5.3. Centrifugal instability of plasma in an inverted coaxial diode with magnetic insulation

The observed phenomenon of accelerated expansion of the cathode plasma across the insulating magnetic field and the associated limitation of the duration of the voltage pulses and microwave radiation in relativistic magnetrons of direct geometry led to RM studies of a fundamentally different configuration – inverted relativistic magnetrons (IRM). These experiments are based on the results of a complex study of the mechanism of plasma crossover of the cathode–anode gap of inverted coaxial diodes with magnetic insulation (ICDMI). It is necessary to note that the inverted magnetrons have practically no losses of the end current, and one can expect an increase in the efficiency of the device.

Since the cathode in such diodes is an external electrode, centrifugal instability can not develop in the cathode plasma, therefore, one can expect an increase in both the duration of the voltage pulse and the duration of the microwave pulse.

Conditionally, ICDMI designs can be divided into two types: without extraction of the electron beam (Fig. 5.3 *a*, *b*) and with the extracted electron beam (Fig. 5.3 *c*, *d*). The first construction was originally used to produce powerful ion beams [12–14] (sometimes called an ion diode) and later used to generate microwave radiation of nanosecond duration in M-type devices [15–17]. In it, unlike in a diode with a an extracted beam, there is no collector of electrons and, accordingly, the possibility of transporting the beam along the magnetic field.

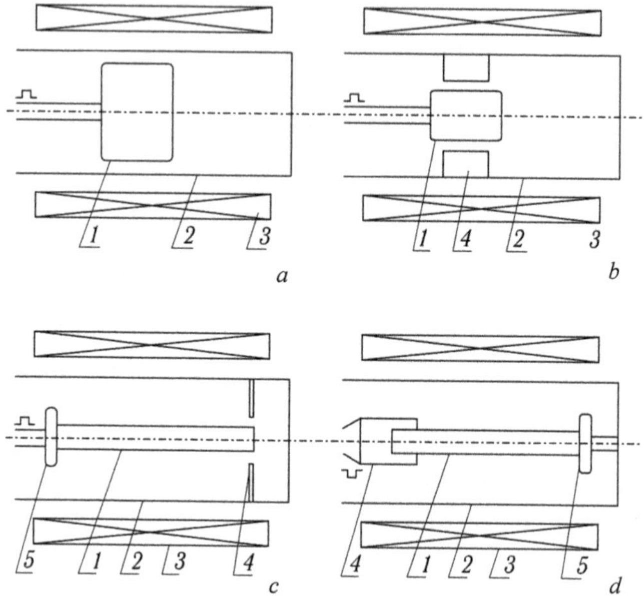

Fig. 5.3. Designs of inverted coaxial diodes with magnetic insulation (*a*, *b* – without electron beam extraction, *c*, *d* – with the extracted electron beam): *1* – anode; *2* – the vacuum chamber; *3* – magnetic system; *4* – cathode; *5* – collector of electrons.

The internal electrode of the diode (anode) receives a high voltage pulse of positive polarity. In the vacuum chamber *2* (Fig. 5.3 *a*) or cathode *4* (Fig. 5.3 *b*) slits are cut to measure the parameters of the ion beam. The magnitude of the magnetic field formed by the magnetic system *3* can greatly exceed the critical value.

The results of experiments with the coaxial diodes, which are a prototype of the magnetron, published in [12–27], boil down to the following:

a) The main factor determining the duration of the voltage pulse in a coaxial diode of both direct and inverse geometry is the magnitude of the electric field strength at the central electrode.
b) For equal values of this strength, the duration of the voltage pulse in the inverted diode is 2–4 times higher. The value of α in formula (5.2.3) is about 1 for ICDMI.

These conclusions are based on the results of studies of inverted coaxial diodes with magnetic insulation, shown in Fig. 5.3 *a*. By varying the polarity of the voltage at the internal electrode, it is possible to realize the geometry of the direct or inverted diodes and to compare the diode time of closing with other conditions being equal (cathode–anode gap, electrode material, vacuum level). The experiments were carried out using the diodes of different curvature (the ratio of the diameter of the outer electrode to the diameter of the inner electrode). This makes it possible, with equal values of the interelectrode gap and voltage, to vary the electric field strength at the electrodes and to evaluate its effect on the time value t_k.

The following two conclusions are based on the results of electron-optical studies of plasma processes in the interelectrode gap of diodes [20]:

c) the breakdown of the inverted diode is caused by the development of an unstable state of plasma formation at the anode and expansion of the anode plasma to the cathode until the interelectrode gap is closed;
d) a noticeable increase in the duration of magnetic insulation in the ICDMI in comparison with the direct diode is explained by the time delay of the plasma layer formation at the anode of the diode.

In [20], electro-optical photographs show the process of formation of a closed anode plasma layer at magnetic fields substantially exceeding the critical value. The authors observed the azimuthal motion of the boundary of the plasma layer with an increase in the time interval from the moment when the accelerator voltage pulse was applied to the internal electrode of the diode before the electro-optical converter was activated. The spatial coincidence of the emission of cathode and anode plasma is also noted, which suggests that there is a positive feedback between the electron and ion fluxes [18].

Further investigations of ICDMI were carried out in the loading regime by an electron beam in accordance with the scheme depicted in Fig. 5.3 *c* [25, 26]. The length of the transport space for the electron beam (cathode–collector distance) was as much as 300 mm and could be varied by approaching the collector. All elements of the inverted coaxial diode were placed in an uniform magnetic field with an induction to 0.7 T. Studies have shown that, for the diodes with an extracted electron beam and also for the diodes without an extracted beam, the duration of the voltage pulse is determined by the magnitude of the electric field at the anode. These facts suggested that the instability leading to the closure of the interelectrode gap develops in the anode plasma.

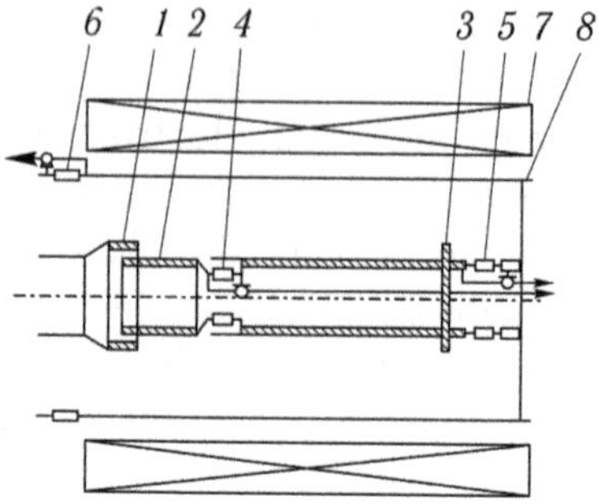

Fig. 5.4. Experimental circuit of ICDMI: *1* – cathode; *2* –anode; *3* – collector of electrons; *4* – shunt of the anode current; *5* – shunt of the full current of the electron beam; *6* – shunt of the accelerator's total current; *7* – magnetic system; *8* – vacuum chamber.

When studying the plasma parameters and determining the mechanism for closing the gap, it was completely natural to study the effect of the anode material on the value of t_k. A special experimental setup was developed and is shown in Fig. 5.4 [24, 25].

A high voltage pulse of negative polarity was applied to the cathode *1* made of stainless steel. The diameter of the cathode *1* was 70 mm, the anode *2* – 50 mm, the collector *3* – 80 mm. The cathode–collector distance exceeded 600 mm, which eliminated the possibility of long-term closure of the diode by the cathode plasma to the collector at the voltage pulse duration characteristic for the experiment (the cathode plasma has a very high propagation velocity along the magnetic field). The anodes made from graphite, aluminium, copper, brass, lead, stainless steel, molybdenum, tungsten and tungsten–rhenium alloy were studied. As follows from the listing, the materials have a wide range of different physical properties. Shielded shunt *4* was used to measure the electron current flowing to the anode across the external insulating magnetic field. In addition, the measuring circuit included a shunt of the collector current *5* and a shunt of the total current of the accelerator *6*, and an ohmic voltage divider. Note that the readings of the shunts *5* and *6* corresponded to each other, indicating that there is no loss of electron current to the surface of the vacuum chamber *8* (diameter 180 mm) and 'parasitic' emission from the cathode holder. Magnetic system *7* formed an uniform magnetic field with an induction up to 2.6 T.

Figure 5.5 *a* shows the dependence of the voltage pulse duration t_k, determined at the 0.1 amplitude level, on the value of the induction of an external magnetic field for anodes from various materials. Figure 5.5 *b* shows the dependence on the magnetic field of the

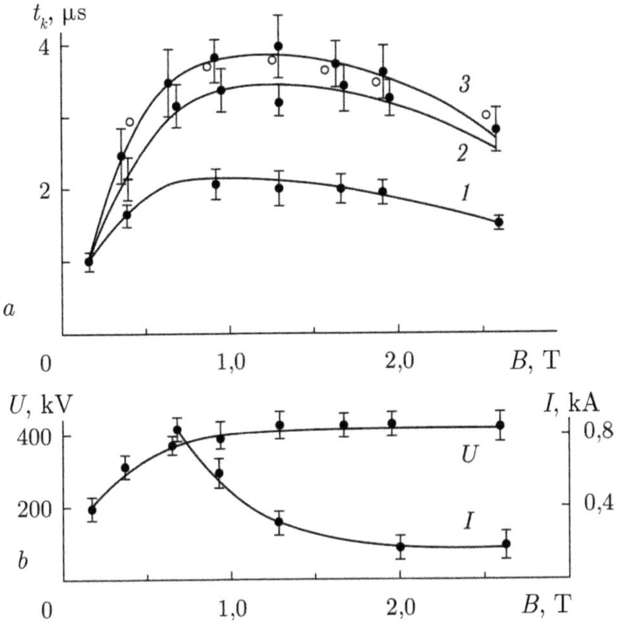

Fig. 5.5. Dependences: *a* – the duration of the voltage pulse; *1* – C, *2* – Cu, *3* – a set of the following materials: brass, stainless steel, Al, Pb, W, Mo, W–Re; ○ – calculated values for stainless steel anodes; *b* – amplitudes of voltage (*U*) and anode current (*I*) on the value of magnetic field induction for stainless steel anodes.

voltage at the anode and the current flowing across the gap typical for all experiments.

For most materials studied, the behaviour of the curves $t_k(B)$ coincides within the limits of the measurement accuracy (curve *3*). Significant differences were observed for the anodes from graphite (curve *1*) and copper (curve *2*), that is, for materials that have lower values of hydrogen desorption energy compared with materials from the set of curve *3* [28]. Table 5.1 shows the maximum values of the voltage pulse duration and the energy of desorption of hydrogen for some of the investigated anode materials.

Table 5.1.

Parameter	Anode material				
	Mo	W	Stainless steel	Cu	C
t_k, μs	4,5	4.6	4.2	3.7	2.4
Energy of hydrogen desorption, kcal/mol	40	46	32	8	2.4

Why was the hydrogen desorption energy selected as the important parameter? This is determined by the data of [12] in which it is shown that ~2/3 of the ion current in the inverted diode is transferred by the protons and only ~1/3 by the ions of the anode material. It can also be noted that the use of hydrogen-containing anode coatings significantly reduces the duration of the voltage pulse in comparison with both the metallic and the graphite anodes [12–14].

To establish the breakdown model of the inter-electrode gap of ICDMI, studies were made of the plasma parameters (electron density N_e and temperature T_e) [27]. The spectroscopic method was chosen for diagnostics, and this is due to the design and accuracy of the method in the range of concentrations 10^{13}–10^{18} cm^{-3}, characteristic for plasma in diodes with explosive emission. In particular, the method of measuring the Stark broadening of hydrogen spectral lines was used to determine N_e and the method of relative intensities of the Fe ion lines for estimating T_e [29].

Investigations of the parameters of electrode plasma were carried out for ICDMI whose construction corresponded to that depicted in Fig. 5.3 *b*. Spectral measurements were made with an ISP-30 quartz spectrograph. The spectra were recorded with a spatial resolution along the radius of the cathode–anode gap, which was achieved by focusing the image of the gap on the input slit of the instrument with an achromatic quartz condenser. The scale of the spatial transformation was 1:10. The spectral lines were recorded on a photographic film. The characteristic curve of the film at the investigated wavelengths of the plasma radiation was determined by means of a pulsed radiation source with a spectrum close to the emission spectrum of an absolutely black body (Podmoshensky source). Photometric measurements of spectral lines were taked with a recording microphotometer. To identify the lines on the film, the reference arc spectrum of iron was also recorded.

The plasma spectrum contained both the lines of ions of the cathode and the anode material and of hydrogen. The plasma temperature was determined from the results of measuring the integrated intensities of several lines I_λ close in length. In the calculation, combinations of the lines Fe (3217.38 Å), Fe (3219.52 Å), Fe (3239.44 Å), Fe (3277.00 Å) were used. It should be noted that the temperature estimate is made assuming the thermodynamic equilibrium of the plasma. The results of estimations of the electron temperature lead to a value of $T_e \sim (0.35\pm0.2)$ eV. However, it is impossible to fully rely on this data because it is not known at what moment to which the measurements of the integral line intensities are related to, and the the diode plasma has a long afterglow. In any case, it can be assume that $T_e < 1$ eV, which is further confirmed by the results of the experiments [30].

Estimates of N_e were obtained by broadening the spectral line H_γ (4340.47 Å). We note that the broadening of the lines of the Balmer series of hydrogen depends weakly on T_e and is mainly determined by N_e. The microphotometer measured the broadening of H_γ at half the maximum intensity. True broadening $\Delta\lambda_{1/2}$ of the line H_γ was determined by subtracting the width of the tool profile, measured on the nearest lines Fe (4337.05 Å) and Fe (4343.26 Å), and the intrinsic broadening was significantly less than the hardware broadening. Measurements of N_e were conducted at three points corresponding to the image of the cathode, the anode and the middle of the interelectrode gap. Using the estimates $T_e < 1$ eV and $\Delta\lambda_{1/2}$ electron concentration calculations were carried out. It was found that at the cathode surface $N_e \sim (4 \pm 2 \cdot 10^{15})$ cm^{-3}, on the anode surface $N_e \sim (2.5 \pm 2 \cdot 10^{15})$ cm^{-3}, in the interelectrode gap $N_e < 10^{15}$ cm^{-3}. The measured value of the plasma density at the anode indicates the possibility of it centrifugal instability at values of $N_e \sim 10^{13} - 10^{14}$ cm^{-3}. We note that such plasma parameters are characteristic for the development of the centrifugal instability of a cathode plasma in a direct diode. In addition, such a plasma density provides sufficient conductivity of the channel for closing the inter-electrode gap of the ICDMI.

In general, the mechanism of closing of the interelectrode gap by the plasma is as follows. The development of the diocotron instability of the electron cloud at the cathode leads to an electron drift across the external magnetic field to the anode and the formation of an anode plasma. The anode plasma is a source of ions moving to the cathode, causing additional electron emission, which, in turn, leads to the development of anode plasma. This is the mechanism of feedback between the electron and ion fluxes. Over time, the formation of closed plasma layers occurs on the surface of the cathode and the anode. The reason for the radial motion of the anode plasma across the insulating magnetic field is the development of the centrifugal instability in the pre-anode plasma layer. The radial drift of the anode plasma (proton component) caused by this instability leads to the closure of the interelectrode gap.

5.4. Inverted coaxial relativistic magnetron at the Stanford University

The possibility of increasing the duration of a voltage pulse in an inverted coaxial diode with magnetic insulation in comparison with a diode of direct geometry, under all other conditions being equal, and a decrease in the end current of losses have caused interest in conducting experiments on the generation of microwave radiation in inverted relativistic magnetron systems. It can be noted that inverted coaxial magnetrons (ICM) allow increasing the number of resonators of the slowing-down system and, consequently, the overall dimensions of the generator, which is promising from the point of view of the tasks being solved by relativistic high-frequency electronics. It is also possible to turn to a shorter wavelength range while maintaining the size of the slowing-down system. Besides, ICM possesses a number of such positive qualities as high power, high electronic efficiency due to an increase in electromagnetic wave slowing-down, and the stability of oscillations.

Since the power flow in the diode can be increased both by decreasing the gap, as well as increasing the electrode area, an obvious alternative is a significant increase in the circumference of the interaction space and, accordingly, the number of resonators. However, frequency separation between the neighbouring modes decreases with increasing number of resonators, and the mode control and stability are the main design problem.

To evaluate the prospects of the approach, a magnetron which is based on a hybrid inverted coaxial design was constructed and is shown in Fig. 5.6 [15–17]. The increased electrode area is achieved by using a coaxial stabilizing resonator between the anode ground system and the cylindrical output waveguide. In this design, the emission area of the grounded cathode was increased to 600 cm^2. The use of graphite as a cathode material provides a rapid formation of an explosive-emission cathode plasma. The use of a coaxial resonator makes it possible not only to increase the radius of the interaction space, but also to ensure the stabilization of the π-mode of the oscillations of the slowing-down system. According to the developers, such a hybrid system is well suited for super-power generation, since the number of resonators of a slowing-down system that stably operates in the π-type is relatively large. This 10-cm magnetron uses 54 resonators. The problem of vacuum microwave breakdown is eliminated by applying a large number (27) of the

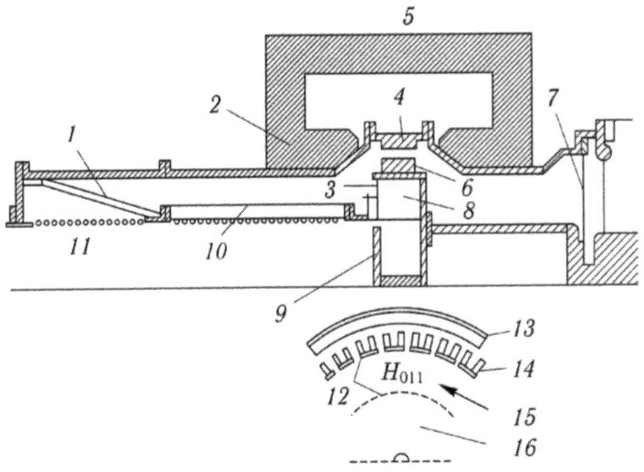

Fig. 5.6. Hybrid inverted coaxial magnetron: *1* – insulating rod; *2* – pole tip; *3* – tuning piston; *4, 13* – cathode; *5* – the magnet; *6* – anode; *7* – power supply; *8, 15* – H_{011}-resonator; *9* –coupling transformer; *10* – filter of wave types; *11* – helical waveguide; *12* – communication slots; *13* – cathode; *14* – slowing-down system; *16* – resonator section of the output waveguide.

slots of the coupling of the slowing-down system to the coaxial resonator. The symmetry of this excitation ensures the H_{011}-mode in the resonator, which, because of its high Q, acts as a frequency and mode stabilizer. Microwave oscillations inside the stabilizing resonator are then coupled to the H_{01}-mode in the output cylindrical waveguide using 24 symmetrically located axial coupling slots in the outer wall. The coupling transformer in the form of a disc allows adjustment of the Q-factor of the H_{011}-type oscillations in the stabilizing cavity and at the same time highly competitive power load modes. The waveguide filter of the wave modes, connected with the coupling transformer, is constructed so as to transmit the H_{01}-mode, but greatly weakens all other H_{0n}-modes, which for $n >$ 2 are prohibitive for an output waveguide with a diameter of 20.3 cm. The spiral winding that connects the filter of types of waves to the external waveguide, works as a waveguide with low losses for microwave power and represents a high-impedance grounding connection for the applied voltage pulse.

The pulse former from the *Febetron* X-ray system was used as a high-voltage single-pulse modulator. It provided a pulse of 600 kV with a half-height duration of 55 ns, an output impedance of 100 ohm, and a peak power of 3.6 GW. The Lucite high-voltage insulator separates the freon insulation of the installation from the vacuum of the magnetron (10^{-5} Torr).

A 1 MW magnetron, tunable in the range from 3.1 to 3.5 GHz, was used to power the excitation circuit. To increase the mode stability during the transient process, an excitation circuit consisting of a line with opposing pins or a strip line was located on the surface of the cathode. The exciting circuit was propagated to the cathode segment at 90° and was covered with a sprayed layer of colloidal graphite. The transition from a standard rectangular waveguide to the exciting circuit was carried out by a comb-shaped waveguide of variable cross section.

The axial magnetic field is produced by a permanent electromagnet with eight pole pieces located radially around the vacuum chamber to provide an azimuthally symmetric field. The measured inhomogeneity of the field in the interaction space was less than 5%. The saturation of steel limits the maximum magnetic field to approximately 2.5 kG. Numerical analysis was used to determine the optimal shape of the pole pieces. The design required a compromise between the conflicting conditions for maximizing the magnetic field in the interaction space and minimizing the electrostatic field along the non-emitting regions of the grounded circuit near the cathode. Careful monitoring of these fields reduces end losses and allows these areas to work as an analogue of conventional end screens.

The time of the growth of the microwave signal between the levels 0.1–0.9 is ~40 ns, which is comparable with the total duration of the current signal at half-height. The total time of approximately 30 ns is comparable with the decay time of the field in the resonator (Q_c/ω, where $\omega = 2\pi f$, $f = 3.2$ GHz, $Q_c = 500$).

The magnetron current and voltage were measured by a calibrated coil and a capacitive divider, respectively, the microwave power was measured using a 60 dB coupler, attenuators, and semiconductor detectors. The frequency of microwave radiation was measured using a solid-state dispersion line and was 3.2 GHz.

Due to the short duration of the voltage pulse (~60 ns), the magnetron effectively operated (1000 MW at an efficiency of 18%) with only external excitation (without external excitation – 100 MW). The authors of this device noted its prospects for long-pulse generation up to 400 ns and further up to 1 μs, but the results of the experiments were not published.

A simpler construction of IRM was investigated in [31, 32] for nanosecond voltage pulse durations. In [31], 8- and 10-cavity blocks of the 10-cm wave band were used. Despite the difficulties in outputting microwave radiation through the cathode slit and reflections from the space-charge spokes, it was possible to obtain microwave pulses with a peak power of ~500 MW at an efficiency of ~15%. In studies [32], the microwave power was extracted from a 12-cavity anode block by means of a wave transformer E_{01} along the axis of the system. Low output parameters of ~100 MW at an efficiency of ~2% are apparently caused by a non-optimal coupling of the transformer with the output waveguide.

5.5. Inverted relativistic magnetrons of the Tomsk Polytechnic University

5.5.1. Design, calculation and 'cold' measurements

The main purpose of the studies described in this section was to study the possibility of obtaining high-energy microwave pulses by increasing their duration. A Luch microsecond HCEA at the TPU was used for these studies (see section 3.1).

The process of investigating inverted relativistic magnetrons includes several stages: construction, calculation of dispersion characteristics, 'cold' measurements and experiment on a high-current electron accelerator.

The design task is to determine the type of the slowing-down system, select the number of resonators, estimate the diameters of the cathode and anode, and the length of the anode block. Based on the conclusions of the theoretical model, calculations by the field theory method, 8- and 10-cavity anode blocks are fabricated (the dimensions of the elements are listed in Table 5.2). The cathode-anode gap, equal to 1 cm, provides good conditions for the capture of electrons in the spokes [33]:

$$r_c - R_1 \approx 2\pi R_1 / N \qquad (5.5.1)$$

and, on the other hand, the sufficient duration of the accelerating voltage pulse (R_1 is the outer radius of the anode block resonators).

The ratio of the width of the lamellae to the width of the cavity gap was chosen equal to unity in order to facilitate the thermal regime of the anode block. The internal diameter of the resonators of the anode blocks was determined when calculating the dispersion characteristics. To develop the surface of the anode block (excluding anode plasma formation and lamella destruction), a 'long anode' system (L ~ 50 mm) was chosen. As in the study of the relativistic magnetrons of direct geometry, three types of resonator systems can

Table 5.2.

Para-meter	Number of resonators								
	$N = 8$				$N = 10$				
Cathode radius	$r_c = 4$ cm				$r_c = 3.4$ cm				
Anode radius	$R_1 = 3$ cm				$R_1 = 2.4$ cm				
Oscillation type number	1	2	3	4 (π)	1	2	3	4	5 (π)
f_{calc}, GHz	1.61	2.04	2.89	3.27	1.35	2.61	3.57	4.08	4.23
β_p	0,657	0.643	0.605	0.513	0.680	0.670	0.600	0.513	0.424

be used: open, half-open (at the end of the anode block a shorting cap is installed) and closed (two caps are installed). Two slits are symmetrically arranged in the cathode section which connect the interaction space with standard waveguides of 72 × 34 mm^2 for the output of microwave radiation.

The results of calculating the spectrum of natural frequencies of oscillations of open anode blocks with 8 and 10 resonators are given in Table 5.2.

As follows from the table, the magnetrons have a good separation from the threshold excitation voltages of the modes calculated using (1.2.10). Thus, the division of this parameter π- and neighbouring (N/2–1) modes for the 8-cavity block is ~12% and for the 10-cavity block ~16%. The anode blocks were made of stainless steel, which, in comparison with other metals, has increased resistance to electron bombardment (see sections 2.3.3 and 5.3), and allow to obtain the longest duration of the voltage pulse (see Fig. 5.5).

The scheme of 'cold' measurements using a frequency meter is shown in Fig. 5.7. The IRM was excited by the sweep frequency generator through one of the slits in the cathode, and in the other slit the signal travelled to a standing wave ratio.The frequency spectrum of the types of oscillations (f_{cold}) and the loaded Q-value (Q_n) were determined in this circuit. The results of measurements of the 8-cavity magnetron are presented in Table 5.5, and the calculated

Table 5.3.

Type of oscillation	Open slowing-down system				Half-open slowing-down system				Closed slowing-down system			
	1	2	3	4 (π)	1	2	3	4 (π)	1	2	3	4 (π)
f_{calc}, GHz	1.61	2.04	2.89	3.27	1.61	2.04	2.89	3.27	3.18	3.63	4.17	4.4
f_{cold}, GHz	–	–	2.81	3.19	–	–	2.81	3.24	3.1	3.57	–	–

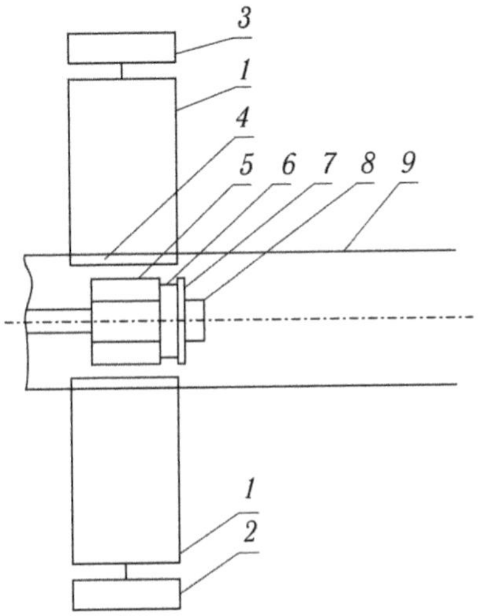

Fig. 5.7. Scheme of 'cold' measurements: *1* – waveguide output of power; *2* – generator; *3* – indicator; *4* – cathode; *5* – anode block; *6* – inserts; *7* – the disk; *8* – matching pin; *9* – drift tube.

values of the frequencies f_{calc} are given for comparison. Three types of slowing-down systems were studied: open, half-open, closed.

It should be noted that the loaded *Q*-factor of IRM in comparison with the previously studied relativistic magnetrons of direct geometry is much smaller. This is due to the use of resonator systems with a large opening angle of the resonators, so we should expect a corresponding expansion of the band of generated frequencies. A close correspondence of the frequencies of the types of oscillations calculated and determined for 'cold' measurements is recorded. This indicates the correctness of the assumptions in the calculation (a limited number of harmonics in the interaction space, the accuracy of calculating the roots of the dispersion equation).

The radial radiation (through cathode coupling slit and waveguide junction) has a drawback due to the fact that an electron cloud is located between the anode block and the cathode, resulting in reflections of the microwave power [31]. In addition, with the use of microsecond voltage pulses, the azimuthal drift of the cathode plasma in crossed electric and magnetic fields can lead to filling of the communication slots and limiting the duration of the microwave pulse. From these considerations, the problem of designing and tuning the 'effective' microwave output 'forward' along the axis of the drift tube, which is a round waveguide with a diameter of 92 mm and a length of 600 mm, was solved at 'cold' measurements. For this purpose, special microwave output devices were used at the H_{01} wave (Fig. 5.8 *b*) and on the wave E_{01} (Fig. 5.8 *c*). Appearance of anode blocks with devices for output of microwave radiation is shown in Fig. 5.8 *d*.

The 'cold' measurements according to the scheme shown in Fig. 5.7 were carried out only for the 8-cavity inverted relativistic magnetron, their results were used in the design of radiation output devices for other anode blocks. Excitation occurred through one

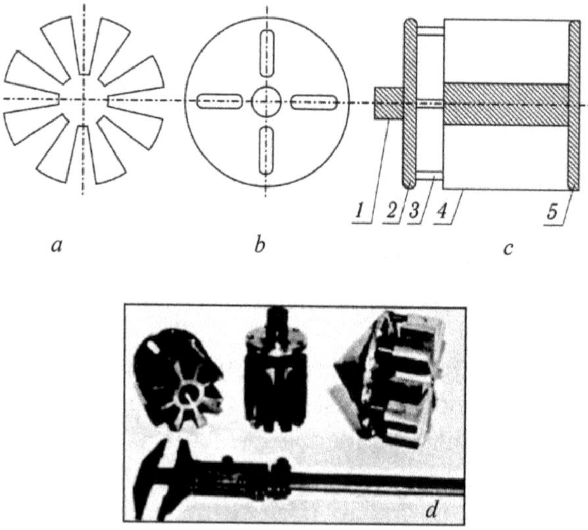

Fig. 5.8. Designs of microwave output devices for IRM: *a* – open end of the anode block; *b* – coupling element on wave H_{01}; *c* – coupling element on wave E_{01}: *1* – matching pin, *2* – disk, *3* – inserts, *4* – anode block, *5* – end cap; *d* – external view of anode blocks with microwave output devices.

of the slits in the cathode. The second communication slot was closed with a metal plug. The radiated power level was estimated by integrating the radiation pattern from the drift tube. The diagram was measured in two mutually perpendicular planes in steps of 5° in the far zone (~1 m). The ratio of the power delivered to the excited relativistic magnetron and the power output from the drift tube determines the attenuation of the signal (expressed in decibels) and serves to evaluate the efficiency of the output device.

Excitation of a waveguide on wave E_{01}. The coupling element on wave E_{01} was constructed using disk *2* (Fig. 5.7 *c*) was used with a diameter equal to the diameter of the anode block of the IRM – 60 mm and thickness 5 mm. A decrease in its diameter increased the attenuation of the signal, and the maximum diameter of the disk was limited by the diameter of the anode block because of fears of loss of the electron current on it during experiments with a high-current electron accelerator. The disc was mounted on four inserts *3* and electrically connected the resonators through one of them, providing additional conditions for the excitation of the π-mode of oscillation. At the opposite end of the anode block there was an end cap *5*. In this case, the maximum of the axial distribution of the high-frequency field is in the region of the coupling element. This makes it possible to increase the efficiency of microwave power transmission from the resonators of the anode block to the waveguide. The inserts were located at the maximum radius of the slots of the anode block where the microwave field is the strongest.

The type of wave in the waveguide was determined by the following methods: a capacitive probe was used to measure the distribution of $E_r(r)$-, $E_r(\varphi)$-, $E_z(r)$-components of the electromagnetic field in the drift tube and its opening. A movable piston in the output waveguide created a resonator and defined the distribution $E_z(z)$ of the formed standing waves, which constitute a resonant dependence with a period of ~8 cm (the estimated value ~7.93 cm). In the far zone (~1 m from the open end of the waveguide), the microwave radiation pattern was recorded in two mutually perpendicular planes.

The results of measurements of the electrical components of the electromagnetic field components indicated good excitation and propagation of wave E_{01} in the waveguide and further in free space.

Excitation of a waveguide on wave H_{01}. The design of the coupling element for outputting the microwave resonators of the anode block for wave H_{01} is an end plate with coupling slots (Fig.

5.7 *b*). Coupling slots open the resonators through one of them, so that the microwave fields in them when the anode block is excited in the π-type of the oscillations turn out to be in the same phase. The maximum value of the coupling was achieved in the case when the dimensions of the slots in the end plate coincided with the size of the resonators. The attenuation for an open anode block was approximately 6 dB at frequency f = 3.2 GHz and for the half-open block 5.4 dB at frequency f = 3.4 GHz.

The type of wave H_{01} was defined as follows: an electric probe measured the E_φ-component of the electric field at the waveguide aperture. With the probe in parallel arrangement to the waveguide axis the E_z-component was not detected, which enables to identify the measured wave with as the *H*-type. A waveguide-coaxial transition in two mutually perpendicular planes with a step of 5° was used in the measurement of the directional pattern of the radiation of the open end of the waveguide in the far zone (distance ~1 m). The change of the polarization plane of the receiving horn by 90° led to a decrease in received power by an amount more than −15 dB. The described results indicate the excitation of a wave in the waveguide and the propagation in free space of the H_{01} wave.

The results of the 'cold' measurements of the inverted relativistic magnetron: the resonant excitation frequency, the attenuation of the microwave signal, and the *Q*-factor of the π-mode for all output devices are summarized in Table 5.4.

A small measured value of the loaded *Q*-factor (~20–40) with the use of effective designs of the microwave output devices indicates a good connection between the magnetron and the waveguide and the possible broadening of the frequency band during IRM operation. The results of the tuning of output devices of inverted relativistic magnetron systems were used in experiments on the generation of microwave radiation with the use of HCEA with a microsecond pulse voltage duration.

5.5.2. Design, calculation and 'cold' measurements of inverted coaxial magnetrons (ICM)

The cylindrical stabilizing resonator with radius r_r of such a magnetron is connected by the coupling slots to the anode block cavities through one ('open' resonators), the other resonator system ('closed') is not directly connected to an external circuit (Fig. 5.9). Interconnected through the interaction space, the 'open' and 'closed' resonators are excited in the antiphase, forming a field of the π-type.

Table 5.4.

Anode block type		Microwave output device			
		Without a coupling element	E_{01}		H_{01}
			Without pins	The pin (diam. 30 mm, l = 15 mm)	$N/2$ coupling slots
Open	f_{cold}, GHz	3.140	3.180	3.180	3.200
	Attenuation, dB	−10.5	−5.4	−3.5	−6
	Q_n	100	47	25	38
Half-open	f_{cold}, GHz	3.225	3.280	3.280	3.440
	Attenuation, dB	−10.5	−4.6	−2.8	−5.4

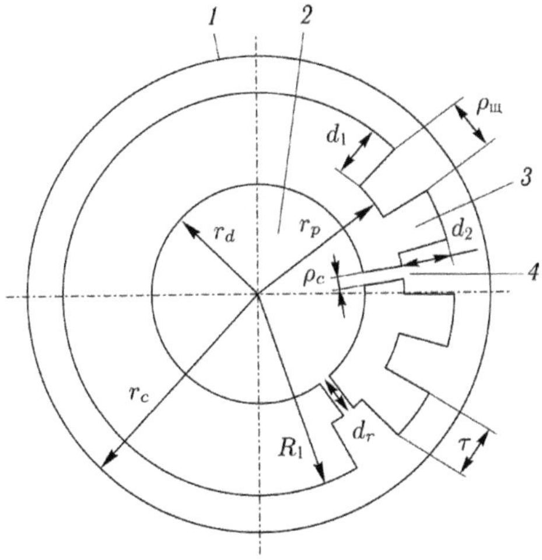

Fig. 5.9. Scheme of the inverted coaxial magnetron: *1* – cathode; *2* – resonator; *3* – anode block; *4* – coupling slot.

Using the internal high-Q coefficient cylindrical resonator operating on oscillations of the H_{011} type, allows effectively stabilize the frequency and type of the oscillations at in the interaction space [34]. Other types of oscillations of the resonant systems of the magnetron are not supported by the type H_{011} oscillations or are supported only very weakly.

The anode block of the investigated magnetron of the 3-cm wavelength band contained 48 resonators. The calculation of the spectrum of the resonant frequencies of the ICM was carried out by the field theory methods. The dispersion equation of the entire coupled system of an inverted coaxial magnetron [35] is obtained by cross-linking the interaction space and the slowing-down system of conductances of the ‘closed’ and ‘open’ resonators, the coupling gap with the stabilizing resonator and the interaction space. The solution of the dispersion equation (obtaining the spectrum of the resonance frequencies of the ICM) was carried out by the method described in section 1.5.

In the cylindrical cavity eigenfrequencies of the *H*-type are arranged on the frequency axis in the following order: H_{111}, H_{211}, H_{011}, H_{311}, H_{121}, H_{221}, H_{321}. Stable operation of the generator requires the same order of following of the modes of oscillations in the cylindrical resonator of the entire couple ICM system. As a result of the calculations and optimization the ICM system with the following geometry was selected [36,37]: R_1 = 30 mm, r_c = 40 mm, d_1 = d_2 = 5 mm, r_p = 25 mm, r_d = 22 mm, d_r = 3 mm, ρ_c = 1 mm, ρ_{rad} = 1.96 mm, τ = 1.96 mm, L_p = 27 mm, L_m = 16 mm, l_{rad} = 20 mm, where L_p is the resonator length, l_{rad} is the length of the coupling slit, L_m is the the length of the lamellae.

The obtained spectrum of the coupled ICM system is located along the frequency axis as follows (Table 5.5):

The possibility of stable operation of the ICM on the π-type of oscillations of the multi-resonator anode block is determined by the separation value of the non-working modes H_{311} and H_{211} from the basic mode H_{011}. The excitation of ‘parasitic’ modes of the stabilizing resonator H_{311} and H_{211} is caused by the types of oscillations of the

Table 5.5.

Type of oscillation	H_{111}	H_{211}	H_{011}	H_{311}	H_{121}	H_{221}	H_{321}
f_{calc}, GHz	7.4	9.32	10.5	11.19	13.0	15.5	23.6

anode block *N*/2 – 3 (f = 11.19 GHz) and *N*/2 – 2 (f = 9.32 GHz), adjacent to the π-type. Because of *N*/2–1 of the slowing-down system of ICM corresponds to H_{111}-mode of oscillation of the stabilizing resonator, which is good separated in frequency. With respect to the excitation voltage, the types *N*/2 – 3 and *N*/2 – 2 are separated from the working type by 24 and 3%, respectively. Since the frequency

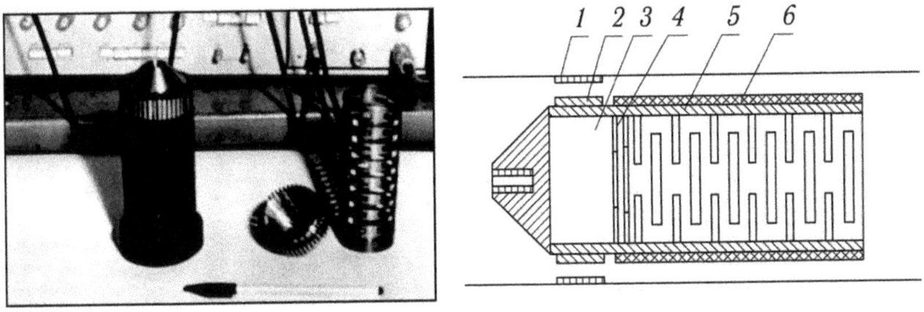

Fig.5.10. External view of an anode block and the construction of an inverted coaxial relativistic magnetron: *1* – cathode; *2* – resonators of the anode block; *3* – cylindrical stabilizing resonator; *4* – annular diaphragm; *5* – filter of the types of waves; *6* – graphite absorber.

separation for ICM with such a number of resonators turned out to be quite satisfactory, the device shown in Fig. 5.10 was manufactured and a full cycle of experimental studies was carried out.

The microwave radiation was outputted from the stabilizing resonator on the wave H_{01} using a circular waveguide with an internal radius, coinciding with the radius of the resonator, that is 22 mm. At the same time, the waveguide performs the function of filtering the types of waves, since along its length ~150 mm there are azimuthal slots cut, preventing the flow of longitudinal currents and propagation of electromagnetic waves with a similar distribution of currents in the walls of the waveguide.

For 'cold' measurements (Fig. 5.11), the anode block with the microwave output device was installed in a coaxial vacuum chamber *3* serving simultaneously as an output waveguide (diameter ~92 mm). On the inner surface of the vacuum chamber there was a cathode *4* with an inner diameter of 80 mm. ICM was excited by 12 capacitive pins installed in the 'closed' resonators evenly along the azimuth. The use of lengths of cable of equal length, connected at one point with the cable from the directional coupler of the generator, ensured the coincidence of the phase of microwave oscillations in the resonators with capacitive pins. Such a method of introducing microwave radiation into the ICM allows symmetric excitation of the anode block resonators and a sufficient power level for the measurement. The connection area of the generator cable and the pins was screened with a metal foil and a 3-cm wide-range diode amplifier, except for excitation of 'parasitic' modes in the output waveguide.

The frequency spectrum of the inverted coaxial magnetron was determined using a sweep frequency generator *5* in the range 8–12

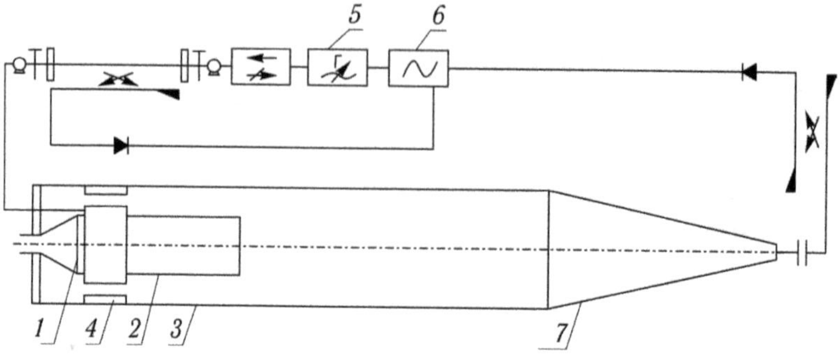

Fig. 5.11. Scheme of 'cold' measurements of an inverted coaxial relativistic magnetron: *1* – anode block; *2* – filter of types of waves; *3* – output waveguide; *4* – cathode; *5* – generator; *6* – indicator; *7* – smooth waveguide transition.

GHz. According to the results of the calculation, three types of oscillations of the cylindrical resonator H_{211}, H_{011}, H_{311} can be excited in the generator band (see Table 5.5). The output radiation was received by a directional coupler of the reflected wave ratio indicator and attenuation *6* using a smooth waveguide transition *7* from round to rectangular with a size of 23 × 10 mm².

The H_{211}- and H_{311}-modes of oscillation were suppressed using graphite ring absorbers installed at the ends of the anode block and overlapping the ends of the coupling slits. They are used because of a more significant sagging of the fields of these types in the end regions, that is, in the direction of the ring absorbers [37]. Indeed, there was a weakening of the output signal level of the types H_{21}, H_{31} (the absorber of the filter of wave types was not used in these measurements), a decrease in the Q-factor of these types due to the growth of losses and, accordingly, the Q-factor growth in the main π-type (Table 5.6).

In the process of 'cold' measurements, the efficiency of various methods of output of microwave radiation from a stabilizing resonator was determined. Measurement results: the spectrum of the modes, the value of the loaded Q-factor and the amount of attenuation of the radiated signal power in decibels are given in Table 5.6.

The installation of the wave type filter resulted in an increase in the attenuation of the output signal level >3 dB for modes with frequencies f = 9.118 and 11.078 GHz, while for the wave H_{01} (f = 10.181 GHz) the microwave radiation level due to the improvement of the coupling of the stabilizing resonator with the output waveguide increased. The use of a graphite absorber installed

Table 5.6.

Oscillation mode	Frequency, GHz		Resonator open		Filter of types of waves		Absorber at the end of resonators		Absorber on the filter and resonators	
	Calculation	'Cold' meas.	Att. dB	Q_n	Att. dB	Q_n	Att. dB	Q_n	Att. dB	Q_n
H_{01}	10.53	10.181	−2	648	0		−0.8	536	−1.2	679
H_{21}	9.32	9.118	−3.5	304	−5.4		−7.9	212	>−10	–
H_{31}	11.19	11.078	−3.7	427	−6.1		−9.3	170	>−10	–

Table 5.7.

Diameter of diaphragm	15 mm	22 mm	32 mm
$U_{microwave}$, rel. units	0.72	1.0	0.53
Q_n	1020	680	430

at the outer surface of the wave type filter led to complete attenuation of the waves H_{21} and H_{31}, while the H_{01} radiation level remained unchanged (Table 5.6).

The possibility of increasing the level of the output radiation by optimizing the diameter of the hole in the annular diaphragm installed at the end of the resonator and connecting it with a filter of the types of waves was studied in 'cold' measurements. On the one hand, the diaphragm must provide high resonance properties of the cylindrical stabilizing resonator, and on the other – its good connection with the filter of types of waves. The dependence of the efficiency of the output of microwave radiation on the internal diameter of the diaphragm is illustrated in Table 5.7.

Estimates of the radiated power level were also carried out by integrating the radiation pattern measured in the far zone (~1.5 m). In this case, the ICM was excited with a tunable 3-cm wavelength generator, with an increased output of microwave power. The radiation from the outgoing waveguide was conducted by means of a cone antenna which was subsequently used in experiments on a high-current electron accelerator. A waveguide-coaxial transition was established in the far zone and the signal from it was fed to the detector, the broadband amplifier, and then to the oscilloscope. The measured radiation pattern is shown in Fig. 5.12. Assessing the level of microwave energy of the ICM by integrating the directivity pattern and directly measured by a smooth transition from a waveguide of circular cross section to a rectangular (position *7*, Fig. 5.11) indicate their correspondence with an accuracy of 10–15%.

The analysis of the results of the 'cold' measurements of the iverted coaxial magnetron makes it possible to note: 1) a good correspondence of the calculated and measured values of the frequency: the difference does not exceed 3%; 2) the division by the magnitude of the slowing-down of the modes $N/2 - 2$, $N/2 - 3$ with the working π-mode of oscillation is

$$\beta_{N/2-2} / \beta_\pi = 0.97,$$
$$\beta_{N/2-3} / \beta_\pi = 1.24.$$

The results of the measurements show the possibility of suppressing the H_{211} type which is most dangerous for stable operation of ICM, (its excitation voltage practically coincides with the excitation voltage of the π-type). The installation of graphite absorbers at the ends of the anode block resonators allows to significantly reduce the power level of the H_{211}-type. The excitation of the H_{311} wave is unlikely because of the significant separation in magnitude of the slowing-down with the π-type.

In comparison with the previously studied relativistic magnetron generators without a stabilizing resonator in the ICM there is a

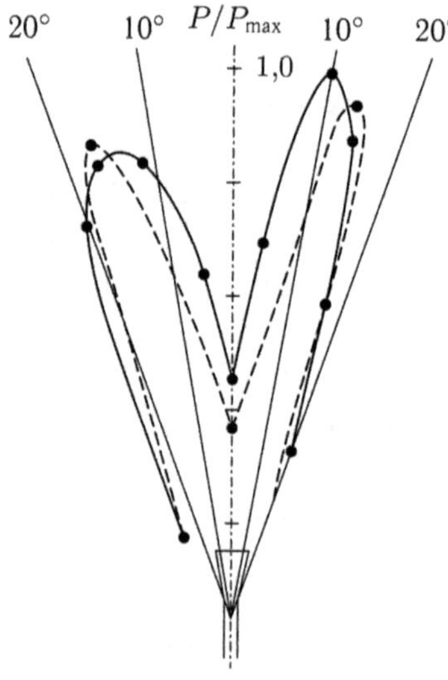

Fig. 5.12. Directivity pattern of the radiation of the ICM in excitation of the H_{01} wave, 'cold' measurements (solid curve), experiments with HCEA (dotted curve).

noticeable increase in the Q-value of the slowing-down system to $Q_n \sim 430\text{–}1020$. This fact assumes the use of voltage pulses of microsecond duration, if an external microwave generator is not used to excite the device [38].

5.6. Experimental investigations of inverted relativistic magnetrons

Experiments on the generation of microwave radiation by inverted relativistic magnetrons were carried out on a high-current electron accelerator 'LUCH' with an input voltage of 300–600 kV, a current of 2–7 kA, pulse duration of 1–1.5 μs according to the scheme shown in Fig. 5.13 [39, 40].

The magnitude of the accelerating voltage was recorded by a resistive voltage divider of the electron gun *1*, the total accelerator current by shunt *2*. Estimates of the level of the generated microwave power were obtained: 1) by integrating the radiation pattern recorded in the far zone; 2) the value of attenuation of the microwave signal between the receiving and transmitting antennas calibrated on 'cold' measurements; 3) using a microwave calorimeter. The results of the three independent methods for determining the microwave power differed by no more than 30%, and the power values given below are the average values of these methods.

Figure 5.14 shows oscillograms of the voltage, current, and signals of the microwave detector when the radiation is 'sideways'-radial: through the slots in the cathode and a smooth waveguide transition,

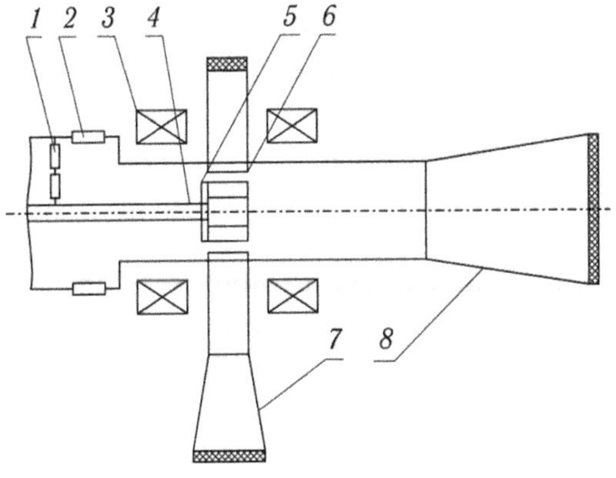

Fig. 5.13. Experimental scheme of IRM research: *1* – resistive voltage divider; *2* – shunt of the total current; *3* – magnetic system; *4* – anode holder; *5* – anode block; *6* – cathode; *7* , *8* – microwave output antennas.

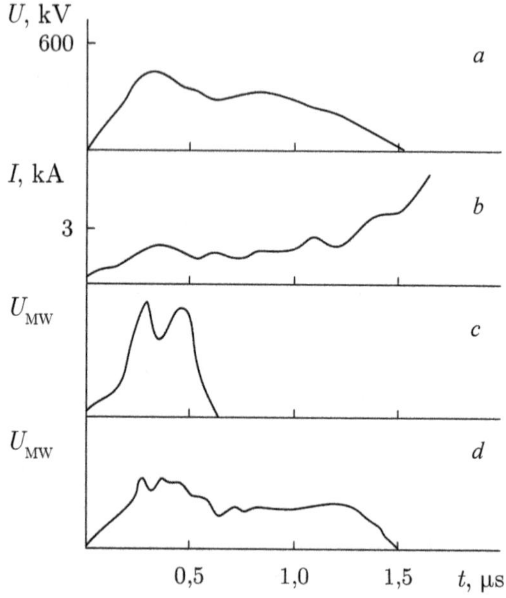

Fig. 5.14. Typical oscillograms of pulses: *a* – voltage; *b* – the total current of the accelerator; *c* – signals of the microwave detector when the radiation is radial; *d* – radiation 'forward'.

similarly to [31]. In addition to low efficiency due to reflections of microwave power from the space charge cloud, measurements showed a limitation of the duration of microwave radiation in comparison with the duration of the voltage pulse. The duration of the microwave pulse 'sideways' is ~0.6 μs with a voltage pulse duration of 1.5 μs, which is apparently due to the azimuthal motion of the cathode plasma and the filling of its coupling slots by it. The duration of radiation 'forward' along the axis of the drift tube is close to t_k (Fig. 5.14). On the basis of this, it was concluded that the output of power 'sideways' was unpromising, therefore further experiments were conducted by the 'forward' output devices described above (5.5.1). The coupling slots in the cathode section were filled with metal inserts to ensure uniform emission from the cathode and the symmetry of the azimuthal distribution of the high-frequency field of the anode block. As can be seen from the oscillograms, an increase in the total current of the accelerator is observed in the region of synchronous magnetic fields, which is caused by the appearance of a high anode current of a magnetron under the influence of powerful microwave fields in the interaction space, the intensity of which, apparently, is comparable with the strength of the statistical electric field.

Calculations of the synchronism ($\beta^{N=8}_{p.\text{meas}} = 0.5 \approx \beta^{N=8}_{p.\text{calc}} = 0.513$, $\beta^{N=10}_{p.\text{meas}} = 0.4 \approx \beta^{N=10}_{p.\text{calc}} = 0.424$) and measurements of the carrier frequency ($f^{N=8}_{\text{meas}} = 3.33$ GHz $\approx f^{N=8}_{\text{calc}} = 3.27$ GHz, $f^{N=10}_{\text{meas}} = 4.22$ GHz $\approx f^{N=10}_{\text{calc}} = 4.438$ GHz) indicating the generation of the π-mode. No other types were observed. As expected, a wide frequency spectrum of the IRM radiation is detected, which is associated with a decrease in the value of the loaded Q-factor (see Table 5.5) in comparison with relativistic magnetrons of direct geometry.

A significant effect on the level of the generated power, as in the ordinary RM, is exerted by the magnitude of the current leaving the interaction space. The growth of the limiting transport current leaving the interaction space with a change in the diameter of the anode holder from 16 to 45 mm resulted in a 1.7-fold decrease in the microwave power in the experiment with a 10-cavity anode block. Note that the calculated maximum transportation current should have increased by approximately 3.5 times. However, the absence of a collector on the anode holder leads to the return of a part of the electrons back to the interaction space and to their involvement in energy exchange with the microwave field of the anode block, similar to the use of drift tubes for relativistic magnetrons of direct geometry [41].

According to the data in Table 5.8, it is possible to compare the efficiency of the power output using different versions of microwave devices.

As can be seen from Table 5.8, the best power parameters are provided by the RMs with an output device at wave E_{01}, and optimization of the coupling element will further increase the output parameters of the device.

Since the loss current in the inverted magnetron systems is minimal and practically the entire recorded current is the operating current of the device, it is interesting to compare the different theoretical models to determine the magnitude of the anode current and the power generated by the device. To do this, we use the relations (2.1.31) and (2.2.50) to calculate the high-frequency electric field strength through the anode current of electrons. The value of the anode current for the calculations was determined as follows. At a voltage $U \sim 600$ kV, the measured value of the total current of the accelerator was $I \sim 7$ kA. Measurements of voltage and current for a smooth metal anode with a diameter that coincides with the

Table 5.8.

Radiation power, MW	Method of output of microwave radiation					
	Open anode block	Slots in the plate (H_{01})	Coupling element on wave type E_{01}			
			Without pin, inserts length 15 mm	Without pin, inserts 10 mm long	Pin, inserts length 15 mm	Pin, inserts 10 mm in length
	59	34	30	50	120	160

diameter of the anode block with the corresponding parameters of the accelerator and the magnetic field showed that the current at the anode is ~1 kA (this current is caused by the action of the diocotron instability of the electron cloud at the cathode). Thus, the anode (operating) current of the IRM is defined as the difference between the total current of the accelerator and the current to the anode under the action of the diocotron instability and is equal to 6 kA.

The intensity of the high-frequency electric field calculated from (2.1.31) and (2.2.50) was $\tilde{E} \sim 270$ kV/cm. Estimates of the microwave power level in the interaction space of the inverted relativistic magnetron were carried out by the formula (2.1.38) and showed a value of ~490 MW.

The level of generated microwave power was also calculated by the model of energy balance [42]:

$$P = \frac{fV(\Delta B)^2}{4Q\cdot 10^7}\left(1 - \frac{f}{2Qf_c}\right), \tag{5.6.1}$$

where f is the microwave frequency; f_c is the cyclotron frequency; V is the volume occupied by the high-frequency field;

$$\Delta B = (B_{\max} - B_{cr})/2,$$

$B_{\max}$ is the value of the magnetic field at which the power is maximum. Substituting the experimental data, we obtain a power value of about 400 MW. As follows from the calculations carried out for different theoretical models, the output powers of the IRM correspond quite accurately to each other, but differ significantly from the values recorded in the experiment. This fact can be explained by the insufficient degree of optimization of the microwave output device.

The above experiments were carried out at the output voltage of the accelerator accelerator of ~700 kV. Scaled up to 800 kV made it possible to obtain at the output of the 10-cavity IRM microwave power pulses P ~ 200 MW and duration ~ 1 μs at an efficiency of ~7.5%.

Higher performance was obtained by using 12-cavity anode blocks with an external resonator diameter of 90 mm. The inner diameter of the cavity is 28 mm, the ratio of the aperture width of the resonator to the width of the lamella $\varkappa$ was selected 2:1 and 1:1. The cathode was the inner surface of the output waveguide with a diameter of 110 mm. The coupling elements on the wave H_{01} and E_{01} were used for the output of microwave radiation from the resonators. The results of studies of IRM on a high-current electron accelerator are given in Table 5.9.

The maximum power of the microwave pulse for these generators was P ~ 350 MW for a 12-cavity IRM with an output device at a wave type of E_{01} (pin l = 50mm, diameter 65 mm and the disc that provides the best coupling). The duration of the microwave pulses was $\tau_{microwave}$ ~ 0.6–0.7 μs at U ~ 400 kV, I ~ 8.5 kA, t_k ~ 1.0 μs (see Fig. 5.15).

As can be seen from Table 5.9, in comparison with the 8- and 10-cavity anode blocks, the voltage pulse duration was reduced due to the presence of a high anode current on the surface of the slowing-down system due to the development of high-power microwave fields in the interaction space. As a consequence, the time interval for forming the plasma layer on the surface of the anode block

Table 5.9

N	$\varkappa$	f_{meas}, GHz	U, kV	I, kA	t_k, μs	P, MW	Type of wave	$\tau_{microwave}$ μs	η, %
8	2	3.300	500	6.0	1.6	120	E_{01}	1.2	4
10	1	4.480	600	6.0	1.3	200 34	E_{01} H_{01}	1.0	5
12	1	2.600	400	8.5	1.0	350 40	E_{01} H_{01}	0.7	10
12	2	2.600	400	8.5	1.0	300 60	E_{01} H_{01}	0.6	9

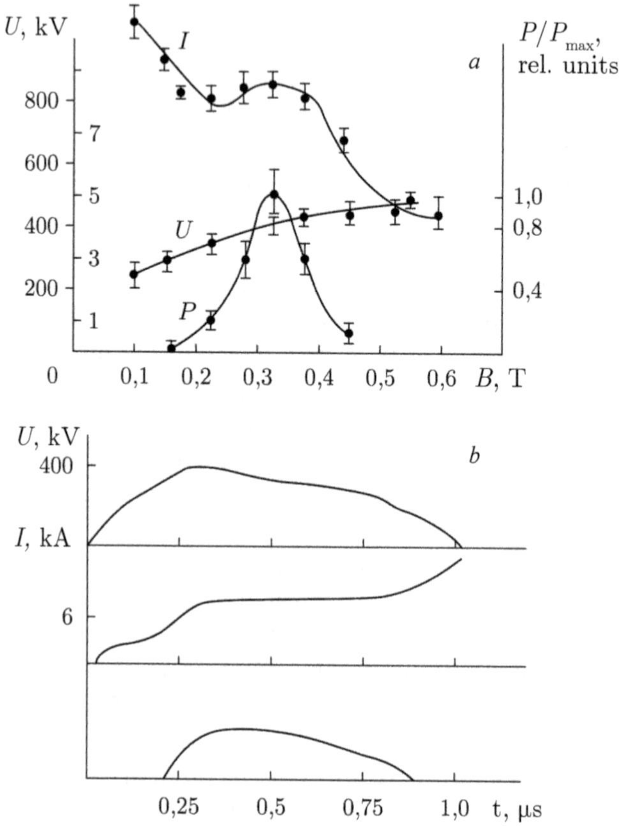

Fig. 5.15. Dependences of microwave power (*P*), voltage (*U*) and the total current of the accelerator (*I*) for the 12-cavity IRM on the value of the magnetic field induction, P_{max} = 350 MW (*a*). Typical oscillograms of voltage pulses, total current and microwave detector signals (*b*).

decreased, and also the duration of the voltage pulse and microwave radiation decreased.

In comparison with RMs of direct geometry, an increase of 2–3 times of the energy content of microwave pulses generated by IRM with corresponding parameters of the charging voltage of the HCEA is recorded. Thus, for equal values of the charging voltage of the HCEA 'Luch' studies were conducted on relativistic magnetrons of inverted geometry ($P \sim 160$ MW, $\tau_{microwave} \sim 1$ μs) and direct geometry ($P \sim 300$ MW, $\tau_{microwave} \sim 0.2$–$0.25$ μs). Calorimetric measurements of the energy of microwave pulses are: for IRM ~ 100 J, which exceeds more than twice the corresponding data for the direct geometry magnetron, ~45 J. Both for direct [1] and for inverted relativistic magnetrons, the energy parameters of microwave pulses are observed to increase with the use of electron beams of higher energy. This fact demonstrates the prospects of using IRM for obtaining microwave pulses of increased energy content.

5.7. Experimental investigations of the ICM

The design of ICM (inverted coaxial magnetron) is described in section 5.5.2 and [36.37]. The scheme for the diagnostics of microwave radiation parameters included: a cryogenic detector on 'hot' carriers, a tunable bandpass filter of the 3-cm wavelength range. The level of generated power was estimated by integrating the radiation pattern of microwave radiation in the far zone (~1.5 m) in two mutually perpendicular planes; on the calibrated magnitude of the attenuation between the antennas and the microwave calorimeter. The ICM was used with a filter of the types of waves, with graphite absorbers, with an annular diaphragm with an internal diameter of 15, 22 and 32 mm (Q-factor values are given in Table 5.7).

The generation of the ICM is characterized by a long time of rising of the microwave oscillations, which can be related to the inertia of the oscillatory system of the ICM. Indeed, estimates of the rise time to a level of 0.9 from the maximum value in accordance with the well-known formula were

$$\tau_0 = 4.6\frac{Q}{f}, \tag{5.7.1}$$

that for a quality factor of the order of 600–700, the values close to the measured values are ~0.5 μs.

Figure 5.12 shows the directivity pattern of the radiation, determined during 'cold' measurements and in the microwave generation mode on a high-current electron accelerator. One can note their coincidence, which allows us to assert that excitation in the output waveguide and propagation in the free space of the H_{01} wave are possible. Correspondence of β_{rad} and β_{calc} shows the excitation of an inverted coaxial magnetron at the working π-mode of oscillations. The output power level was ~35 MW with a pulse duration of ~0.6 μs at the base with the parameters of the HCEA: voltage $U \sim 330$ kV, current $I \sim 4.5$ kA, duration $t_k \sim 1.2$ μs. An increase in the parameters of the accelerator to $U \sim 400$ kV, $I \sim 5.5$ kA, $t_k \sim 1.0$ μs resulted in an increase in power up to

~70 MW with a duration of 0.7 μs. Just as in the case of 'cold' measurements, diaphragms of different internal diameters were used in experiments on HCEA. Using a diaphragm with a diameter of 15 mm led to an increase in the rise time of oscillations to 0.7 μs. At the same time, a decrease in the output power level by 2–3 times is recorded, which, as the study of the surface of the diaphragm has shown, is caused by microwave breakdowns at the coupling hole of the coaxial resonator with a filter of the types of waves. The increase in the diameter of the aperture of the diaphragm is more advantageous at higher levels of the accelerating voltage. Reduction of the loaded Q-factor (see Table 5.7) leads to a faster increase in the microwave field and the power generated by ICM. At the parameters of the accelerator: voltage $U \sim 400$ kV, current $I \sim 5.5$ kA, duration $t_k \sim 1.0$ μs the parameters of microwave radiation of ICM were 100 MW with a duration of ~0.7 μs.

After the experiments with the limiting values of the charging voltage of the HCEA the inner surface of a cylindrical resonator contained traces of microwave breakdown in the region of some coupling slots. This, apparently, explains the limitation of the microwave power level generated by the ICM.

In general, the results of the ICM studies of the 3-cm wavelength range have shown the possibility of its effective operation with an output power of $P \sim 100$ MW with the sub-microsecond duration of the microwave radiation. Limitations of the duration of the microwave pulse are caused by a long transient time of oscillations due to the high Q of the cylindrical resonator. The duration of the microwave pulse can be increased by using an external exciting microwave generator to reduce the rise time of the oscillations to values of ~10–30 ns.

Thus, experiments with the inverted relativistic magnetrons have shown the possibility of their effective work in the emission of microwave radiation 'forward' along the axis of the drift tube. It can be noted that the duration of the microwave pulse in the IRM is close to the duration of the voltage pulse, although the pulse shape is not uniform, which is probably due to the formation of anode plasma on the anode block lamellae. The increase in duration is explained by the fact that there is no radial motion of the cathode plasma in the inverted magnetron diode and during the time interval determined by the vacuum conditions of the anode material, there is not anode plasma absence or its concentration does not exceed the critical value for the absorption of microwave radiation [43]. The expansion of the band of the emitted signal is experimentally observed up to 10–16%, which is determined by the low value of the loaded *Q*-factor of the resonant systems of IRM. In the region of synchronous magnetic fields there is an increase in the current sampling of the slowing-down structure, which indicates the high strength of the microwave fields in the block (comparable to the electric field strength formed by the accelerator). A significant effect on the amount of radiated power is exerted by the type of microwave output device, which needs further optimization in cold measurements.

The results of ICM studies of a 3-cm wavelength range have shown the possibility of its operation with an output power of ~100 MW for a submicrosecond microwave duration. The limitations of the duration of the microwave pulse are caused by a long transient time of oscillations due to the high *Q* of the resonator system. The increase in the duration can be achieved by using an external exciting microwave generator.

In experiments, there was an inverse dependence of the duration of the microwave pulse and the power level of the device, which imposes certain limitations on the output parameters of the IRM. This circumstance forces one to turn to another promising construction – a relativistic magnetron with external injection (RMI) of high-current relativistic electron beams. The following sections of the monograph are devoted to an analysis of the operation of such a device and the results of experimental studies.

5.8. Relativistic magnetrons with external injection of an electron beam

5.8.1. Physical preconditions for the creation of RMI

The causes of the limitation of the duration of voltage pulses and microwave radiation in the relativistic magnetrons of direct geometry due to the accelerated expansion of the cathode plasma in high-power microwave fields of the anode block can be eliminated in a magnetron-type device with external injection of a high-current relativistic electron beam. In a similar device, the explosive-emission cathode is taken out of space interaction, its role is played by an electron beam injected by an additional diode (Fig. 5.16).

The device design was proposed by experts of the Institute of Applied Physics and demonstrated its efficiency in [44]: $P \sim 100$ MW, $\eta \sim 4\%$, $\tau_{\text{microwave}} \sim 25$ ns. A complete cycle of experimental

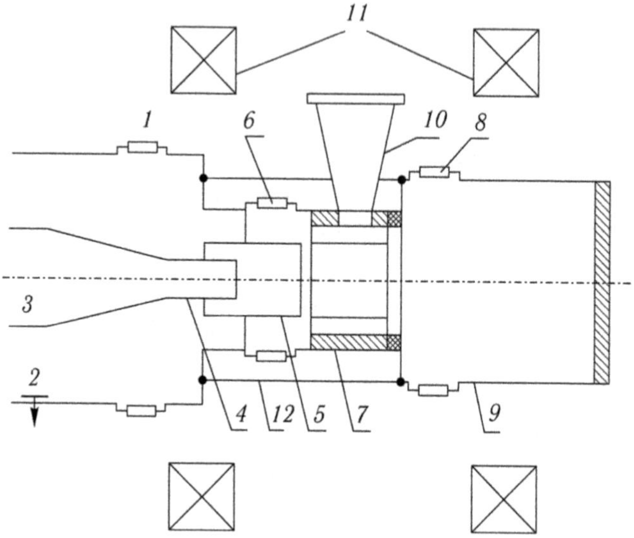

Fig. 5.16. Experimental scheme of RMI investigations: *1*, *6*, *8* – shunts of full, anode, and end currents; *2* – capacitive voltage divider; *3* – cathode holder; *4* – cathode; *5* – anode of an additional diode; *7* – anode block; *9* – drift tube; *10* – output of microwave power; *11* – magnetic system; *12* – reverse current liners.

studies of an RM with external injection of an electron beam was performed at the TPU [45, 46].

When an electron beam is injected into the anode block, its space charge creates, together with an external magnetic field, crossed fields that ensure the synchronization of the azimuthal drift of electrons with the high-frequency field. The synchronism condition in this case has the form [45]

$$\frac{mc^2}{eH}\frac{\gamma-\gamma_B}{d}\approx\frac{2\pi R}{\lambda n}, \tag{5.8.1}$$

where $\gamma = 1 + eU/mc^2$ is the total electron energy; $\gamma_B = \sqrt{2\gamma + 0.25} - 0.5$ is the kinetic energy of the electrons.

Unlike the well-known formula for the synchronism condition, here the intensity of the electric field in the interaction space is created not by the applied voltage to the cathode, but by the potential of the injected electron beam.

The mechanism of the interaction of electrons with the microwave field of the slowing-down system remains the same as in the magnetron with the cathode in the interaction space. However, the electron beam in the interaction space can be in different states, both in the high-energy state ($\gamma^{1/3} < \gamma_B < \gamma$), and in a state of a virtual cathode ($\gamma_B = 1$). In the case of the high-energy state of the beam the minimum value of its kinetic energy is $\gamma_{B\,\min} = \gamma^{1/3}$, therefore the maximum possible electronic efficiency without consideration of the cyclotron rotation energy is limited by the quantity

$$\eta = 1 - \frac{\gamma_B - 1}{\gamma - 1} = \frac{\gamma - \gamma^{1/3}}{\gamma - 1}. \qquad (5.8.2)$$

The state with a virtual cathode (VC) is characterized by a minimum kinetic energy $\gamma_B = 1$, and the maximum possible electron efficiency tends to the efficiency of an RM with a cathode in the interaction space.

Both states of the electron beam have their own advantages. The high-energy state avoids the deposition of electrons on the surface of the slowing-down system inherent in the M-type generators if the length of the anode block is chosen such that the time of flight of electrons along it is less than the time of their drift to the surface:

$$L \leqslant V_z \int_{r_c}^{R} \frac{dr}{V_r},$$

where V_z, V_r are the longitudinal and radial velocities of the electrons. In this case, the reasons for limiting the pulse duration are associated with the formation of anode plasma and the destruction of the anode blocks can be completely eliminated.

Estimates of the efficiency according to the formula (5.8.2) show that at typical voltages of the order of 1 MV of high-current electron accelerators (which corresponds to $\gamma = 3$), the magnetron with transit electron loses by efficiency in comparison with the magnetron with a cathode in the interaction space by ~20–30%. However, as seen from (5.8.2), to increase the voltage on the additional cathode the efficiency of the device increases as the total electron energy γ increases faster than their kinetic energy γ_B. This allows us to define a relativistic magnetron with external injection of an electron beam as a purely relativistic device.

From the point of view of obtaining high values of efficiency, it is preferable to use the stationary state of an electron beam with a virtual cathode. To realize such a state, the current I_{inj} injected the interaction space must exceed by the generation of microwave radiation:

$$I_{\text{inj}} > I_{\text{lim}} + I_{\text{an}}, \tag{5.8.3}$$

where

$$I_{\text{inj}} = \frac{mc^3}{e} \frac{(\gamma - \gamma_B)\sqrt{1 - 1/\gamma_b^2}}{2\ln(R_d / r_c)}; \tag{5.8.4}$$

$$I_{\text{lim}} = \frac{mc^3}{e} \frac{\left(\gamma^{2/3} - 1\right)^{3/2}}{2\ln(R / r_b)}, \tag{5.8.5}$$

quantity I_{an} is determined by the relation (2.1.31) or (2.2.50); R_d is the internal radius of the anode of the additional diode; $r_b \approx r_c$ is the mean radius of the electron beam. The fulfillment of condition (5.8.3) means that the diameter of the additional anode must be much smaller than the diameter of the multi-cavity anode block. Therefore, the synchronism condition (5.8.1) may be incompatible with the magnetic insulation condition of the additional diode, which imposes restrictions on the amount of the injected current. We also note the following peculiarity of the formation of a virtual cathode under the indicated conditions: the distribution of the minima of the potential occupies the small axial length of the drift space. This allows us to conditionally call the virtual cathode being formed 'narrow'. Under these conditions, the selection of electrons is carried out on some local surface of the anode block, which does not exclude the possibility of the formation of an anode plasma. The removal of the restrictions associated with the need to form a current in the diode, significantly exceeding the limiting transport current in the anode block, and distributing the flow evenly across the anode block uniformly can be achieved by forming an 'extended' VC. To form it, it is sufficient that the injection current exceeds the sum of the anode current and the limiting transport current in the drift tube $I_{\text{lim.tube}}$:

$$I_{\text{inj}} > I_{\text{lim.tube}} + I_{\text{an}}, \tag{5.8.6}$$

where the

$$I_{\text{lim.tube}} = \frac{mc^3}{e} \frac{\left(\gamma^{2/3} - 1\right)^{3/2}}{2\ln(R_{\text{tube}} / r_b)}. \tag{5.8.7}$$

To obtain the state of a beam with a virtual cathode in the interaction space, the beam potential at the input to it must be close to the anode potential. To fix the potential, an electrode transparent to electrons is used: a foil, a grid or a diaphragm whose inner radius is close to the outer radius of the beam: $r_\Delta = r_b + \Delta$, where Δ is the beam thickness. At the initial time, the beam passes freely along the anode block and begins to be reflected in the region of the drift tube. The reflected part of the beam together with the injected current creates an interaction space that is sufficient for reflection already in this area. Over time, the virtual cathode moves rather close to the plane that fixes the potential. Numerical simulation [47] for this case shows that the distribution of the minimum of the potential has an 'extended' character along the entire length of the anode block – an 'extended' VC is formed.

The most significant question related to the operation of a magnetron with external injection is whether the beam potential remains constant when interacting with a high-frequency field? In other words, is the principle of synchronism retained when taking of some of the electrons in the anode current as a result of the formation of electron spokes of the space charge? As the analytical calculations show, the electron beam in the drift region is stratified with respect to azimuthal velocities. The outer fraction of the beam can be captured to the anode, while the inner part of the beam and the incorrect phased electrons maintain the VC potential constant. The escape of electrons to the anode does not lead to the disappearance of the VC, since they are rapidly replaced by electrons of the direct or reflected flux.

The brief analysis of the operation of a relativistic magnetron with external injection indicates the prospects of its use, first, for ultra-high power installations and, secondly, in the mode of microsecond voltage pulses for the following reasons. In the interaction space of the device the energy losses due to incomplete matching of the impedances of the accelerator and the generator are eliminated. It is also possible to increase the duration of the microwave pulse in such devices in comparison with relativistic magnetrons of direct geometry, since the rate of expansion of the outer boundary of the electron beam is somewhat different from the radial motion of the cathode plasma, and one can expect an increase in the time interval during which the synchronism conditions will be satisfied.

5.8.2. Experimental studies of RMI of nanosecond duration

Experimental studies of a relativistic magnetron with external injection of an electron beam were carried out on Tonus-1 and Luch high-current electron accelerators of the TPU. The experimental setup is shown in Fig. 5.16. The experiments were conducted using anode blocks of relativistic magnetrons of direct geometry with 6 and 12 resonators of the 10-cm wavelength band.

The electron beam of the tubular section was formed in a coaxial additional diode with magnetic insulation consisting of a cathode *4* and an anode *5* and injected into the interaction space (anode block *7*) and further into a drift tube *9* of large diameter. Installation parameters: the total current of the accelerator, anode and collector currents were measured by shunts, voltage by the capacitive divider *2*. The shunt of the end current *8* fixing the current leaving the interaction space is separated from the anode block by the insulator and connected by the reverse current liners *12* to the shunt output of the anode current *6*. Thus, shunt *6* recorded only the current flowing to the anode block. The magnetic field was created by two coils *11* forming a Helmholtz pair, and could reach a value of 1.5 T. The end current was also measured by the Faraday cylinder installed in the drift tube. A diaphragm fixing the potential of the electron beam was installed at the end of the anode block facing the additional diode (Fig. 5.16).

To select the optimal power and efficiency modes of operation, studies were carried out of a 6-cavity relativistic magnetron with an external injection on the Tonus-1 HCEA at a voltage of up to 1 MW, a current of ~20 kA with a duration of 60 ns.

The parameters of the electron beam and, consequently, its own electric field varied during the experiments by changing the diameters of the cathode, the anode, the drift tube, and also by installing various diaphragms at the input of the interaction space. The results of the investigations are summarized in Table 5.10, which also indicates the power of microwave pulses from a relativistic magnetron with a cathode in the interaction space at the same charge voltage of the accelerator.

In all the experiments, the microwave generator worked on π- type oscillations, as evidenced by the coincidence of the calculated (3146 MHz) and measured (3075 MHz) values of the radiation frequency. The fact that the RM was working on a potential equal to the cathode potential is shown by measurements of a synchronous magnetic field. This can be used to make a conclusion regarding the formation of a virtual cathode generator in the interaction space.

The results of the experiments are presented in diagrams reflecting the dependence of the efficiency of the generator on the ratio between

Table 5.10.

No.	Anode diameter, mm	Cathode diameter, mm	Diameter of the drift tube, mm	Anode block type	Presence of diaphragm	Power, MW
Relativistic magnetron with external injection of the electron beam						
1	40	16	180	Closed	–	690
2	40	16	180	Closed	+	1300
3	40	16	180	Half-open	–	570
4	40	16	180	Half-open	+	690
5	40	16	92	Closed	–	290
6	40	16	92	Closed	+	510
7	40	16	92	Half-open	–	560
8	40	22	92	Closed	–	440
9	30	16	180	Closed	–	570
10	30	16	180	Closed	+	1050
11	30	16	180	Half-open	–	570
12	30	16	180	Half-open	+	930
13	30	16	92	Closed	–	570
14	30	16	92	Closed	+	1000
15	30	16	92	Half-open	–	510
16	30	16	92	Half-open	+	570
17	60	16	60	Closed	–	50
Relativistic magnetron with a cathode in interaction space						
18	43	16	180	Closed	–	1900
19	43	16	180	Half-open	–	800

the current injected by the additional diode and the limiting transport current in the anode block (Fig. 5.17–5.19).

The quantities I_{inj} and $I_{lim.m}$ were calculated using the relations (5.8.4) and (5.8.5) for the measured values of voltage on the diode. The excess of the injected current above the limiting transport current characterizes the value of the accumulated electron charge in the interaction space, which performs the function of a negative electrode.

The numbers of the points plotted on the diagrams correspond to the experiment numbers in Table 5.10. The efficiency of the generator is determined by the formula

$$\eta = \frac{P}{UI_{\text{inj}}}. \tag{5.8.8}$$

The value of the total current of the accelerator, measured by shunt *1* (see Fig. 5.16), was not used in estimating the efficiency. This is due to the fact that when forming a beam with a VC in an additional diode, a significant reverse current was observed to the surface of the electron gun, which distorts the signal from the shunt. The diagrams were plotted using the magnitude of efficiency % rather than the maximum power level and is determined by the fact that the wave impedance of the forming line of the Tonus-1 HCEA is $2_\rho \sim 24$ ohm and reducing the diode impedance $Z = U/I_{\text{inj}}$ lead to the mismatch of the diode with the line and, consequently, to a decrease in the power of the injected beam. Thus, in a series of experiments No. 1–7 (see Table 5.10), the power of the beam was ~8 · 10^9 W, and in the series No. 9–16 ~5 · 10^9 W.

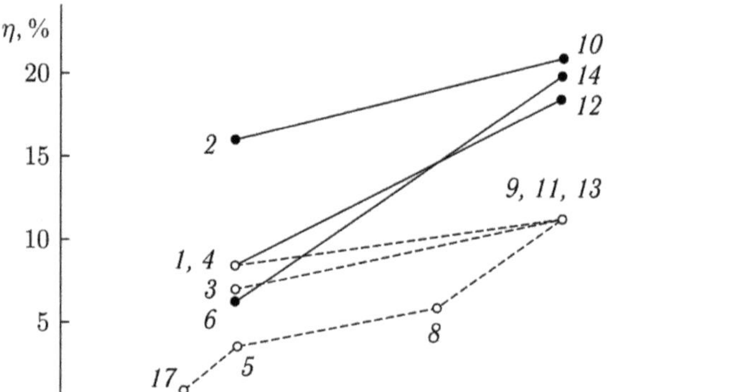

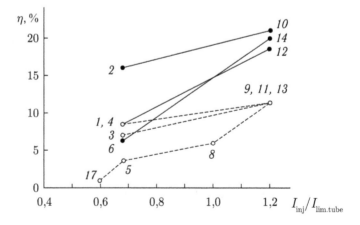

Fig. 5.17. Diagram of the dependence of the efficiency of RMI on the ratio of the current injected by the diode to the limiting transport current in the anode block by the use of the input diaphragm (solid line) and the absence (dotted curve) of the input diaphragm.

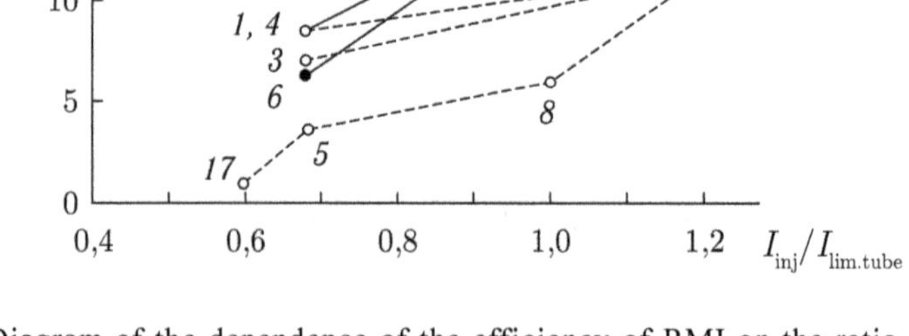

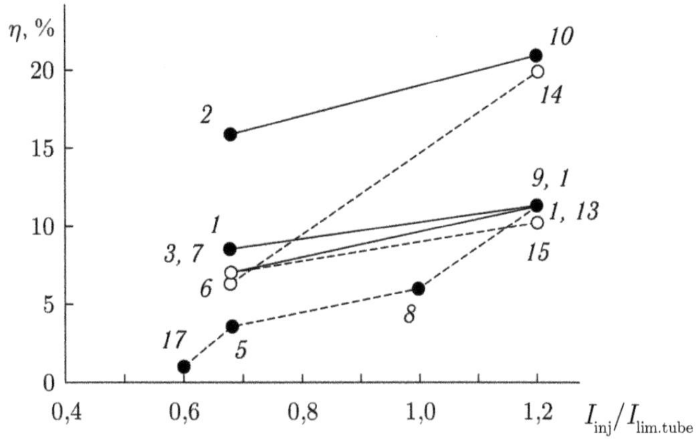

Fig. 5.18. The diagram of the dependence of the efficiency of the RMI on the ratio of the current injected by the diode to the limiting transport current in the anode block with diameters of the drift tube 180 mm (solid) and 92 mm (dotted curve).

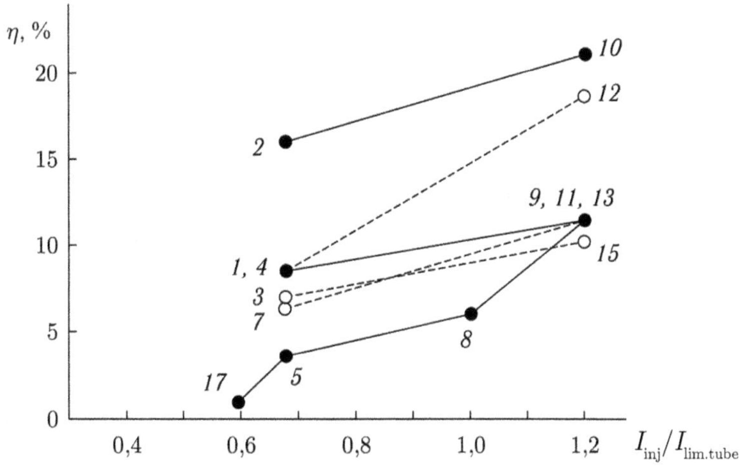

Fig. 5.19. The diagram of the RMI efficiency as a function of the ratio of the current injected by the diode to the limiting transport current in the anode block for the closed (solid) and half-open (dashed curve) anode blocks

Analysis of the presented diagrams (Figs. 5.17–5.19) allows the following conclusions to be drawn. In RMs with external injection of the beam, an increase in the efficiency is achieved:

1) with an increase in the ratio $I_{inj}/I_{lim.tube}$;

2) fixing the potential at entry to the slowing-down system, i.e. by the formation of an 'extended' VC.

When carrying out experimental studies of the device in these modes, the repeatability of the shape and amplitude of the generated microwave pulses in different starts of the HCEA was high.

The beam parameters in the interaction space depend not only on the injector, but also on the transport conditions of the beam in the drift tube. If, for example, the diameter of the drift tube in the investigated generator is increased, then an increasing part of the beam will be reflected and returned back to the interaction space. Such beams are practically not applicable in conventional O-type instruments, except for vircators [48]. Here such conditions are most favourable. To illustrate this, drift tubes with a diameter of 92 and 180 mm were used in a magnetron with external injection of the beam (Fig. 5.18).

A significant influence of the output parameters of the device continues to be exerted by the type of the resonator system, namely the axial distribution of the high-frequency field, as illustrated in Fig. 5.19. The spatial coincidence of this distribution with the distribution of the electric field strength of the beam (virtual cathode) ensures high output parameters of the generator.

Experiment No. 17 investigated the work of a magnetron with external injection beam in high-energy state. The geometry of the additional diode and the diameter of the drift tube were chosen so that the injection current became:

$$I_{\text{inj}} < I_{\text{lim.tube}},\ I_{\text{lim}}, \tag{5.8.9}$$

and the efficiency of the device decreased to 0.6% in accordance with the formula (5.8.2). The magnitude of the synchronous magnetic field, connected with the decrease in the beam potential (equation (5.8.1)), also decreased.

Studies using a 12-cavity anode block in a mode with a potential fixation at the input of the slowing-down system showed that the generator operated on the π-type oscillations with a power of $1.3 \cdot 10^9$ W (individual pulses up to $1.75 \cdot 10^9$ W).

5.8.3. Experimental studies of RMI of microsecond duration

The experiments with microsecond duration of the injected beam were conducted with 6- and 12-cavity anode blocks. As noted above, the desire to obtain high values of efficiency in the RMI determines the choice of such an additional diode geometry that ensures the formation of a virtual cathode in the interaction space. However, during the initial experiments it was found that this process is accompanied by a reduction in the duration of the voltage pulse and reduces the energy parameters of the beam. Therefore, it became necessary to study the mechanism of this phenomenon and develop special measures to prevent its occurrence. As a result of the studies [49–51], it was found that the shortening of the duration of the voltage and current pulses in the coaxial diodes with magnetic insulation during the formation of a virtual cathode is caused by reflection of the beam electrons from it, which fall into the region of the cathode–anode gap and settle on the surface of the additional anode. To reduce the number of reflected electrons and, consequently, to increase the duration of the current pulse, one can: a) choose the distance between the cathode and the injection plane; b) installation of metallic diaphragms in the injection plane of the beam in the interaction space of the device.

The scheme of experiments with relativistic magnetrons with external injection of a microsecond electron beam was analogous to that described above (see Fig. 5.16). In addition, the current injected by the diode and the anode current on the slowing-down system were recorded. The magnitude of the signals for plotting the voltage and current dependences on the value of the magnetic field induction was determined at the time when the voltage was applied to the maximum. This time corresponds to the maximum microwave power generated by the device. To limit the end current from the interaction space, a drift tube 180 mm in diameter was used in all experiments.

In the first series of experiments, the diameters of the additional anodes were chosen close (or coincident) with the inner diameter of the anode block. The cathode–anode block distance was 200 mm. The output power of the generator when operating in the π-mode was ~100 MW, the duration of the microwave pulse on the base is 0.15–0.2 μs. The dependences of voltage, anode, total and end currents, as well as microwave power, on the magnitude of the magnetic field induction are shown in Fig. 5.20. It is characteristic that an anode current growth is observed in the region of synchronous magnetic fields. The time-synchronized oscillograms of the signals and their dependence on the magnitude of the magnetic field induction are shown in Fig. 5.21. As can be seen from the oscillograms, the voltage pulse has a gently sloping front of about 0.25 μs, and after its termination, the generation of microwave radiation begins, since the synchronism conditions begin to be fulfilled. The termination

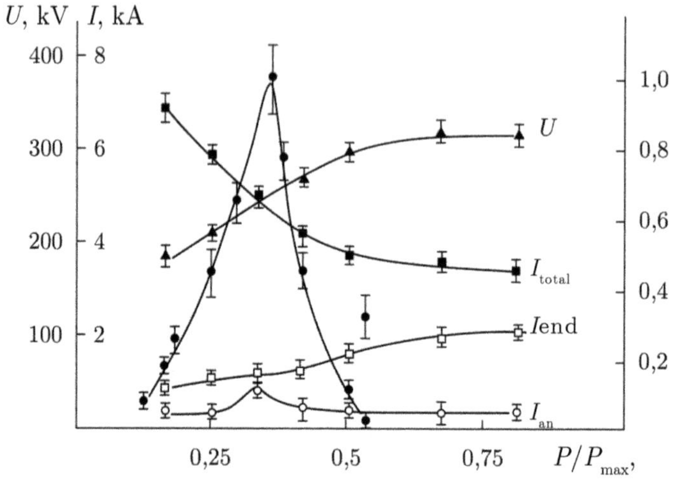

Fig. 5.20. Dependences of microwave power (P), voltage (U), anode (I_{an}), total (I_{total}) and end (I_{end}) currents on the value of magnetic induction fields for a 6-cavity RMI, P_{max} = 100 MW.

of the voltage pulse is characterized by a rapid increase in the total current and its transition to a short-circuit regime, connected with the current flow through the interelectrode gap shorted by the plasma. The oscillograms of the anode current show an increase in the signal synchronized with the increase in the microwave pulse. Moreover, a small value of the anode current indicates a low value of the output power of the RMI (100 MW).

For a 12-resonator anode block with an increased interaction space a power of 100 MW with a pulse duration of 0.35–0.4 μs was obtained

In all experiments it is noted that the duration of the output microwave pulse $\tau_{microwave}$ is shorter than the duration of the voltage pulse. The process of limiting the duration of the microwave pulse in the RMI with both 6- and 12-cavity anode blocks is as follows.

The transverse expansion of the outer boundary of the cathode plasma leads to an increase in the diameter of the beam and an increase, according to (5.8.5) and (5.8.7), of the limiting currents of beam transport on the anode block and the drift tube. A virtual cathode located at the edge of the anode block (this is evidenced by a higher level of generated power in the half-open anode block in comparison with the closed one) during the voltage pulse moves to the drift tube, reducing the value of the electrostatic field in the region of maximum energy exchange. This process results

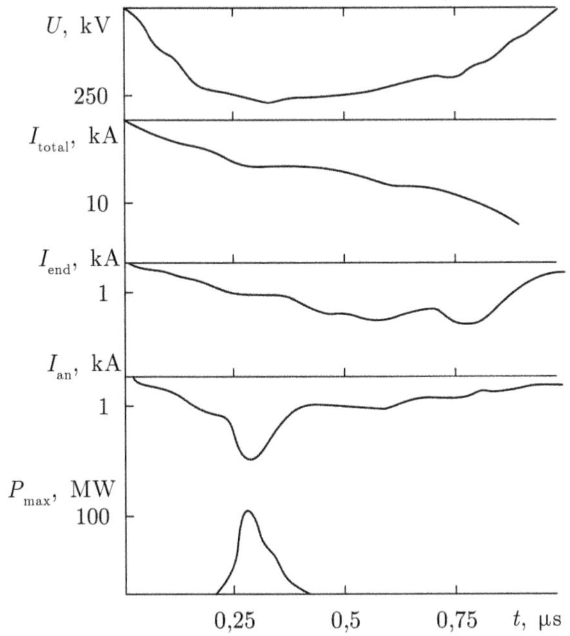

Fig. 5.21. Typical oscillograms of voltage, total current, end current, anode current, and the microwave detector signal.

in a 'smooth' decrease in the power of the generated microwave oscillations (Fig. 5.21). It can be seen that the envelope of the microwave signal has a short rise time, which is related both to the parameters of the slowing-down system and to the time of accumulation in the interaction space of a sufficient charge (the time of formation of the VC). Estimates of these two processes show that the time of rise of oscillations in the resonator system is ~15 ns (5.7.1) with the value of the loaded $Q_n \sim 100$; the time interval associated with the formation of a virtual cathode can be estimated using the results of numerical simulations of the injection of a beam with the current higher than the limiting current into the vacuum chamber [47]. The time of accumulation of a charge is equal to $t = 8\ R/c = 8$ ns. Assuming that the process of disruption of the microwave oscillations is determined by the disappearance of the virtual cathode due to the current selection to the anode block (since I_{an} has a characteristic peak in Fig. 5.21), the decline of the microwave pulse time would be characterized by the time corresponding to the duration of the leading front. However, since the rear edge of the pulse of microwave radiation is much longer, then the proposed mechanism for moving the virtual cathode into the region of the drift tube appears more plausible.

The proposed mechanism for shortening the duration of the microwave pulse is experimentally confirmed by introducing into the drift tube a Faraday cylinder, which acts as an electron collector. As the cylinder approached the end of the anode block, the collector current increased, that is, the current selection from the VC was reached. This, in turn, led to a change in the position of the VC, a decrease in the synchronous magnetic field, and a drop in the power of the microwave oscillations. Figure 5.22 shows the dependence of the power of the microwave pulse and the collector current on the position of the cylinder relative to the end of the anode block. At a distance of ~5cm, when the current is the largest, the generated microwave signal corresponds to the noise level.

The case of an additional anode with a diameter smaller than the diameter of the anode block was studied experimentally. With this geometry, the injected current is exceeded by the sum of the anode current and the limiting transport current in the anode block. However, the expected increase in output parameters device did not happen. This is due to the specifics of the operation of the Luch HCEA on a lowresistance load. The use of a diode with an anode diameter of 50 mm and a cathode diameter of 40 mm for a 12-cavity anode block led to a sharp decrease in the voltage amplitude due

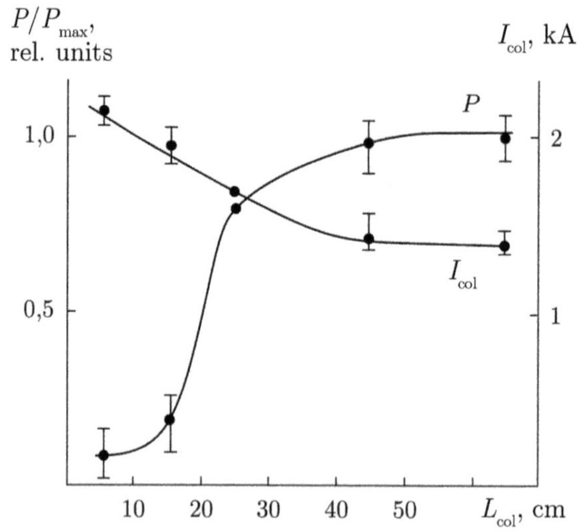

Fig. 5.22. Dependences of the microwave power of the RMI (P) and the collector current (I_{col}) on the distance (L_{col}) between the Faraday cylinder and the end of the anode block of the relativistic magnetron with external injection of the electron beam.

to an increase in losses at the internal resistance of the accelerator and a decrease in the duration of the voltage pulse to 0.5 μs with a microwave radiation power of 60 MW with a microwave pulse duration of up to 0.1 μs.

The fixation of the potential at the entrance to the interaction space caused the opposite of the expected effect: the power decreased to ~10 MW. Probable causes are apparently related, as in relativistic O-type generators, to a decrease in the beam potential in the interaction region due to the compensating action of the diaphragm plasma, whose velocity along the magnetic field can be about 10^7–10^8 cm/s. Also, it may not be possible to avoid the electrons entering the surface of the diaphragm and reducing the amount of this injected current.

One way to increase the output parameters of the device is to increase the power of the injected beam. In this case, the difference in the injected and end currents increases, and in the interaction space a larger space charge is accumulated. The oscillograms of the envelopes of the microwave pulses at three values of the charge voltage of the HCEA (which correspond to the different beam power), presented in Fig. 5.23, show an increase in the efficiency of the device with an increase in the power of the injected electron beam while maintaining the duration of the microwave oscillations. The maximum RMI parameters in this mode are: microwave pulse power ~200 MW, duration ~0.45 μs, efficiency ~9%.

To compare the efficiency of relativistic magnetrons with external injection and a magnetron with a cathode in the interaction space, the results of calorimetric measurements can be used for the microsecond pulse duration of the voltage. Thus, for a 12-resonator RMI the energy of the microwave pulse was 60 J (P ~ 200 MW, $\tau_{\text{microwave}}$ ~ 0.45 μs) and for the magnetron with the cathode in the interaction space 45 J (P ~ 300 MW, $\tau_{\text{microwave}}$ ~ 0.2–0.25 μs). There is a noticeable increase in the energy content of the microwave pulse, while we note the following

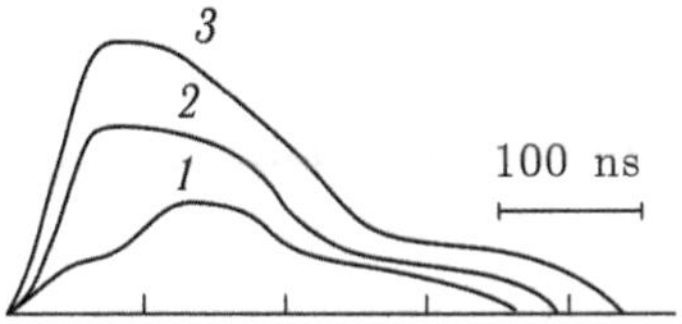

Fig. 5.23. Oscillograms of signals from microwave detectors: curve *1* – P_{beam} = 800 MW, P = 50 MW; curve *2* – P_{beam} = 1400 MW, P = 100 MW; curve *3* – P_{beam} = 2200 MW, P = 200 MW.

circumstance. An increase in the power of the beam in the application of external injection leads to an increase in its efficiency. However, with the design of the high-current electron accelerator used in the experiments, it was impossible to realize the advantages of an additional diode design from the point of view of the efficiency. An analysis of the experimental results showed that the low output parameters of the RMI of microsecond duration are limited by the capabilities of the available power source. For experiments with RMI, a source with an increased power of the injected beam and having a low internal resistance is required to minimize energy losses.

In conclusion, we note that the effective use of relativistic magnetrons with external injection of an electron beam is promising from the standpoint of increasing the life of anode blocks, first, because the energy of repeated pulses is scattered in an additional diode. Secondly, when microsecond power supplies are used, the short-circuit current flows not in the magnetron diode, but between the cathode and the anode of the additional diode. The effective application of RMI imposes a requirement on the design of additional diodes: the current generated by them must significantly exceed the limiting transport current in the anode block. High currents can be realized by using the configuration of an additional diode with the compression of an electron beam from a large radius to a small radius [52]. The construction of a relativistic magnetron with external injection can also be performed in inverse geometry. The variant of RMI combining the last two sentences was proposed in [53].

A series of studies of relativistic magnetrons with external injection of high-current relativistic electron beams of nanosecond and microsecond duration allows us to draw the following conclusions. Generation of microwave radiation in RMI or nano or microsecond duration (with the exception of experiment No. 17 – Table 5.10) occurs on a virtual cathode whose potential coincides in magnitude with the cathode potential. Setting the input diaphragm changes the location of the virtual cathode in the interaction space. The type of the resonator system, namely the axial distribution of the high-frequency field, has a significant effect on the microwave power generated by the device. The spatial coincidence of the maximum of this distribution with the position of the virtual cathode provides a higher level of generated microwave power. The presence of destabilizing factors (selection of current to the anode block during the generation of microwave radiation, the presence of high-frequency fields, expansion of the beam diameter during the time of the voltage pulse) does not lead to the disappearance of the virtual cathode. The output parameters of the RMI can be improved in the following ways:

a) reduction of the end current when using drift tubes with a large diameter;
b) the use of anode blocks of large internal diameter and the magnitude of the slowing-down down of the electromagnetic wave on the working mode of oscillations;
c) alignment of the axial distribution of the high-frequency in the anode block, while maintaining sufficiently high values of the quality factor;
d) using low-impedance HCEA and increased capacity of the injected beam.

This chapter presents the results of studies of inverted relativistic magnetrons and relativistic magnetrons with external injection of an electron beam of nanosecond and microsecond duration.

The use of the inverted RMs is promising for obtaining microwave pulses of a microsecond range of durations when microwave output devices are used along the instrument axis. In the inverted coaxial diode with magnetic insulation used by such devices, the duration of the voltage pulse is 2–4 times greater than in a direct diode with the same strengths of electric fields at the diode electrodes and other equal conditions. The closing of the cathode–anode gap of an inverted coaxial diode with magnetic insulation is caused by the radial motion of the anode plasma under the influence of the centrifugal instability developing in it.

The use of relativistic magnetrons with external injection of an electron beam is advisable for obtaining microwave pulses of high power, since it allows increasing the life of the anode block. Effective generation of microwave radiation in these systems occurs when a stationary state of an electron beam with a virtual cathode is used, which is achieved by using diaphragms at the input of the anode block, limiting the current leaving the interaction space of the device, increasing the ratio of the injected current to the limiting transport current in the anode block.

References

1. Didenko A.N., et al., Pis'ma Zh. Teor. Fiz. 1978. Vol. 4. No. 14. P. 823–826.
2. Glazer I.Z., et al., Pis'ma Zh. Teor. Fiz. 1980. TV 6. No. 1. P. 44–49.
3. Ballard W.P., A relativistic magnetron with thermionic cathode. Suipr. Report 1981.

No. 840. 142 p.
4. Ilic D.B., et al., in: Proc. IEEE Int.Conf. onPlasmaScience.Monterey, 1978. P. 286.
5. Ballard W.P., et al., Abstr. IEEE Int. Conf. on Plasma Science. Montreal, 1979. P. 45.
6. Ballard W.P., et al., J. Appl. Phys. 1982. V. 53. No. 11. P. 7580-7591.
7. Bugaev S.P., et al., Fizika plazmy. 1983. 9. 9. No. 6. P. 1287–1291.
8. Bugaev S.P., Kim A.A., A breakdown of a coaxial diode across a magnetic field. Preprint of the Siberian Branch of the Academy of Sciences of the USSR. Tomsk, 1987. 12 pp.
9. Bakshayev Yu.L., Blinov P.I., Zh. Teor. Fiz., 1983. V. 53. No. 9. P. 1882–1884.
10. Glazer I.Z,. et al., Zh. Teor. Fiz. 1980. V. 50. No. 6. P. 1323–1326.
11. Bugaev S.P., et al., Zh. Teor. Fiz. 1980. V.50. No.11. 2463.
12. Luckhardt S.C., Fleischman H.E., Appl. Phys. Lett. 1977. V. 30. No. 4. P. 182–186.
13. Luckhardt S.C., et al., Observation of superdense megavolt ion beams in a long-pulse magnetically insulated diode, Cornell University Report FRL-1. Ithaca, 1977. 10 p.
14. Bakshayev Y.L.,et al., Production of ion beams of microsecond duration in a diode with magnetic insulation. Supplement to Proc.. 3 Symposium on High-Current Electronics. Tomsk, 1978. pp. 16–18.
15. Black W.M., et al., in: Proc. Int. Electron Devices Meet. Washington, 1979. P. 175–178.
16. Parker R.K., et al., Abstr. IEEE Int. Conf. on Plasma Science. Montreal, 1979. P. 44.
17. Black W.M., et al., Proc. Int. Electron Devices Meet. Washington, 1980. P. 100.
18. Bakshayev Yu.L., Fizika plazmy. 1979. V. 5. No. 5. P. 1041–1043.
19. Blinov P.I., et al., *ibid.* 1982. Vol. 8. No. 5. S. 958–962.
20. Vintizenko I.I., et al., Pis'ma Zh. Teor. Fiz. 1986. T. 12. No. 8. P. 449–453.
21. Vintizenko I.I., et al., Proc. 6th Symp. om High-current electronics. Tomsk, 1986. Part 2. P. 9–11.
22. Dolgachev G.I., et al., Preprint IAE 3908/7, IAE. 1984. 25 p.
23. Basmanov A.B., et al., Fizika plazmy. 1986. V. 12. No. 11. P. 1319–1323.
24. Vintizenko I.I., et al., Zh. Teor. Fiz. 1988. Vol. 58. No. 6. P. 1171–1173.
25. Vintizenko I.I., et al., *ibid.* 1988. Vol. 58. No. 8. P. 1584–1586.
26. Vintizenko I.I., Sulakshin A.S., Breakdown of the interelectrode gap and the formation of a high-current electron beam in an inverted coaxial diode with magnetic insulation. Dep. in VINITI No. 5492-B89, 1989. 71 p.
27. Vintizenko I.I., Sulakshin A.S., Sulakshin S.S., Fizika plazmy. 1990. V. 16. No. 4. P. 504-507.
28. Cherepnin N.V., Sorption phenomena in vacuum technology, Moscow, Sov. radio, 1973.
29. Bekefi J., Plasma in lasers. Moscow, Energoatomizdat, 1982.
30. Pal H., Hammer D., Phys. Rev. Lett. 1983. V. 50. No. 10. P. 732–734.
31. Close R.A., et al., J. Appl. Phys. 1983. V. 54. No. 7. P. 4147–4151.
32. Erli L.M., et al., Pribory Nauch. Issled. 1986. No. 9. C. 86–96.
33. Magnetrons of centimeter range: Trans. from English. Ed. S.A. Zusmanovsky. – Moscow, Sov. radio, 1950.
34. Shmatko E.I., Radiotekhnika. 1981. Vol. 17. No. 58. P. 115–121.
35. Shlifer E.D., et al., Elektronnaya tekhnika. Ser. 1. Elekronika SVCh. 1968. No. 7. P. 55–64.
36. Vintizenko I.I., et al., Proc. 7th Symp. High-current electronics. Part 1. Tomsk, 1988. P. 197–199.
37. Vintizenko I.I., et al., Radiotekhnika i elektronika. 1990. V. 35. No. 4. P. 899–901.

38. Shlifer E.D., Debelov D.E., Elektronnaya tekhnika. Ser. 1. Elekronika SVCh. 1971. No. 6. P. 3–8.
39. Vintizenko I.I., et al., Pis'ma Zh. Teor. Fiz. 1987. V. 13. No. 10. P. 620–623.
40. Vintizenko I.I., et al., Radiotekhnika i elektronika. 1990. V. 35. No. 9. P. 2004–2006.
41. Sulakshin A.S. Restriction of current leakage from the interaction space, Zh. Teor. Fiz. 1983. V. 53. No. 11. P. 2266–2268.
42. Sze H., et al., IEEE Trans. on Plasma Science. 1987. V. PS–15. No. 3. P. 327-334.
43. Golant V.E., Zh. Teor. Fiz. 1960. V. 30. No. 11. P. 1265–1320.
44. Fuks M.I. Investigation of electron currents in systems with magnetic insulation: Dissertation, Gorky, 1983.
45. Vintizenko I.I., et al., in: Relativistic high-frequency electronics. IAP of the USSR Academy of Sciences. Gorky, 1988. Issue. 5. P. 125–140.
46. Chernogalova L.F., et al., Proc. 6 Int. Conf. on High-Power Particle Beams. Kobe, 1986. P. 573–576.
47. Berezin Yu.A., et al., Numerical simulation of injection of a powerful electron beam into a vacuum a chamber with a strong magnetic field. Novosibirsk, Preprint of the Institute of Applied and Theoretical Mechanics of the SB AS USSR, 1979.
48. Rukhadze A.A., Radiotekhnika i elektronika. 1992. Vol. 37. No. 3. P. 385–396.
49. Vintizenko I.I., et al., Proc. 6th Symp. High-current electronics Tomsk, 1986. Part 2. P. 29–31.
50. Vintizenko I.I., et al., Zh. Teor. Fiz. 1987. Vol. 57. No. 3. P. 605–607.
51. Vintizenko I.I., et al., Fizika plazmy. 1988. V. 14. No. 2. P. 246–248.
52. Author Cert. No. 1342335 USSR. Relativistic device of M-type. Declared 11.05 85, Vintizenko I.I., Sulakshin A.S., Fomenko G.P.
53. Author Cert. No. 1690507 USSR. Inverted relativistic magnetron. Declared 03.05.89, Vintizenko I.I., Sulakshin A.S., Fomenko G.P.

6

Relativistic Magnetrons with External Coupling of Resonators

The relativistic magnetron, like its classical analogue, refers to instruments of the so-called resonance type with short-term interaction of the active medium (electron beam) with the electromagnetic field. The oscillatory system of the magnetron is a distributed electrodynamic structure and is multimodal. Oscillations of various types (modes) differ in eigenfrequencies and are characterized by spatial distributions of high-frequency fields in the resonator system. This property, as already noted, is the main cause of instability in the operation of generators, expressed in transitions between types of competing oscillations, frequency jumps and power. Modal instability is particularly pronounced when perturbations of the electrical modes of the device occur: in the pulsed mode, when working on unmatched loads, etc. Another powerful factor in the instability of relativistic magnetrons is the non-stationarity of plasma in the near-cathode region. The expansion of the cathode perturbs the synchronism condition and causes a significant frequency drift during the pulse.

The resonator system of a magnetron can be considered as a set of local generating elements interconnected via an electron beam. Prior knowledge of the distributions of high frequency fields (modes of oscillations) and their spatial symmetry allows us to introduce 'internal' links between the selected elements that simulate the real (electronic) stability mechanism of oscillations. From a physical point of view, such a model is equivalent to a system of interconnected self-oscillators, which opens up new possibilities for modifying the

properties of magnetron generators. The introduction of additional, controlled external links into their resonator system provides an additional channel of influence on the stability of the coherent regime.

6.1. Controlling couplings in the oscillatory system of a magnetron

As noted at the beginning of the first chapter, methods are used to separate the types of oscillations in frequencies and to stabilize the main working mode of oscillations in classical magnetrons based on the introduction of additional elements in the form of a straps or high-quality resonators into resonant systems of generators. For relativistic short pulse devices, these techniques are not effective. Thus, the RM anodes usually have a long length, and the high-frequency fields, which vary along the axis of the device according to the hyperbolic law, are rapidly weakening. The effect of the straps located at the ends of the anode block in this case will be weakened. The high-*Q* resonators assembled with the anode cavity of the magnetron, generally speaking, do not directly affect the distribution of the high-frequency field in the interaction space. It is known that the stabilizing effect of an external circuit is a property of one of the types of oscillations that arise in the system. From this point of view, the introduction of an additional resonator into the oscillatory system makes its frequency spectrum even more saturated, and the range of parameters is more dimensional. Since the supply voltage in the pulsed mode varies over a wide range, the excitation conditions can be satisfied for different modes of oscillation [9]. The use of high-*Q* resonators in the pulsed mode is also limited by their natural inertia (see sections 5.5.2 and 5.7). Finally, a high level of generated power can lead to breakdowns in the straps and in nodes of connections with additional resonators.

It can be assumed that a more effective solution of the stabilization problem will be the introduction into the self-oscillator circuit of the magnetron of such couplings that would provide a strong and direct effect on the stability of oscillations with a given phasing. Obviously, this effect can be obtained from the connections introduced directly into the active region of the device. Such a connection – it can be called active – will provide additional interaction of oscillations in different parts of the active space. Then, with the appropriate adjustment of the communication channel, it is possible to raise the degree of stability of the working mode of oscillations.

Control couplings in principle can be introduced into any devices with a developed resonant structure. The magnetron is the most convenient for such a modification. Indeed, the self-oscillating system of the magnetron, taking into account the local character of the energy exchange between the electron beam and the electromagnetic field of the resonators, can be represented as a set of generating elements – oscillatory subsystems, connected through an electron flux. If we formally withdraw from consideration a specific physical mechanism of these connections and take into account only the fact of their action, then we see a transition to a qualitatively different model – a system of interconnected autogenerators. The transition to this model makes it possible to attract methods of the theory of oscillations and mutual synchronization for analyzing and modifying the properties of the generator, in particular those sections in which the questions of the influence of mutual links on the stability of coherent motions are considered. Within the framework of these tools and concepts, a new approach to solving the problems of stabilizing the process of generation of a relativistic magnetron and the microwave output from its resonator system is further discussed. An example of such an approach can serve as the schemes of a vircator and a relativistic backward-wave tube with additional resonance elements in the interaction region [10, 11].

The modified scheme of the relativistic magnetron constructed on the basis of these ideas [12, 13] is shown in Fig. 6.1. Opposite resonators of the anode block *1* of the magnetron have two waveguide outputs of microwave power *2* and are connected to each other by the waveguide channel *3* with a radiating antenna *4*, which is a common load for the resonators.

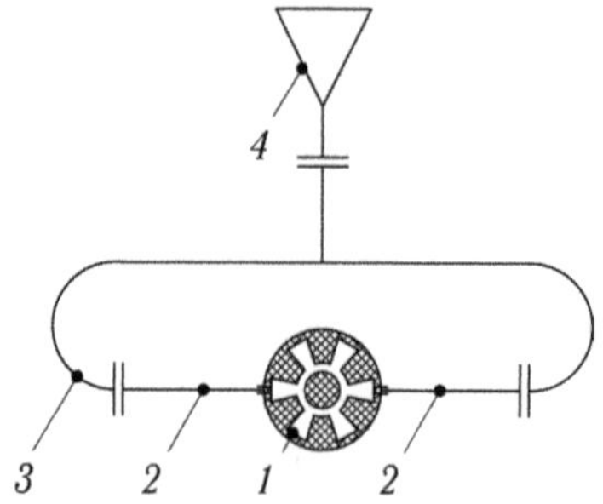

Fig. 6.1. Modified scheme of the relativistic magnetron.

The mechanism of influence of external couplings is as follows. The presence of a communication channel leads to the interaction of oscillations in the resonators. In this case, the phase of the signal arriving at the input of the resonator is determined by the total electrical length of the communication channel. For a magnetron operating on the π-type with a number of resonators satisfying the the condition $N/2$ – an odd number, the oscillations of the opposite resonators, when measured in azimuth, are antiphase, and with respect to the output waveguides, are in phase. For a magnetron with a number of resonators, satisfying the condition $N/2$ – an even number, the oscillations of the opposite resonators, when measured in azimuth, are in phase, and with respect to the output waveguides – they are antiphase. For other types of oscillations, the distribution of the high-frequency field is different from the distribution of the π-type. If the length of the communication channel is chosen so that the signals at its input and output are in phase with the oscillations of the resonators, then one should expect the stabilization of this mode of oscillation and the possible suppression of all other modes. Due to the symmetry of the fields in the interaction space, it is also possible to predict an improvement in the mechanism of energy exchange of electrons with a microwave field. The use of a common load-radiator in the communication channel makes it possible to summarize the oscillations of the resonators in it and thereby ensure a more efficient extraction of the energy from the relativistic magnetron, while at the same time increasing the stability of its operation.

Depending on the phase and impedance properties of the communication channel, as well as on its electrical symmetry, a series of various designs of the generator are viewed. Thus, for a magnetron with a number of resonators satisfying the condition $N/2$ – an even number, the coupling channel length must be close to

$$(2q-1)\lambda_w/2$$

(here q is a positive integer, λ_w is the working wavelength in the waveguide). For a magnetron with the number resonators $N/2$ – an odd number, the channel lengths should be close to $q\lambda_w$. If, in this case, the tee connection is in-phase (waveguide *H*-tee), and the channel is symmetric, then the microwave power radiated by the π-type resonators will be summed in the total load-radiator; in the case of excitation of antiphase oscillations ($2\pi/3$-mode), the power will be subtracted. When using an antiphase tee connection (*E*-tee) for output of the π-type power, the channel must be antisymmetric. Further, the communication channel may comprise a system of loads-radiators; in this case, in order to simultaneously stabilize the working mode of oscillations and the given spatial phasing, it is necessary to specify a quite definite order of placement of the radiators along the channel. In the output of microwave energy through the slit radiators, depending on the location of the slits on a wide or narrow wall of the waveguide, the configurations of the communication channel also appear. Finally, it is possible to connect not only the opposite resonators to an external channel, but also any other pairs.

6.2. The construction of a relativistic magnetron with an external coupling of resonators

In this section, we describe RM structures with external coupling of resonators. The results of experimental studies of which will be presented in sections 6.4–6.9. The main purpose of the experiments was to study the influence of the communication channel parameters on the energy and spectral characteristics of microwave radiation; analysis of the possibilities of stabilizing the operating conditions of the magnetron and distributed output of power from the resonant system. Relativistic magnetrons of direct geometry with 6- and 8-cavity anode blocks, which have power outputs from opposite resonators, were used. The most detailed investigation was made of a 6-cavity magnetron with various versions of the waveguide channel of resonator coupling; the characteristics of microwave radiation were determined in wide intervals of electric and magnetic fields. This made it possible, through a consistent comparison of the results, to obtain complete and reliable information about the oscillatory processes in the instrument.

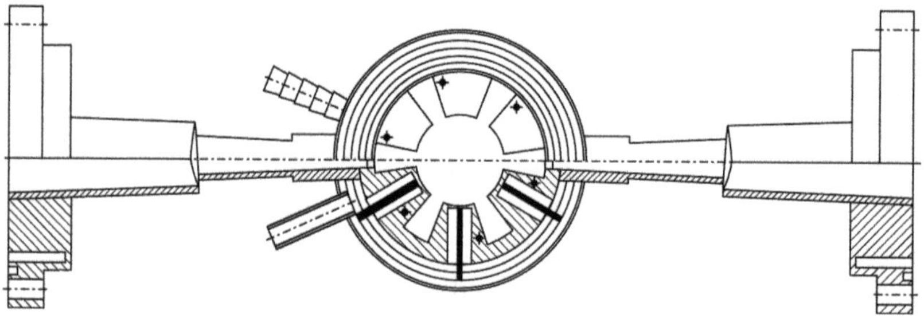

Fig. 6.2. Relativistic magnetron with two power outputs.

The construction of the six-resonator relativistic magnetron is shown in Fig. 6.2. Power outputs are made in the form of narrow slots in two opposite resonators and smooth waveguide transitions to the standard section of 72 × 34 mm^2. The magnetron has the following parameters. The dimensions of the anode block: inner diameter – 43 mm, outer diameter – 86 mm, length of the anode block 72 mm. The working mode of oscillations of the magnetron is π; waveguide power outputs are excited in this phase.

The magnetron uses a graphite cathode that withstands high temperatures and has a low threshold for the formation of explosive electron emission. The diameter of the cathode is 21.5 mm, the length of the cathode is 110 mm. The part of the cathode protruding from the anode block into the drift tube has a smooth rounding at the end.

The external coupling channel of magnetrons is made of segments of a copper rectangular waveguide with a cross section of 72 × 34 mm^2; One or three radiators can be included in the channel. The position of the radiators and the length of the communication channel can vary. The construction of the communication channel with one radiating load is shown in Fig. 6.3.

The position of the *H*-tee in relation to the axis of electrical symmetry of the communication channel axis was ensured by two special symmetric inserts *5*. The output of the waveguide *H*-tee was connected to a radiating pyramidal horn antenna *6* for the output of the microwave oscillations from the system and equipment for measuring the parameters of microwave radiation. The inserts *4* are paired in the right and left arms of the tract and are used to change

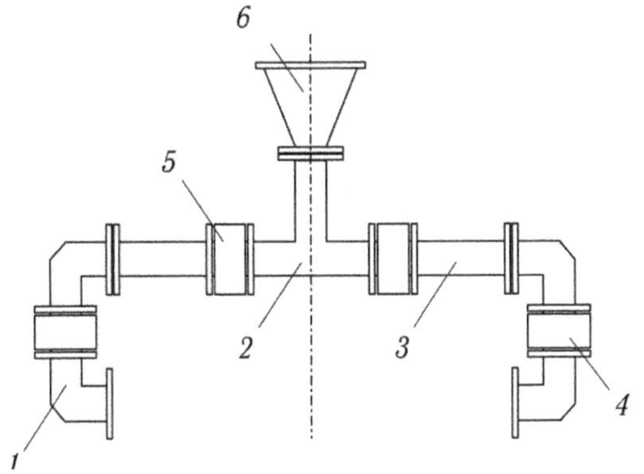

Fig. 6.3. The communication channel of magnetron resonators with one radiator: *1* – waveguide *H*-bends; *2* – *H*-tee; *3* – segments of waveguides; *4*, *5* – set of inserts; *6* – antenna.

the length of the communication channel. A set of inserts made it possible to change discretely the total length of the communication channel L_{in} in steps of 10 mm. The interval for changing the length of the communication channel was from 2810 to 3235 mm, that is, it overlapped two and a half waveguide wavelengths ($\lambda_w \approx 168$ mm).

The symmetric inserts *5* were connected between the waveguides *3* and the tee. They had the same length of 42 mm each, which corresponds to a quarter of the waveguide wavelength. In the case when the symmetric inserts were located on both sides of the tee, a symmetric communication channel configuration was realized; in this case the tee *2* was on the axis of the electrical symmetry of the waveguide channel. Such inclusion of the tee provides a summation of the microwave oscillations during in-phase excitation of the channel. An antisymmetric scheme was formed when both symmetric inserts were included in one arm of the channel and the tee shifted relative to the axis of the electrical symmetry of the communication channel by $0{,}25\lambda_w$. This scheme for in-phase excitation provides the subtraction of microwave oscillations.

The construction of the communication channel with the distributed power output is shown in Fig. 6.4. The output of the microwave oscillation energy from the system was implemented by regular pyramidal horn antennas *4*, *5*, *6*, connected to each of the three *H*-tees *2*. The lengths of the inserts *3* are such that the central radiator *5* is displaced relative to the axis of the electrical symmetry of the communication channel by $0.25\lambda_w$. Thus, the circuit is antisymmetric relative to the central load. Since the radiating antennas have significant geometric dimensions, they are spaced

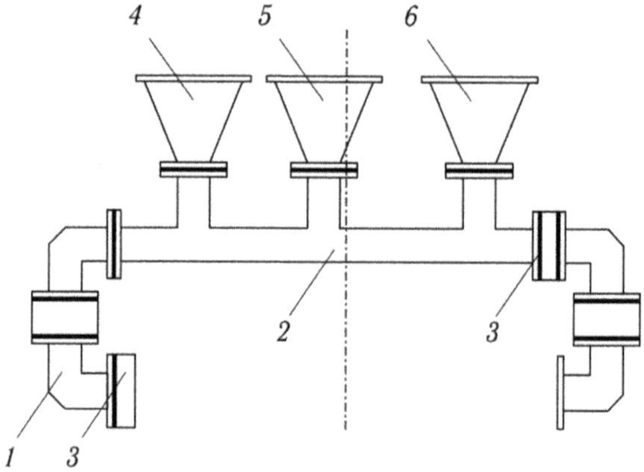

Fig. 6.4. The communication channel of the magnetron resonators with three radiators: *1* – waveguide *H*-bends; *2* – *H*-tee; *3* – waveguide inserts; *4*, *5*, *6* – antennas.

Fig. 6.5. Experimental setup.

along the channel: the distance between the side radiators *4*, *6* is 3.5 λ_w, and the distance between one of the side radiators and the central radiator is 1.25 λ_w. The specified communication channel configuration for the π-type of the 6-cavity magnetron (the channel is excited in phase) ensures in-phase arrival of resonator oscillations in each of the side radiators. It should be borne in mind that the oscillations in the exit arms of the side tees relative to each other are in antiphase.

The general view of the experimental setup is shown in Fig. 6.5. The magnetron was powered from the LIA 04/6, its characteristics are given in paragraph 4.7.2.

6.3. Features of measuring the characteristics of microwave radiation of a relativistic magnetron

To measure the spectral characteristics of the microwave radiation of a relativistic magnetron, two waveguide directional couplers were sequentially connected to the output waveguide channel (see section 4.7.1). The microwave pulse from the output of one of the couplers was detected; the signal of this channel was the reference one. The second detector was connected to the second coupler through a bandpass filter. The filter was tuned in the range 2550–3250 MHz; the bandpass of the filter on level –3 dB was 25–35 MHz. The duration of the transient process in the filter is characterized by a constant time $\tau_k = 2Q/\omega \approx 10$ ns.

As is known (see, for example, [1]), the nature of the transient process in a narrow-band filter is determined by detuning the carrier frequency of the radiopulse from the resonant frequency of the filter. With the proximity of these frequencies, the process has an

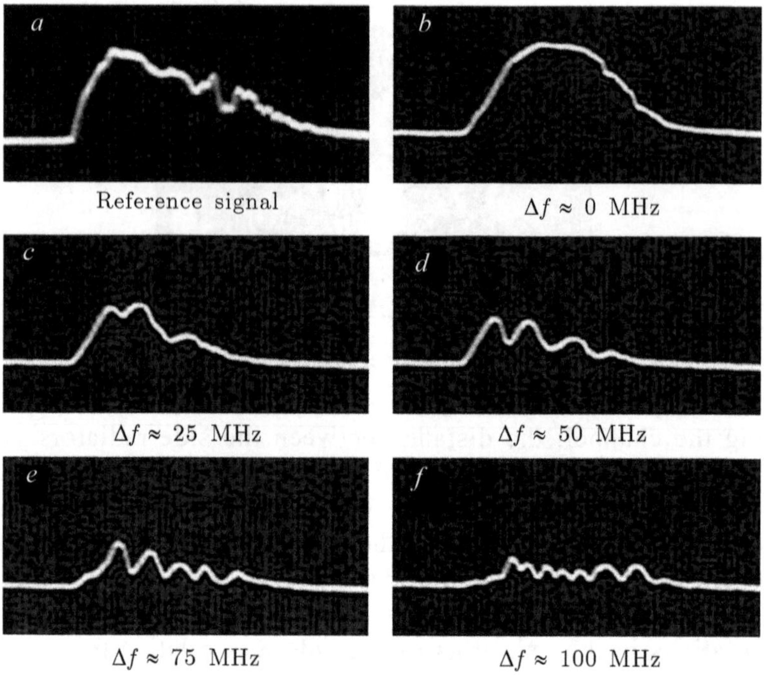

Fig. 6.6. Signals of the detector at different detunings Δf of the filter from the generation frequency of the magnetron.

aperiodic character, and beats are produced with a sufficiently large detuning. Figure 6.6 shows the oscillograms of the pulses transmitted through the filter and of the detected pulses of the magnetron at various detunings $\Delta f = |f_{res} - f_0|$ of the resonant frequency f_{res} of the filter from the frequency f_0 of the magnetron generation. The oscillograms reflect the known fact of decreasing the period of beats with increasing frequency difference of the signal and the resonant frequency. As can be seen, if these frequencies coincide, the signal at the output of the filter does not contain beats and has a smoother form in comparison with the reference signal. A more gentle front of the output signal of the filter indicates the aperiodic nature of the transient process. Thus, the beats have the least influence on the accuracy of measurements near the central part of the spectrum. It can also be assumed that the magnitude of the response of the filter in different time sections of the pulse, which is more than 10 ns, will be proportional to the spectral density of radiation in these sections. The absence of beats in the response indicates that the microwave pulse spectrum is concentrated in the pass band of the filter. At sufficiently large detunings ($\Delta f \approx$ 50–75 MHz), the amplitude of the beats becomes comparable to the pulse amplitude, which limits the accuracy of measurements at the edges of the spectrum. Despite this, the spectral measurements of the experiments described below were carried out over the entire range of filter tuning. This made it possible to control the possibility of the appearance of magnetron generation on other types of oscillations in a sufficiently wide frequency interval.

The procedure for measuring the current emission spectrum was as follows. The resonant frequency of the filter f_{res} was tuned discretely with a step of 5 MHz in the range 2550–3250 MHz. At each filter position, up to 10 pulses of the detectors were recorded. Then, for the selected time t_i, the ratio of the average power levels of the signal after the filter P_f and the reference signal P_0 [2] was calculated for each filter setting:

$$S(t_i, \Delta t) = \frac{P_f(t_i, \Delta t)}{P_0(t_i, \Delta t)}.$$

Averaging was carried out on the signal levels of the pulse in the range Δt near t_i. These data were used to construct spectral diagrams related to the selected time periods of the pulses and reflecting the time dynamics of the spectrum. The normalization at the power

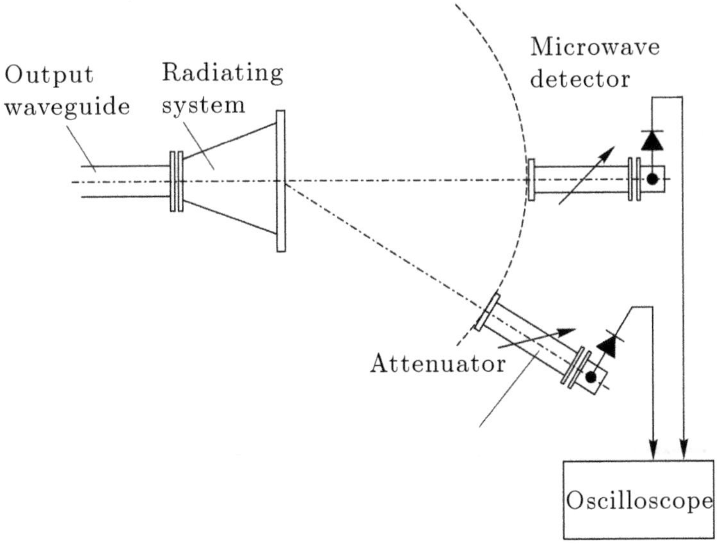

Fig. 6.7. Scheme of measuring the spatial distribution of microwave radiation.

levels of the reference signal P_0 allows to significantly reduce the influence of the amplitude spread between pulses, and also to take into account the change in the power level during the pulse.

A diagram of the spatial distribution of the microwave radiation was measured for a magnetron with a distributed power output. One of the detectors (see Fig.6.7) was the reference one and was located on the axis of the radiating system, the second one was moving along a circle centred on the center of the aperture. The radiation diagram was constructed from the power ratio of the signal measured by the moving detector to the power of the reference detector. The profile of the radiation pattern, especially in the direction of 'zeros', characterizes the phase relationships in the system, and the depth of the zeros is their stability.

6.4. A six-resonator relativistic magnetron with uncoupled resonators

In this case, the power outputs of the two opposite resonators are connected to separate horn antennas through the channel. The characteristics of a magnetron without external coupling of resonators are base characteristics and can be identified in subsequent experiments. The characteristics of the 6-resonator magnetron are given below.

At a preliminary stage, to determine the energetically optimal operating conditions of the magnetron, measurements of the microwave radiation parameters were performed with a change in the magnetic field induction in the range 0.34–0.54 T (Fig. 6.8). The maximum power of microwave radiation from each output falls on the interval of the magnetic field of 0.48–0.52 T and is on the

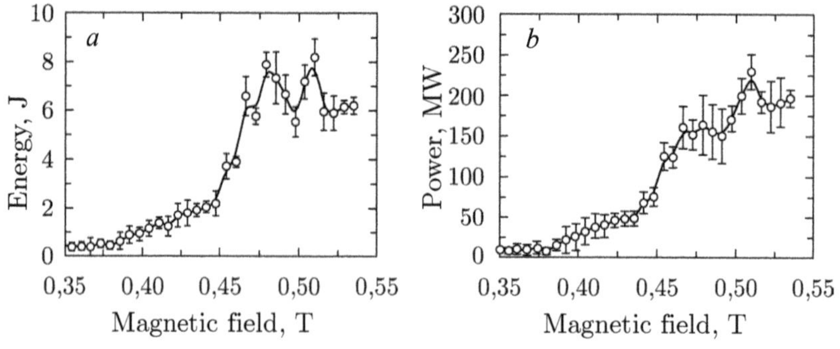

Fig. 6.8. Dependence of the energy and power of microwave radiation on the magnetic field.

average 180 MW. The maximum energy in the pulse (from each output) also occurs at the given values of the magnetic field and varies near ~7.5 J. Noticeably, the generation begins at a magnetic field induction value of ~0.44 T and exists up to the upper limit of the change in the magnetic field. In the generation region, the total current of the magnetron and the cathode–anode voltage level varied within the limits of 3.5–2 kA and 300–400 kV, respectively; the efficiency of a magnetron at an optimum magnetic field reaches ~20%.

The envelopes of the pulses of microwave radiation from two detectors (Fig. 6.9 *a*) have a rectangular shape with a short front edge. The maximum amplitude of the oscillations is 20 ns from the beginning of the pulse, then it recedes. The pulse duration at levels –3 dB and –10 dB is ~20 ns and ~85 ns, respectively. The work of the magnetron is characterized by a very significant spread of amplitudes (up to ±20%) and pulse shapes. This is clearly seen

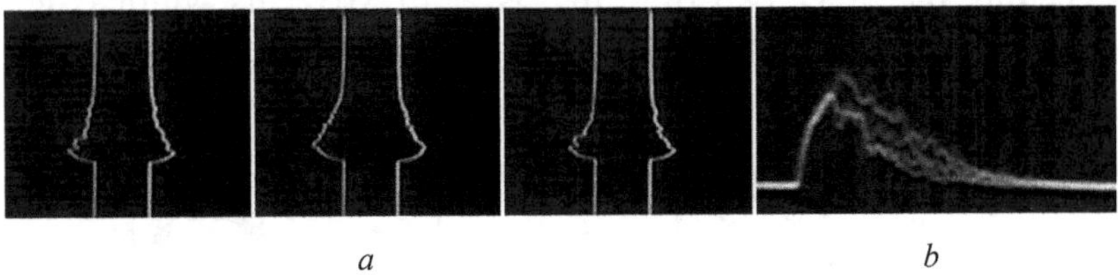

Fig. 6.9. Oscillograms of microwave pulses from a magnetron with uncoupled outputs: *a* – typical pulse shape; *b* – ten coincident pulses.

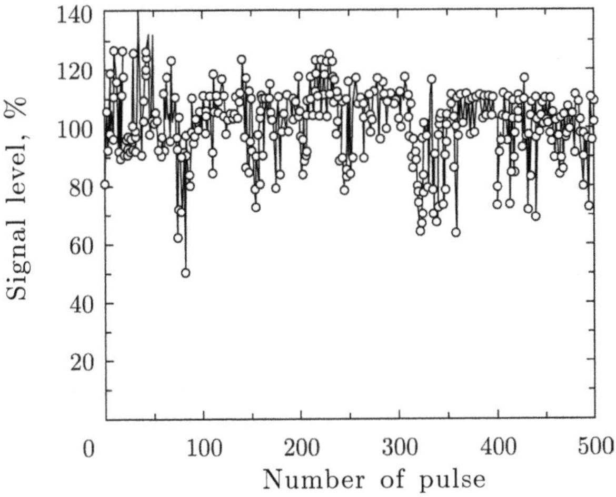

Fig. 6.10. Dynamics of the microwave pulse amplitude of a magnetron with uncoupled outputs in a continuous series of pulses.

when recording ten coincident pulses (Fig. 6.8 *b*), as well as from a continuous series of 500 pulses (Fig. 6.10).

The spectral characteristics of the magnetron were measured at the induction of a magnetic field of 0.5 T in four time parts of the pulse: at the leading edge ($\Delta t \approx$ 5–20 ns), at the point of maximum power ($\Delta t \approx$ 20–35 ns) and at two points on the edge of the pulse $\Delta t \approx$ 40–55 ns and 60–75 ns). These spectra are given to the 0 dB level and are shown in Fig. 6.11.

As can be seen, the radiation spectrum undergoes considerable changes during the pulse. At the point of maximum power it resembles the spectrum at the leading edge (Fig. 6.11 *a*, *b*) and is characterized by the absence of clearly expressed frequency components; the width of the spectra in the level –3 dB reaches 80–100 MHz with the centre around 2780 MHz. The spectrum is formed only in the second half of the pulse: at $\Delta t \approx$ 40–50 ns (Fig. 6.11 *c*) there are three fairly well-expressed maxima, spaced from each other by approximately 70 MHz. The frequency of the central maximum, exceeding the other two by 5 dB, is 2740 MHz; its width is about 40 MHz. This spectral component corresponds to different types of

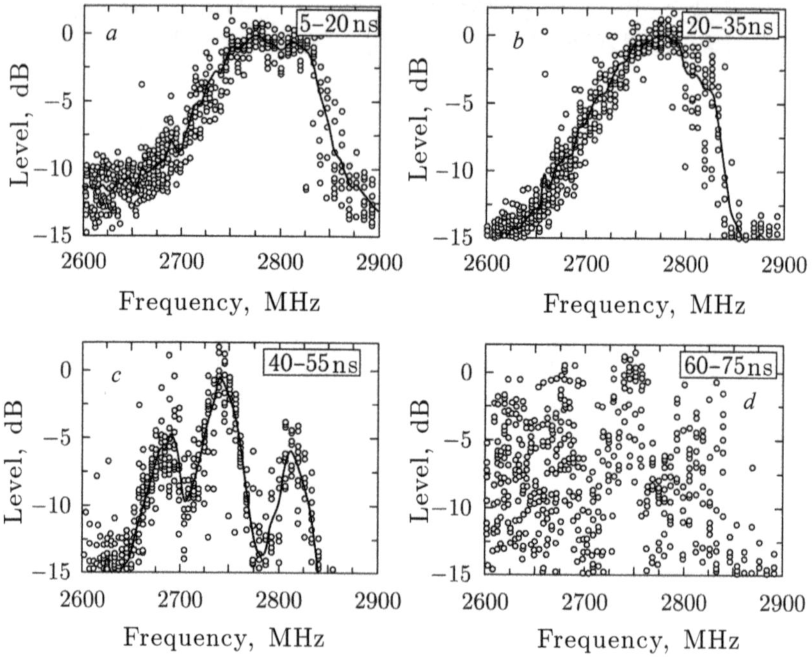

Fig. 6.11. The frequency spectrum of a magnetron with uncoupled outputs in different time parts of the pulse.

oscillations of the magnetron. The small amplitude of the microwave signal in this time interval, as well as the marked instability of the amplitude and shape of the pulses are apparently due to the competition of the modes of oscillation. In the final stage of the pulse, at $\Delta t \approx$ 60–75 ns (Fig. 6.11 *d*), the regular components are practically lost in the spectrum and noise is significantly increased. In experiments the frequency shift during the pulse is fixed. The frequency deviation is explained by the radial motion of the cathode plasma, which leads to a change in the resonance properties of the oscillating system of the magnetron, as well as the electric field strength between the cathode and the anode. This issue was discussed in detail in Chapter 1.

6.5. Relativistic magnetron with a symmetric coupling of resonators

The symmetrical scheme of the communication channel of the magnetron is shown in Fig. 6.11 [2–6]. The outputs of the two opposite resonators of the anode block of the 6-cavity magnetron *1* are connected by a waveguide channel *3* with a common load – a radiating horn antenna *4* located on the axis of the electrical symmetry of the channel. The length of the channel was varied by means of waveguide inserts.

In the preliminary experiment, the length of the channel L_w was discretely increased by 2.5 λ_w with steps of 25 mm. At each step, the spectral and energy parameters of the radiation were measured. A cyclic change in the measured parameters was observed. After that more detailed measurements were made in the region 16.5–17.7 λ_w. In this case, the channel length changed with a smaller step (10 mm) and at each step a full procedure of spectral measurements was performed in the whole range of filter tuning (2550–3250 MHz), the energy characteristics of the radiation were determined, and the parameters of the current and voltage in the pulse were monitored. Spectral studies, as in the case of a magnetron with uncoupled resonators, were carried out at the induction of the magnetic field of 0.5 T.

The power of the pulses of the microwave radiation of the magnetron in the common load varied from 300 to 475 MW (Fig. 6.12 *a*), and the average energy from 11 to 20 J (Fig. 6.12 *b*). The spectral measurements can be determined both by the shift of the position of the frequency maximum (Fig. 6.13) and the change in the

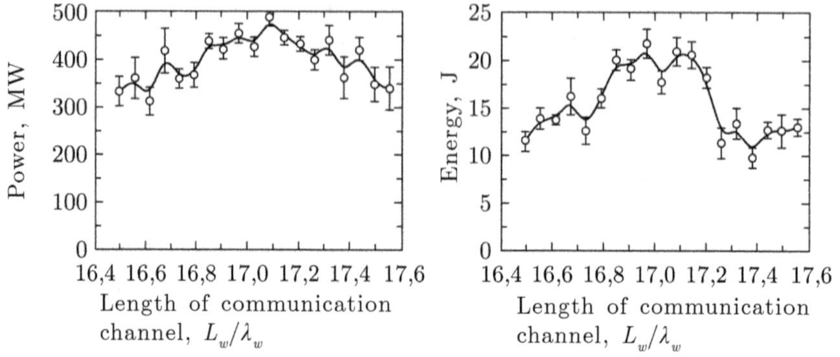

Fig. 6.12. Dependence of the power and the microwave energy of the magnetron on communication channel length L_w.

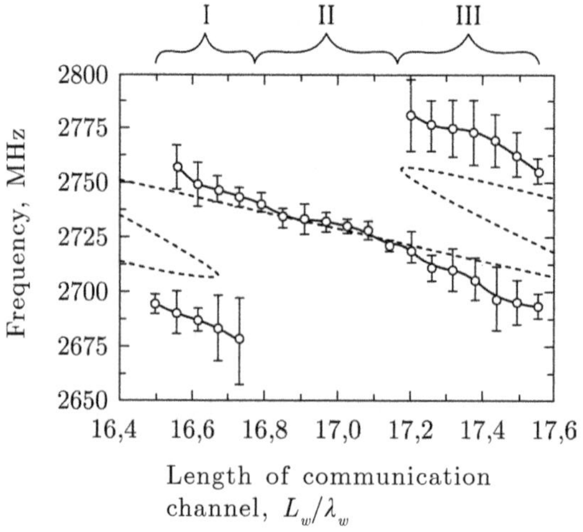

Fig. 6.13. The change in the position of the frequency maximum in dependence on the length of the communication channel L_w.

spectral width of radiation (Fig. 6.14). It was found that in certain parts of the investigated channel length interval, a second frequency maximum appears in the spectrum.

Based on the results of the measurements, it is possible to distinguish three areas for tuning the channel length – they are marked in Fig. 6.13. The region 16.8 λ_w–17.2 λ_w (region II) is characterized by a higher energy efficiency of the magnetron: the power level reaches 400–450 MW, and the energy of the pulse is 20 J (against 180 MW and 7.5 J for each output for the magnetron with uncoupled outputs). The oscillation spectrum in this region is characterized by a single frequency and the narrowest band: the

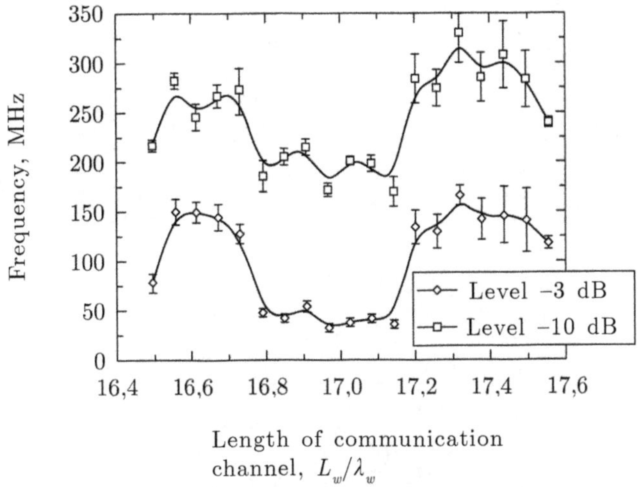

Fig. 6.14. The change in the width of the frequency spectrum in dependence of the length of the comuunication channel L_w.

level – 3 dB – about 30–40 MHz (Fig. 6.14). Outside of this region (in domains I and III) an abrupt expansion of the radiation band to 150 MHz (level –3 dB) appears, additional frequency maxima and the energy characteristics of the pulses are reduced: up to 350 MW and 11 J (see Fig. 6.12). It is characteristic that the centres of the regions are located at a distance of approximately 85 mm from each other, which is 0.5 wavelengths in the waveguide.

However, there are also noticeable differences. So, the power reduction occurs more slowly, and in the region of the maximum (at 25 ns) there is a flat part with a duration of about 20 ns. The pulse duration at the level of –3 dB is ~45 ns, which is 25 ns longer than in the case of uncoupled outputs. At the level of –10 dB the pulse duration remained unchanged, of the order of 85 ns. The process of generation is characterized by more constant shape of the pulse signal (Fig. 6.15 *b*) and the radiation power is also constant from pulse to pulse. The amplitude instability in a continuous series of 500 pulses (Fig. 6.16) does not exceed ±5%, which is much less than in the case of a magnetron with uncoupled outputs (see Fig. 6.10).

Figure 6.15 shows typical envelopes of microwave pulses from two detectors in region II (optimal tuning). As in the case of a magnetron with uncoupled outputs, they have a shape close to a triangular one, with a steep leading edge (see Fig. 6.9).

The spectral characteristics of the radiation of the magnetron were measured for the induction of a magnetic field of 0.5 T, also

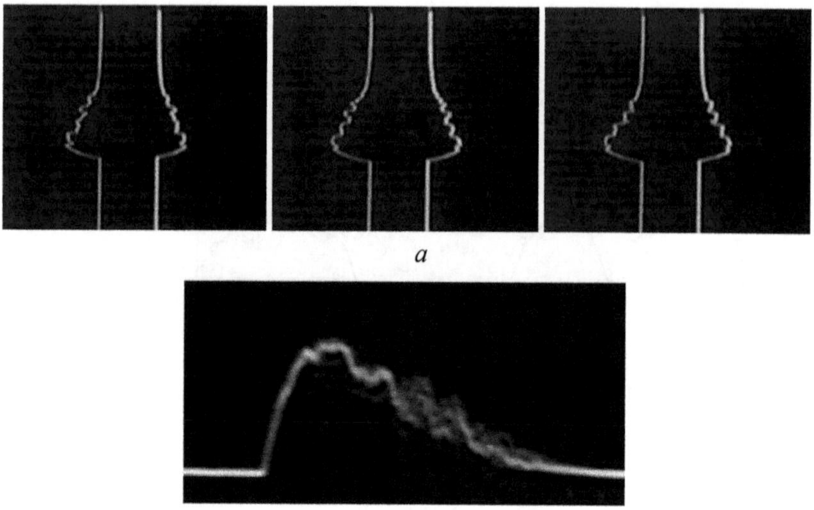

Fig. 6.15. Oscillograms of microwave radiation pulses of a magnetron with coupled resonators in the region of optimal tuning: *a* – typical pulse shape; *b* – ten coincident pulses.

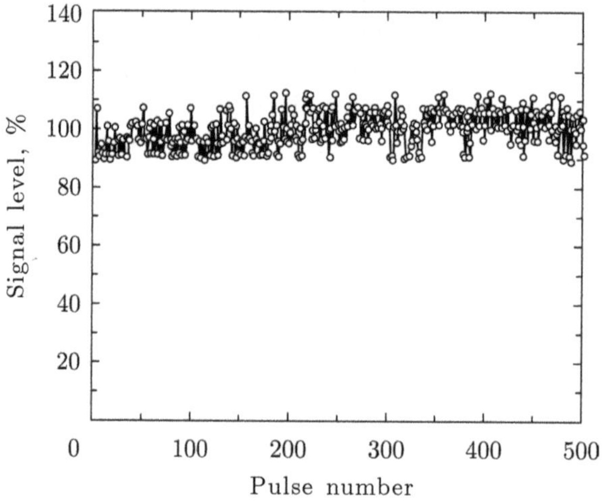

Fig. 6.16. Dynamics of the microwave pulse amplitude of a magnetron with coupled resonators in the region of optimal tuning in a continuous series of pulses.

in the course of four temporal parts of the pulse. All the spectra presented are given to the 0 dB level. The spectrum at the leading edge of the pulse (Fig. 6.17 *a*) resembles the analogous spectrum of a magnetron with uncoupled outputs (see Fig. 6.10 *a*), but its width in the level is 3 dB somewhat smaller 65 MHz. The spectrum clearly shows a maximum of the frequency of which is close to 2730 MHz.

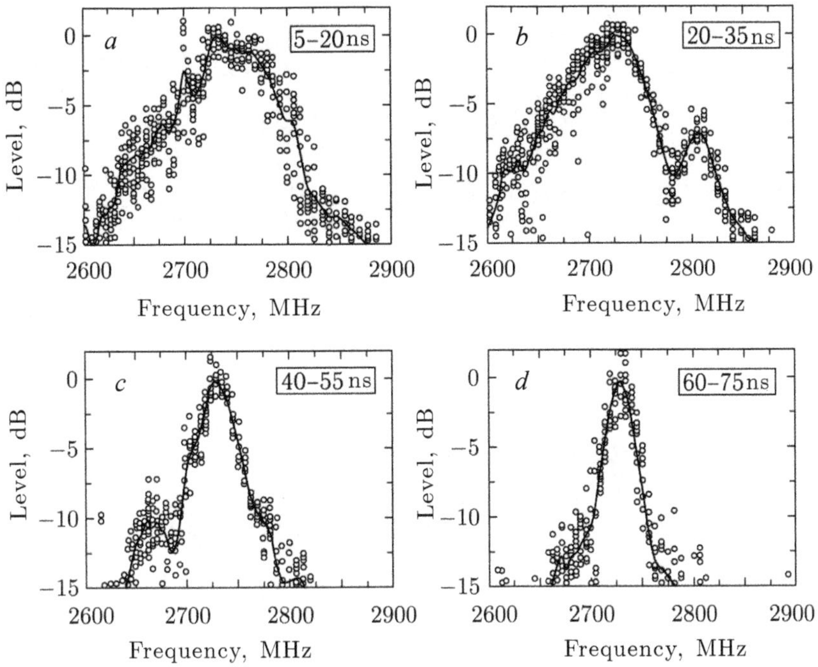

Fig. 6.17. The frequency spectrum of a magnetron with coupled resonators for optimal tuning in different pulse time parts.

The pulse spectrum at the maximum power point (Fig. 6.17 *b*) is already noticeably different. By the middle of pulse near $\Delta t \approx$ 40–55 ns (Fig. 6.17 *a*) the spectrum, while preserving its position, becomes more narrowband. The width of the spectrum at the level –3 dB is 25–30 MHz. The shape of the radiation spectrum and the position of the maximum remain unchanged right up to the very end of the pulse (Fig. 6.17 *d*). The presence of heterogeneities at the edges of the spectrum at the level of –10 dB is due to the beating that occurs when the filter is detuned.

The work of the magnetron in the region of the channel lengths of the couplings I and III (non-optimal tuning) is characterized by poor energy characteristics and a wide spectrum of the oscillations. The maximum power levels here are lower and do not exceed 340 MW and 13 J, respectively. The shape of the pulses of the microwave radiation has changed noticeably (Fig. 6.18); their duration at level –3 dB and –10 dB decreased and now stands at ~30 ns and ~65 ns, respectively. The stability of the pulses is lower: the average power spread in the continuous series reaches ±30% (Fig. 6.19).

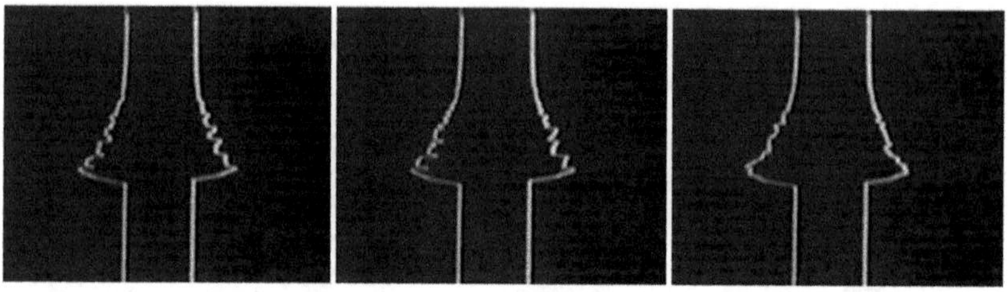

Fig. 6.18. Oscillograms of microwave pulses from a magnetron with coupled resonators in the area of non-optimal tuning.

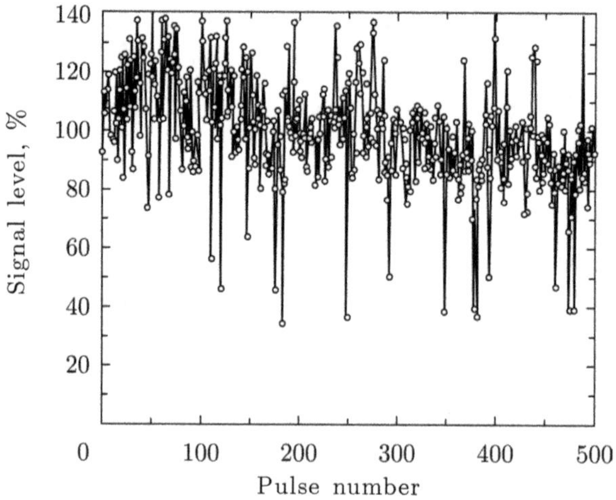

Fig. 6.19. Dynamics of the microwave pulse amplitude of a magnetron with coupled resonators in the non-optimal tuning region.

The greatest changes, as has been said, underwent the spectral characteristics of the radiation. Figure 6.20 shows the spectra recorded for the channel lengths 16.6 λ_w (Fig. 6.20 *a*) and 16.7 λ_w (Fig. 6.20 *b*). On the diagrams, two spectral maxima are clearly visible. The width of each of them is about 60 MHz, the distance between them is ~70 MHz. Analysis of the current spectrum showed that the 'bifurcation' of the spectrum begins to occur only to the middle of the pulse after the maximum power. Further, the separation of the spectrum intensifies and near the end of the pulse it becomes most contrast. It is characteristic that the frequency difference between the spectral maxima remains constant throughout the pulse, i.e. ~70 MHz. The frequencies of the maxima in contrast to the case with uncoupled outputs, where their displacement occurred, are practically unchanged. This emphasizes the effect of

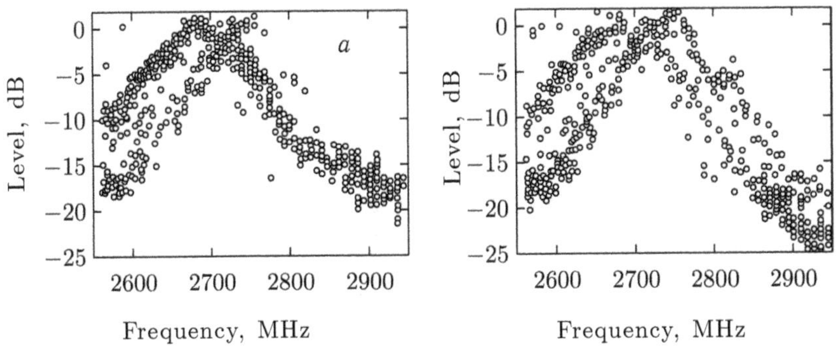

Fig. 6.20. The frequency spectrum of the magnetron in the areas of non-optimal length of the communication channel: *a* – 16.6 λ_w; *b* – 16.7 λ_w.

the communication channel on the generation process, which leads to a strong pulling of the frequency of the microwave oscillations of the magnetron.

Comparing the results of studies of magnetrons with coupled and uncoupled outputs, it can be concluded that the interaction of resonator oscillations through an external communication channel has a significant effect on the characteristics of the generation process. This influence depends on the length of the communication channel and can be either stabilizing (region II) or destabilizing (regions I and III). In the case of optimal selection of the channel length (region II), this interaction, despite a significant delay (~17 periods, i.e. ~6.2 ns), causes the formation of the radiation spectrum and the absence of frequency drift during the pulse. An increase in the stability of the time and energy characteristics of the radiation should also be associated with this mechanism.

A more detailed analysis of the experimental data was carried out to clarify the reasons for the spectral instability of radiation in the tuning regions I and III. First of all, it was taken into account that all pulses from these regions have a fairly high peak power level 300–340 MW. This allowed us to assume that in the magnetron there are oscillations of one phasing, namely, the π-type.

At the same time, pulses from the continuous series forming part of the spectrum with one maximum have a noticeable difference in shape from the pulses that form the spectrum with another maximum. In other words, from a pulse to a pulse, a random excitation of a magnetron occurs at one of the frequencies, and the observed 'bifurcation' of the spectrum is actually the result of the superimposition of two spectrograms, each of which has a single maximum. Thus, the frequency instability takes place in the RM at the specified settings: the main π-type of oscillations splits into two competing subtypes. The marked stability of the frequencies of competing oscillations is a consequence of a strong frequency pulling.

The appearance of subtypes of the basic mode of oscillation, as well as the delay in the frequency of radiation, is explained by the mismatch of the common load. In the experimental diagram in Fig. 6.13, the dashed line shows the calculated dependence of the resonant frequencies of the equivalent magnetron circuit for the case of an underloaded communication channel. (the dependence was calculated for load resistance R_{load} = 200 Ohm, this corresponds to the dynamic mismatch of the channel with a standing wave ratio K_{st} = 2, which is close to the characteristics of real waveguide tee). The qualitative agreement between the experimental and calculated areas frequency ambiguity confirms their direct simplified relationship with the channel length. The mismatch of the frequency of the subtypes can be explained by the simplified type of the the computational model of the magnetron.

6.6. Relativistic magnetron with an antisymmetric channel of coupling of resonators

The scheme of the experiment is shown in Fig. 6.21. In this case, the central radiator *2* is displaced relative to the axis of electrical symmetry of the communication channel *1* by $\lambda_w/4$. In this case, if the magnetron (6-resonator magnetron is considered here) operates on the π-type, the oscillations of its resonators are subracted in

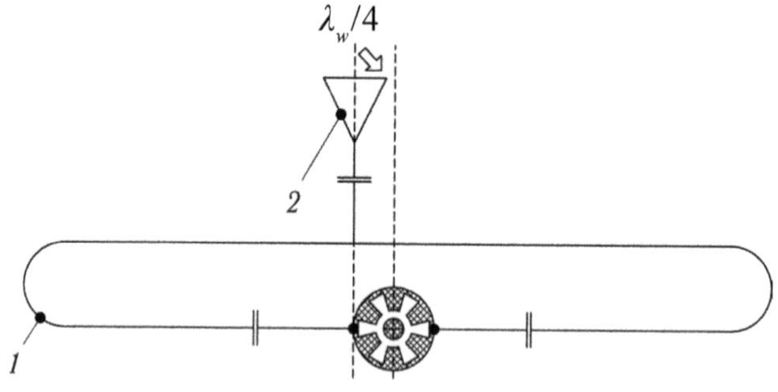

Fig. 6.21. Antisymmetric scheme of the modified relativistic magnetron: *1* – communication channel; *2* – antenna.

the common radiator, and in the case of $2\pi/3$-type excitation– the radiation is summed.

The results of an experimental study of the antisymmetric scheme are as follows. The output power of microwave radiation does not exceed 20 MW, and their energy is 1 J, which is much lower than the power and energy of a magnetron with a symmetric communication channel. When measuring the spectral characteristics of the radiation of a magnetron, their dependence on the length of the communication channel was observed. The behaviour of the spectrum here is similar to the behaviour of a spectrum in a magnetron with a symmetric communication channel.

Thus, the low power level fixed in this experiment is a consequence of a deep subtraction of antiphase oscillations in the load-radiator. This unambiguously indicates the existence and uniqueness of the working π-mode of the oscillations in the system.

It should be noted that the operating of the magnetron in the antisymmetric scheme differs significantly from that for the symmetric version (when the resonator powers are equal, the oscillations in the common load are subtracted and the magnetron operates without load). With this, in particular, we can relate the observed shift in the generation frequency, which in this experiment was about 2.8 GHz. Despite this, there is a well-developed process of generating microwave oscillations in a magnetron. This is also indicated by the coincidence of the generator power parameters (magnetron current and accelerator voltage) with similar parameters for a magnetron with uncoupled outputs.

A special experiment made it possible to find out that with the optimal tuning of the connections, the stability of the main (in-phase) type of oscillations was strengthened and weakened for competing (antiphase) types.

Calculations performed in the design of the anode magnetron block (see section 4.3) showed that for a given electric field strength, the excitation conditions for the π-mode are performed in the magnetic field range 0.3–0.53 T. Outside this region, the excitation conditions are satisfied for oscillations of the type $2\pi/3$ (0.25–0.3 T) and its (−1)-th harmonic (0.53−0.56 T). On the border of the mode stability intervals (0.3 and 0.53 T), one can expect excitation of the oscillations of both modes. Indeed, when measuring the energy characteristics of a magnetron in the magnetic field, close to 0.53 T, the power levels, an order of magnitude and more than the level of 20 MW, were periodically recorded. These levels were correlated with oscillations of the $2\pi/3$ type, which allowed to formulate the idea of the experiment [7].

In the course of the experiment, at a magnetic field induction value of 0.53 T for each length of the communication channel, 50 shots of the installation were performed throughout the entire range of its variation. The radiation power was determined for each pulse and the relative number of realizations of each mode was calculated from the power level.

Figure 6.22 shows the change in the relative number of realizations of the modes π and $2\pi/3$ (N_π and $N_{2\pi/3}$, respectively) in the series of $N_\Sigma = 50$ pulses, depending on the length of the channel. It was found that in the length region of the tract 16.8 λ_w –17.2 λ_w, the probability of excitation of the π-mode oscillations reaches 87%, and the probability of excitation of $2\pi/3$ modes is 13%. Outside this region, the probability of excitation of the π-mode sharply reduced, so that at the communication channel lengths 16.7 λ_w and 17.3 λ_w which belong to the suboptimal interaction region (regions I and III in Fig. 6.13) the oscillations of the type $2\pi/3$ are already dominant: the probability of excitation reaches 70%.

From the above results, it can be seen that the change in the electric length of the circuit makes it possible to obtain intensification and also weakening of the competing modes. With a non-optimal coupling, the interaction of resonator oscillations leads to suppression of the fundamental π-mode of oscillation of the magnetron. This, in

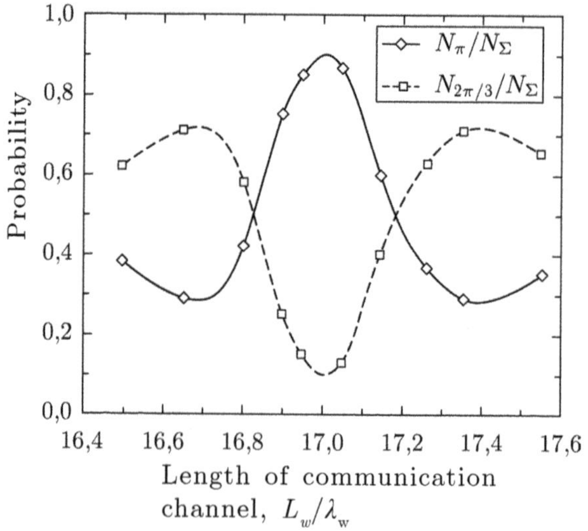

Fig. 6.22. Dependence of the relative number of realizations of the oscillation modes π and $2\pi/3$ on the length of the communication channel.

particular, contributes to the expansion of the generation zone of the mode of $2\pi/3$ oscillations toward smaller magnetic fields. Thus, external interaction has a sufficiently strong selective effect on the existence of oscillations of various modes, which is in accordance with the predictions obtained theoretically.

6.7. Relativistic magnetron with distributed radiation output from the channel of coupling of resonators

For a distributed output of microwave radiation from a relativistic magnetron, a system of radiators (loads) is introduced into the channel of external coupling of resonators. The circuit of the communication channel, providing the necessary amplitude–phase distribution of microwave oscillations to the radiators and stabilization of the operating mode of the magnetron, was constructed according to the rules stated in section 6.1.

The source scheme based on the 6-cavity magnetron is shown in Fig. 6.23 [8–11]. The length of the communication channel of resonators *1* was chosen based on the results of a study of a system with one radiator. It was 17 λ_w for the working π-mode of oscillation and corresponded to the optimum tuning of the channel. The central radiator (antenna) *2* is offset relative to the axis of the electrical symmetry of the communication channel by $\lambda_w/4$; the left *3* and right *4* side radiators are located at a distance of 5 $\lambda_w/4$ and 9 $\lambda_w/4$ with respect to the central radiator. The excitation mode of the circuit for the π-mode is in-phase. The circuit is asymmetric with respect to the central radiator, so that in the operating mode the oscillations of the resonators in this radiator are subtracted. Due to the geometric

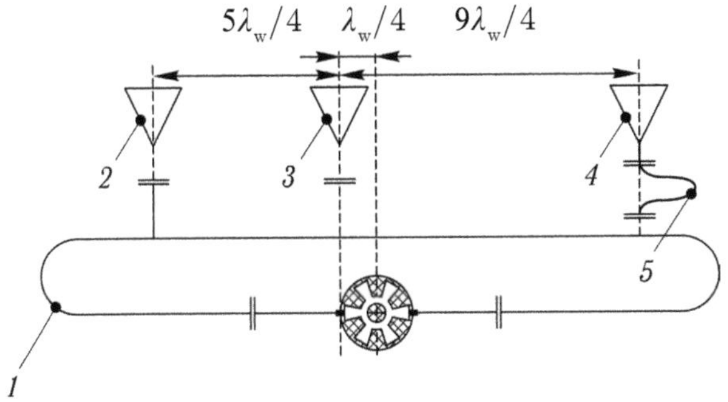

Fig. 6.23. Scheme of the magnetron with three radiators: *1* – communication channel; *2*, *3*, *4* – antennas; *5* – additional insert.

dimensions of the antennas, the side radiators are spaced apart from each other. The difference of the line segments to each radiator is a multiple of λ_w, so the resonator signals enter each of the side radiators in phase, that is, they are summed. The distance between the side radiators is 3.5 λ_w, and are therefore excited in the antiphase. For in-phase excitation of the radiators the phase difference was compensated by the additional waveguide insert *5*.

The registered power level of microwave pulses in the channel of the central radiator does not exceed 25 MW. The power of the radiation from the side antennas is 180 MW. The spectra of the emitted signal are analogous to the spectra of a tuned symmetrical system with summation of power. In this case, the spectrum has one maximum at 2720 MHz; frequency deviation is absent. These data allow us to conclude that the magnetron is stable on the π-type of oscillations.

The spatial distribution of the radiation was measured by the method described in section 8.2; the radiation diagram is shown in Fig. 6.24. The radiation intensity is maximal in the direction normal to the radiators. The radiation intensity is sufficiently low in the direction of the minima of the diagram (below –13 dB). It is important to note that this level is repeated in a continuous series of pulses. This indicates a high stability of the amplitude–phase relations in the RM. The radiation pattern calculated for the two pyramidal horn antennas (dotted line in Fig. 6.24) is in qualitative agreement with the experiment.

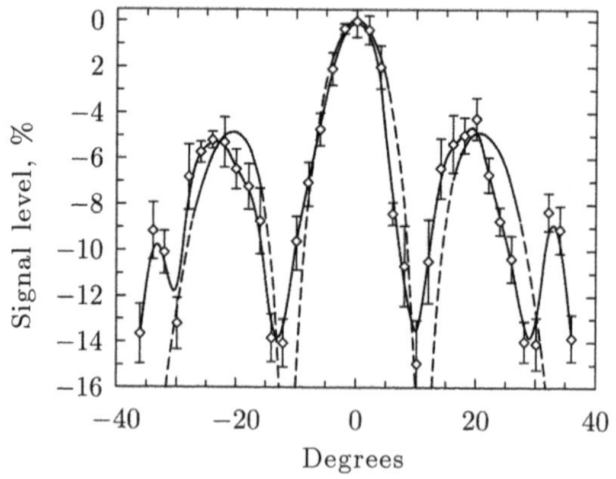

Fig.6.24. Diagram of the spatial distribution of radiation.

6.8. Characteristics of an eight-cavity relativistic magnetron with coupled resonators

A relativistic magnetron with an eight-cavity anode block can provide a higher efficiency of the source. Excitation of the working π-mode of oscillations in this magnetron should occur at higher synchronous magnetic fields; the radius of the cyclotron rotation of the electrons decreases, which should lead to an increase in the electronic efficiency of the device. The anode block of the magnetron had geometrical dimensions identical with the 6-cavity device (cathode and anode radii, internal radius of the resonators). This provided the same load for the power source, which allowed a correct comparison of the results of the experiments. The eight-cavity magnetron was investigated in combination with a communication channel containing one radiator whose symmetrical and antisymmetrical arrangement with respect to the axis of the electrical symmetry of the channel made it possible to obtain both summation and subtraction of oscillations at the output of the system, as well as selective excitation of various modes of the resonance system.

The use of systems with an increased number of resonators sharply increases the competition of modes of oscillation. Table 4.1 shows the results of calculating the wavelengths of the modes and the slowing-down of the electromagnetic wave for the working and nearest ($N/2-1$) and its (-1) and ($+1$) spatial harmonics of the oscillation modes, as well as the results of 'cold' measurements (highlighted figures). As can be seen, the separation of the oscillation types on the basis of the phase velocity for the 6-cavity anode block is $\beta_\pi/\beta_{2\pi/3} \approx 27\%$, in relation to its ($-1$)-th harmonic $-\beta_\pi / \beta_{2\pi/3}^{-1} \approx 31\ \%$. The separation of the types of the oscillations for the 8-cavity anodic block is approximately one-third smaller and equal to $\beta_\pi/\beta_{3\pi/4} \approx 21\%$ and $\beta_\pi / \beta_{3\pi/4}^{-1} \approx 24\ \%$. From the table it also follows that the slowing of the working mode of oscillations in the 8-cavity anode block in comparison with the 6-cavity anode block is approximately 26% higher, therefore one can expect from this block a higher efficiency of operation.

Figure 6.25 shows the dependences of the output power of 6- and 8-cavity relativistic magnetrons with one output on the value of the charging voltage of the primary storage of the LIA. As it turned out, the 6-cavity RM demonstrates higher efficiency (~20%) with a maximum output power of 300 MW (power pulse parameters: charging voltage U_{C0} = 800 V, cathode–anode voltage 340 kV, current

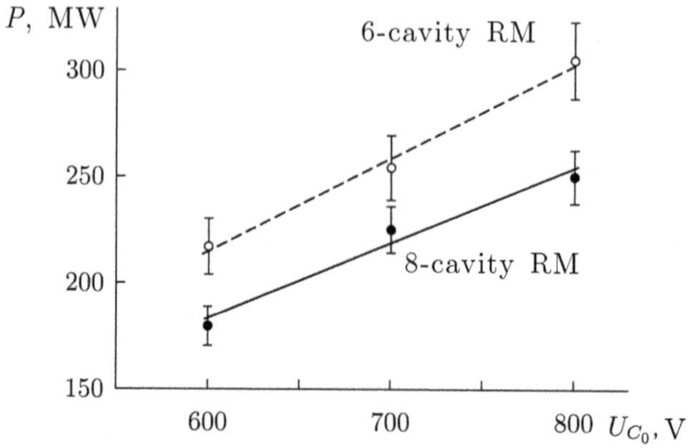

Fig. 6.25. Dependences of the output power of the RMs with one waveguide power output on the value of the charging voltage of the primary storage device of LIA 04/6.

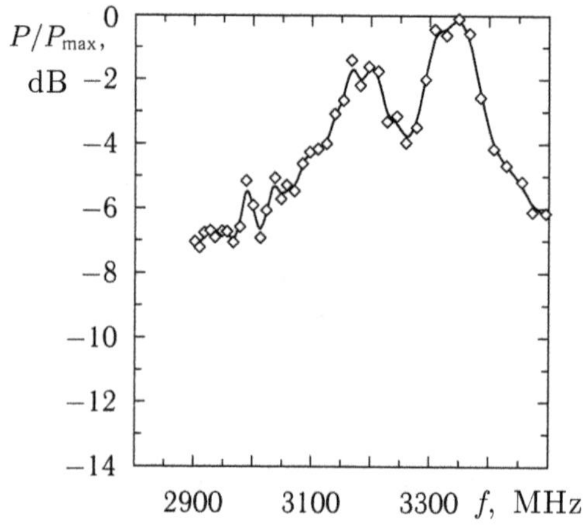

Fig. 6.26. The frequency spectrum of an 8-cavity relativistic magnetron.

4.3 kA). At the same power source the RM with an 8-cavity anode block has an output power of ~250 MW.

Figure 6.26 shows a typical frequency spectrum of an 8-cavity RM. The spectrum differs by a large width (~ 200 MHz) and has two maxima corresponding to two types of oscillations having close frequencies. The competition of these types of oscillations can be observed from a set of oscillograms, shown in Fig. 6.27 and recorded at different central frequencies of the bandpass filter. On the leading front the higher amplitude has the form of oscillations with a shorter wavelength (higher frequency); then during an impulse the amplitudes of the signals corresponding to the two modes of oscillation are equalized, and to the end of the pulse the higher amplitude is already acquired by the mode of oscillations with the longer wavelength. Taking into account the calculation data, we can conclude that the π-type of oscillations predominates on the front, and by the end of the pulse there is the 3π/4-type having the greater wavelength. Simultaneous existence of two types of oscillations and competition between them lead to a decrease in the operational efficiency of the RM and the widening of the spectrum of oscillations.

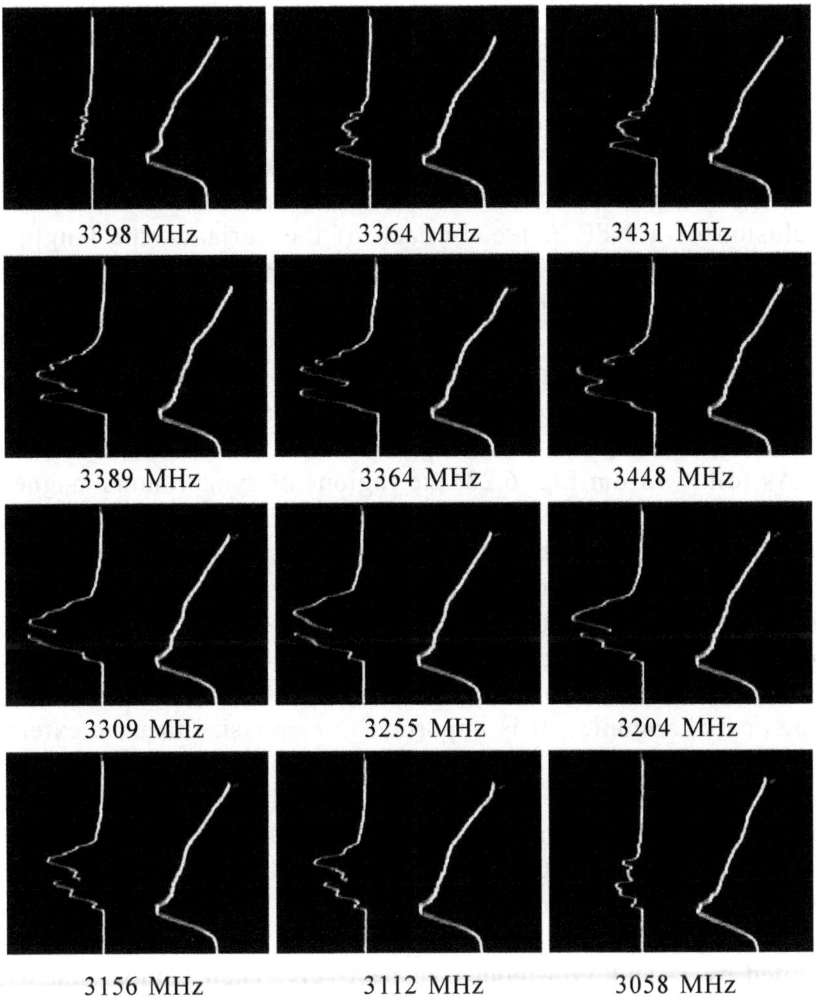

Fig. 6.27. Oscillograms of the signals of the microwave detector with the bandpass filter installed in front of it (on the left) and voltage pulses for the 8-cavity relativistic magnetron (right).

For a magnetron with a number of resonators satisfying the condition $N/2$ – even number, the oscillations of the opposite resonators for the π-type in the azimuth are in phase and in relation to the output waveguides are opposite to phase. Therefore, the signals of the given outputs of the magnetron are summed in the common load in the antisymmetric scheme and subtracted in the symmetric one. For the 3π/4-type, on the contrary, the oscillations of the opposite resonators are in phase. Consequently, the output signals of the resonators for the 3π/4-type are subtracted in the common load of the antisymmetric scheme and are summed up in the symmetric scheme. These simple rules allow in the course of the experiment to uniquely identify the mode of oscillation. The symmetrical and antisymmetric versions of the waveguide channel are realized with the help of a set of waveguide inserts and *H*-tee (see section 8.1).

Both variants of the 8-cavity RM with an external communication channel were investigated: for a symmetric and antisymmetric inclusion of a 3-dB *H*-tee. In each of the variants, the length of the waveguide channel was selected separately with the help of a set of inserts by the criterion of the maximum level of the output power and the stability of the radiation spectrum. After that, with the optimal channel settings, the energy and spectral characteristics were measured. In these experiments, the channel length was of the order of 13 waveguide wavelengths.

As follows from Fig. 6.28, the regions of synchronous magnetic fields differ markedly. In the antisymmetric scheme, the excitation of the π-mode of oscillation (it corresponds to the maximum power level in the load-radiator) is achieved at high magnetic field values, since this mode has the greatest slowing-down. The maximum power in a symmetrical circuit obviously corresponds to in-phase oscillations of the resonators, that is, to the excitation of the $3\pi/4$-type or its harmonics. It is important to emphasize that the external communication channel eliminates the competition of these modes, ensuring their separate existence in each of the circuits in a wide range of values of the magnetic field.

The excitation of various modes is also illustrated by the results of spectral measurements. Figure 6.29 shows the radiation spectra of the 8-cavity RM, taken for the antisymmetric and symmetric variants, excited by π- and $3\pi/4$-modes, respectively. Their comparison with the spectrum of a magnetron with one output (Fig. 6.26) shows that with the help of an external coupling of resonators it is possible not only to eliminate the phenomenon of competition types of

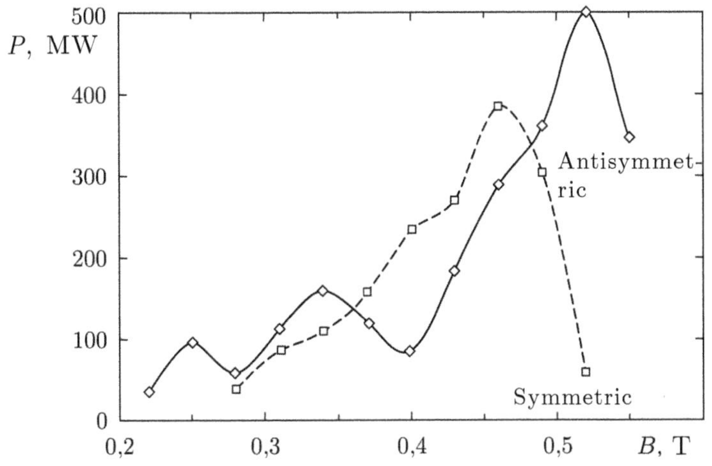

Fig. 6.28. The dependence of the power of the 8-cavity RM on the value of the induction of the magnetic field with an external coupling channel for the antisymmetric positioning of the *H*-tee ($P_{max} \approx 500$ MW) and symmetric positioning ($P_{max} \approx 390$ MW).

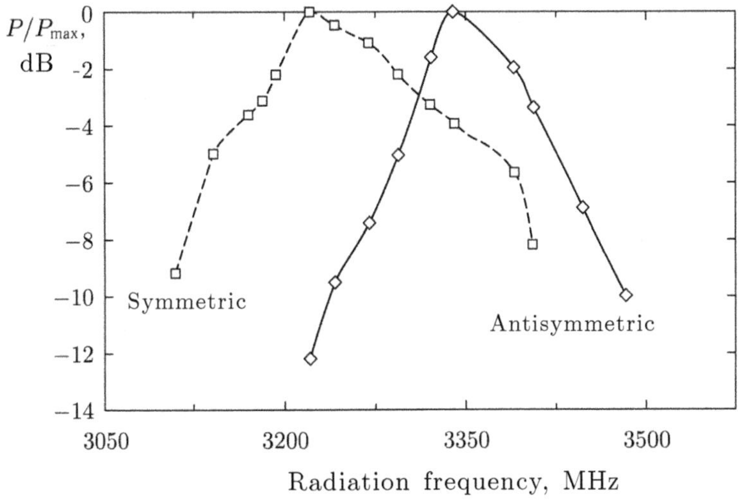

Fig. 6.29. The frequency spectrum of the 8-cavity RM in the antisymmetric (π-mode) and symmetric positioning of the *H*-tee waveguide ($3\pi/4$-mode of oscillations).

oscillations, but also significantly increase the degree of coherence of RM radiation and, thereby, narrow the emission spectrum to 60 MHz and increase the output power to 500 MW.

Thus, a direct confirmation obtained from the analysis of the phenomenological model of the magnetron structure is found on the

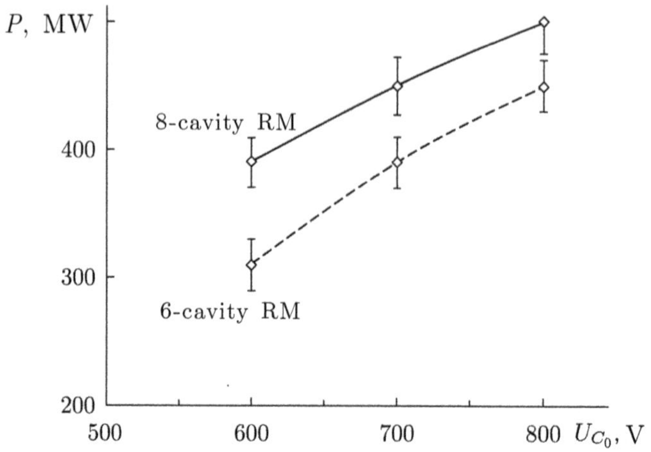

Fig. 6.30. Dependence of the output power of the 8- and 6-cavity RMs with the external communication channel on the value of the charging voltage of the primary storage of LIA 04/6.

possibility of increasing or decreasing (up to suppression) the power and stability of the modes.

Figure 6.30 shows the dependence of the output power of the RM on the charging voltage of the primary storage of LIA 04/6. Unlike the devices with one output (see Fig. 6.25), an 8-cavity RM with integrated outputs delivers 12% higher power than a similar 6-cavity RM.

It should be emphasized that the effect of the external coupling channel of the resonators of the RM on the generation processes depends on its phase properties. Optimal conditions for the interaction of resonator oscillations are realized in a certain region of the electrical length of the channel; experience shows that the width of this region is 30–50°. In each specific case, the spectral and energy studies described here are needed to determine the tuning region and its localization. The length of the external communication channel in the considered devices is large and causes a significant delay in the interaction signal: up to 10–20 oscillation periods. Such a delay leads to a corresponding delay in the process of coherent interaction of the oscillations. However, after this, due to a strong coupling, there is a rapid formation and stabilization of the radiation spectrum.

The distributed radiation from the communication channel can be made in the form of a waveguide–slit antenna array. Scheme of a microwave source based on an 8-cavity RM and the appearance

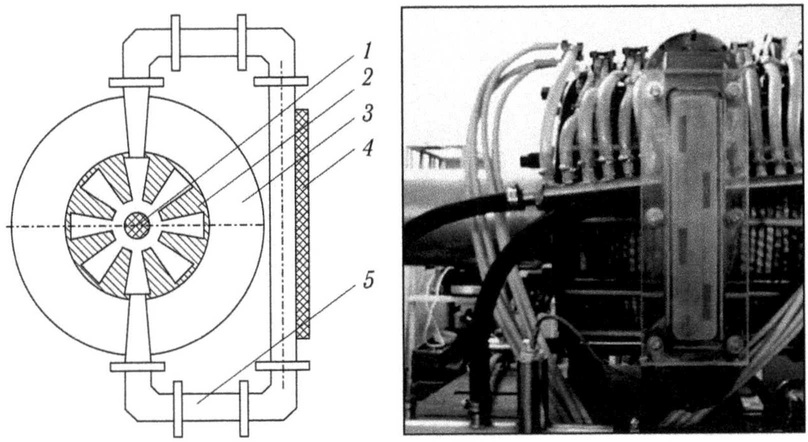

Fig. 6.31. Scheme of the experiment and the appearance of the antenna array: *1* – cathode; *2* – anode block; *3* – magnetic system; *4* – antenna array; *5* – waveguide inserts.

of a waveguide array with four slits are shown in Fig. 6.31. The in-phase excitation of adjacent slits located at a distance $\lambda_{in}/2$ is provided by opposite transverse currents, which is equivalent to an additional phase shift of 180° [12]. With antiphase oscillations of the resonators (π - type), all the slits are excited in the phase, and their radiation is focused in a plane passing through the longitudinal axis of the waveguide and in the direction normal to the wide wall of the waveguide. The slits are closed with a dielectric window made of plexiglass with a vacuum seal.

In this experiment [13, 14], the directional radiation characteristics were also measured with a successive change in the length of the communication channel by means of waveguide inserts. As expected, the optimum channel length region fully corresponds to the experiment with the *H*-tee and the horn antenna. Figure 6.32 shows a radiation diagram. Its comparison with the calculated (dotted line) unambiguously proves the existence in RM of π - mode. The length of the communication channel also has a significant effect on the stability of the amplitude of the output pulses (Fig. 6.33). In the process of the investigation, a large number (100 for each insertion) of measurements of the amplitude of the microwave signals was performed, and the root-mean-square (rms) was calculated from their values. As can be seen, with an optimal length of the communication channel, a more stable generation of the microwave radiation of the relativistic magnetron is observed.

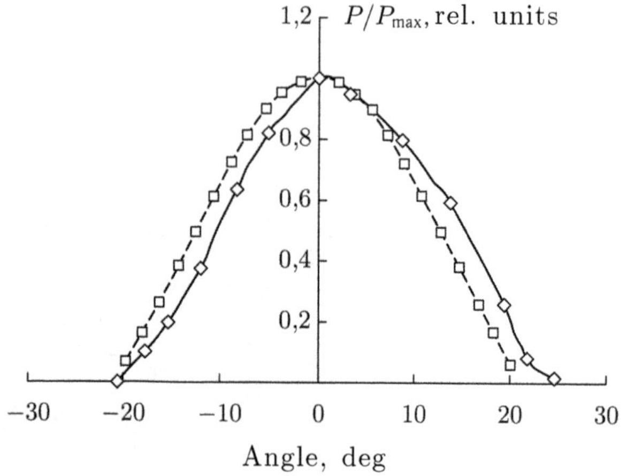

Fig. 6.32. Radiation directional diagram. Dashed line – calculated.

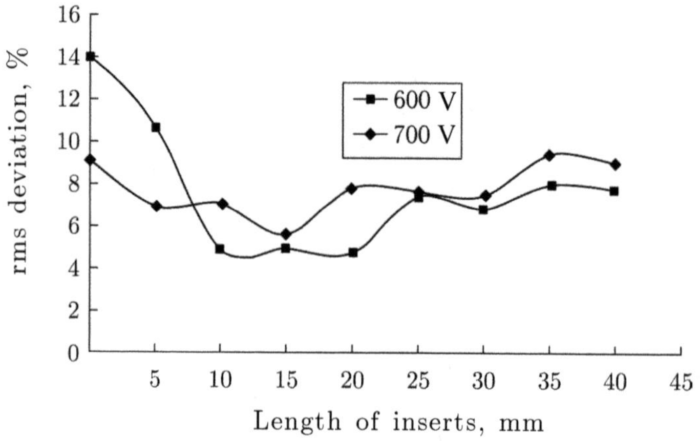

Fig. 6.33. Dependence of the rms deviation of the signals of the microwave detector, installed on the axis of the antenna array, on the length of the inserts at two values of the charge voltage of the primary storage of LIA 04/6.

The output power of the magnetron using a waveguide–array was 260 MW, and the power flux density at the axis of the array at a distance of about 2.7 meters was 150 W/cm^2.

As an indication of the systems of interconnected auto-generators, to increase the stability of the relativistic magnetron with a short waveguide communication channel, it is necessary to introduce a special load into the channel. The effect of this load can be explained by considering the losses introduced by it into the oscillatory system. From this point of view, the central load for non-working modes of

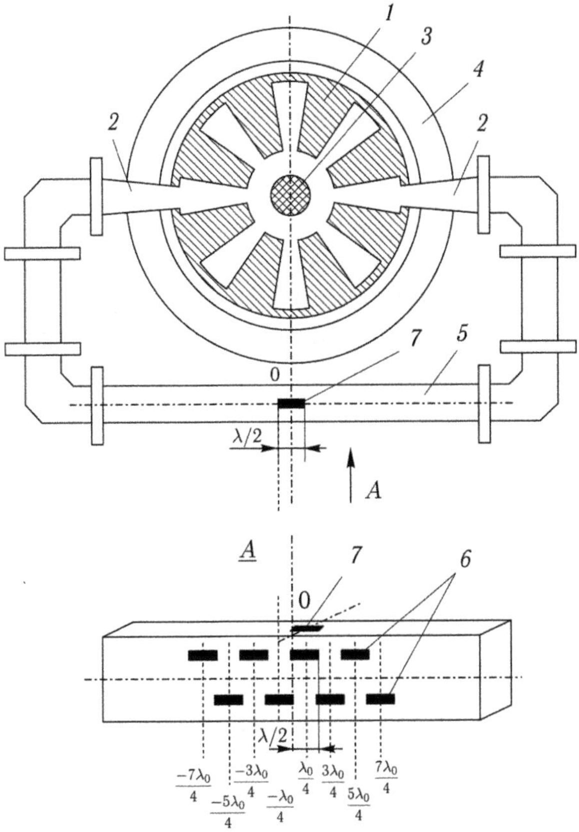

Fig. 6.34. Relativistic 8-cavity magnetron with a system of slit radiators in the waveguide channel of coupling of resonators: *1* – multiresonator anode block; *2* – waveguide power outputs; *3* – cathode; *4* – magnetic system; *5* – waveguide communication channel; *6* – radiating slits; *7* – central load gap.

the magnetron introduces additional losses, reduces their *Q*-factor and ensures fast damping. In addition, as already noted, recording the level of radiated power from the central load makes it possible to control the adjustment of the instrument.

Figure 6.34 shows a constructive version of the introduction of an additional load due to a slot in the narrow wall of the waveguide communication channel of the eight-cavity RM [15]. The antiphase oscillations of opposite resonators (π-type) create a voltage node on the axis of symmetry of the channel; therefore, a slot located on this axis is not excited. The distributed power output from the channel is realized similar to the construction in Fig. 6.31 through a system of slits in a wide wall; the slits are arranged in such a way that their in-phase excitation is ensured.

6.9. Dynamics of the radiation frequency of a relativistic magnetron

Narrow-band resonance filters are traditionally used to estimate the frequency spectrum of relativistic devices. However, with a sufficiently dense spectral composition of the oscillations, these filters do not provide the required resolution. An increase in the quality factor of the filters does not give the desired results, since an increase in the duration of the transient processes alters the form of the response of the filter. A reliable accuracy in estimating the level of the spectrum can be achieved only in the narrow resonance band. Finally, because of the integral nature of the response of the filter, this method can give only approximate information on the time dynamics of the spectral density of the signal (on the current spectrum). To extract this useful information, laborious processing of the filter responses is required at various settings for different time slices of the generation pulse. The highest accuracy is obtained by direct recording of a pulsed microwave process and its further digital Fourier analysis. To realize this possibility, expensive oscillographic equipment is required. In the measurements described below [34, 35], a frequency meter for the microwave pulses, developed at the TPU, was used.

In studies, the behaviour of the frequency of a relativistic magnetron during a pulse was determined for several cases:

1. RM with two uncoupled power outputs, each of which is loaded on its pyramidal radiating antenna.
2. RM with two outputs connected by an external communication channel and a symmetrically installed radiating antenna.
3. RM with two outputs connected by an external communication channel and an antisymmetrically installed radiating antenna.
4. RM with two outputs connected by an external communication channel with an installed waveguide–slit array.

A measuring instrument for the frequency composition of microwave pulses. The instrument [36] is used together with a digital oscilloscope with a bandwidth (350–500) MHz and any compatible computer. The functional diagram of the meter is shown in Fig. 6.35.

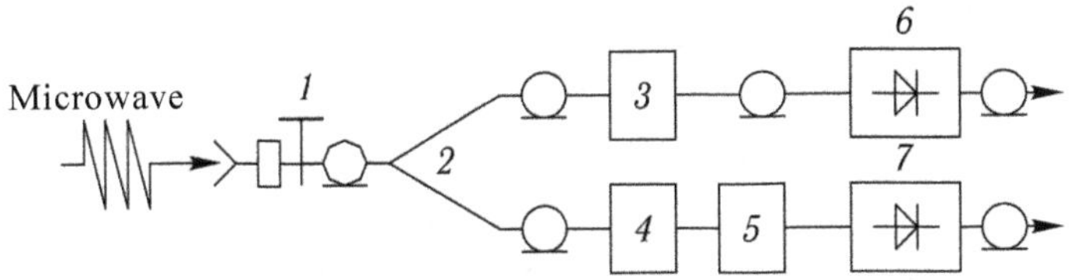

Fig. 6.35. Functional diagram of the frequency meter: *1* – coaxial–waveguide transition; *2* – tee; *3, 4* – fixed attenuator; *5* – the beyond-cutoff attenuator; *6, 7* – detector head.

The meter consists of the coaxial–waveguide transition *1* with the waveguide section 72 × 34 mm^2, a coaxial matched broadband tee *2*, two broadband fixed attenuators, an 'beyond-cutoff' attenuator, and two broadband coaxial detectors with semiconductor microwave diodes. Microwave power goes to the antenna and then through the matched tee *2*, spreads through two measuring channels. In one of the channels, called 'linear', after a fixed attenuator, a coaxial detector head with a semiconductor microwave diode is installed. In the second channel, called 'non-linear', between the fixed attenuator and the coaxial detector, which is similar to the detector in the 'beyond-cutoff' channel, there is an 'over-limit' attenuator in which the amount of attenuation of which varies from frequency according to a certain law. Detected signals from two measuring channels are fed to different channels of the oscilloscope. For both channels there are calibration curves for the dependence of the output voltage level on the microwave power at the input of the spectrum meter at different frequencies of the microwave signal.

To determine the frequency and the pulse power of the microwave signal at any time, the voltage from the 'linear' U_L and the 'nonlinear' U_N channels is measured. When using the calibration curves of the channels, two relationships are constructed. For the 'linear' channel, the dependence of P_L on the frequency is plotted for a voltage $U = U_L$; for the 'non-linear' channel – the dependence of P_N on the frequency at a voltage $U = U_N$. The intersection point of these curves gives the value of the frequency and power at a defined time. To determine the frequency value during the whole microwave pulse, it is necessary to perform the described procedure for each digitized time point of the signal.

RM with uncoupled outputs. As noted above, with the change in the magnitude of the magnetic field in the range from 0.2 to 0.55 T, excitation of several types of oscillations: $\pi/2$-type with a frequency of ~3040 MHz, $3\pi/4$-type with a frequency of ~3140

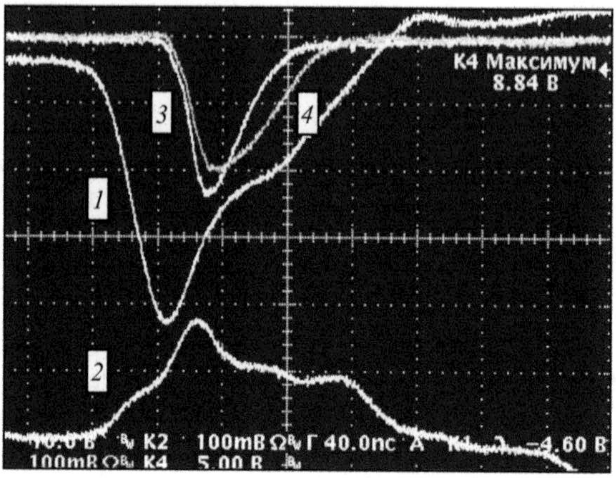

Fig. 6.36. Synchronized oscillograms: *1* – voltage; *2* – current; *3* – detector signal of the 'linear' channel; *4* – 'non-linear' channel.

MHz and π-type with a frequency of ~3300 MHz was observed. In these experiments, it was not possible to initiate a 'pure' π-mode of oscillation at large magnetic fields (0.5–0.55 T); the frequency spectrum contained simultaneously two frequencies corresponding to π- and $3\pi/4$-types. The maximum power from the two magnetron outputs was 230 MW. The synchronized oscillograms of the voltage, current, and signals of the detectors of the 'linear' and 'non-linear' channels are shown in Fig. 6.36.

The dynamics of the radiation frequency of the relativistic magnetron and the microwave power pulse for the $3\pi/4$-type are shown in Fig. 6.37. It can be noted that the frequency is constant for a sufficiently short time interval of about 15 ns, beyond which the rate of change of frequency varies sharply and amounts to ~10 MHz/ns. This behaviour of the frequency can be explained as follows. In the initial period of the supply pulse, the excitation of a relativistic magnetron occurs at high voltages (Fig. 6.37) in the $\pi/2$ mode, which has a small slowing-down of the electromagnetic wave (frequency about 3 GHz). As the anode current increases, the voltage between the cathode and the anode decreases and the conditions for exciting $3\pi/4$-mode oscillation, which has a lower excitation voltage at a fixed magnetic field (frequency about 3140 MHz), start to be fulfilled. At the end of the pulse, the behaviour of the frequency of the RM is determined by plasma effects, namely, an increase in the diameter of the cathode electron layer. An increase in the diameter of the

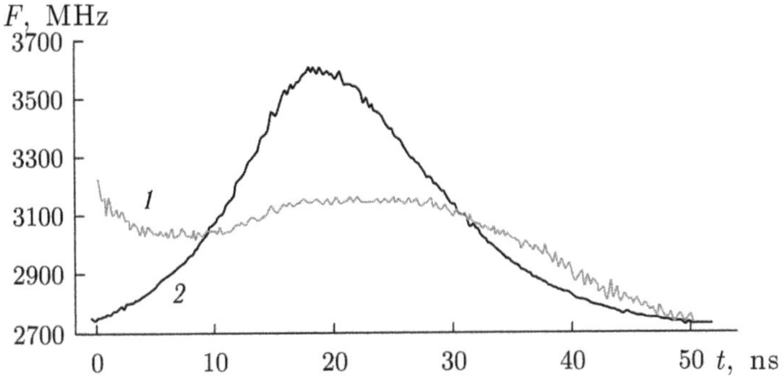

Fig. 6.37. Dynamics of the frequency (*1*) of the radiation of an 8-cavity RM with uncoupled outputs, (*2*) – the shape of the microwave power pulse.

cathode plasma layer is equivalent to an increase in the cathode–anode capacitance, which causes a frequency change in accordance with the approximate formula

$$f_n = f_0 / \sqrt{1 + \frac{C'}{2C\left(1 - \cos(2\pi n / N)\right)}}, \tag{6.9.1}$$

where C is the equivalent capacitance of a single resonator, C' is the concentrated capacitance between one lamella and the cathode, f_0 is the resonant frequency of a single resonator [37]. From these data, we can estimate the speed of radial motion of the cathode plasma in a magnetron diode, which amounted to 4–5 cm/μs. The results obtained are in good agreement with the data of [38], in which a different method of estimation was used.

RM with one load in the communication channel of the resonator. Measurements were made of the radiation frequency of a relativistic magnetron with an external coupling channel of resonators when various modes of oscillation were excited. Figure 6.38 shows the results of measurements in the case of symmetric and antisymmetric connection of the central load and at optimal channel lengths. As can be seen from the figure, the radiation frequencies of the RM differ for these cases by approximately 50 MHz. Taking into account that these frequencies correspond to both the anode block calculation data and the RM spectrum measurements with the help of a bandpass filter, they should be correlated with the π- and $3\pi/4$ modes of oscillation. Thus, the measurements confirm that

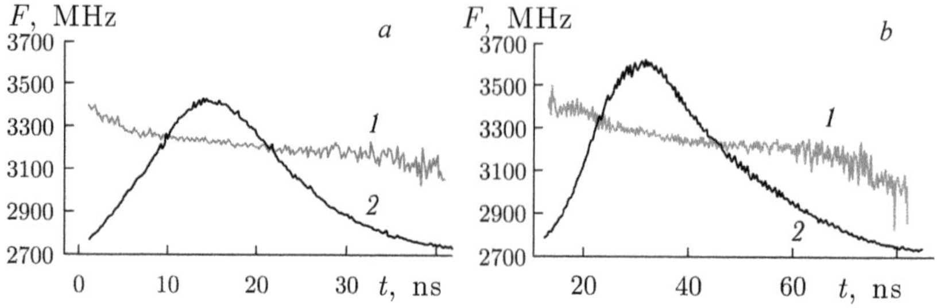

Fig. 6.38. The dynamics of the frequency (1) of the radiation of the 8-cavity RM, (2) is the shape of the microwave power pulse for a symmetric (*a*) and antisymmetric (*b*) installation of the load.

for a symmetric or an antisymmetric configuration of the coupling channel of RM resonators with an appropriate tuning of the channel length it is possible to isolate 'pure' π- and $3\pi/4$-modes. It was also stipulated that the magnitude of the synchronous magnetic field in the excitation of the $3\pi/4$-mode oscillations equal 0.41–0.44 T, below the corresponding value for π-type oscillations (0.5–0.55 T). This indicates different values of the slowing-down of the electromagnetic wave, which also agrees with the calculated estimates.

The experiments showed a characteristic expansion of the region of magnetic fields at which stable microwave generation on a specific type of oscillation was observed in comparison with the RM with uncoupled resonators. The maximum power of the magnetron at symmetric connection of the tee was 290 MW, at antisymmetric installation was 350 MW.

RM with a waveguide–slit array in the coupling channel of the resonators. Figure 6.39 shows the behaviour of the frequency of an RM with a waveguide–slit array. In contrast to the previous experiments, a higher stability of the radiation frequency during the pulse was recorded. This result can be related to the resonant properties of the coupling channel. The resonance nature of the load, as is well known, can stabilize the frequency of auto-oscillations due to the phenomenon of pulling. These effects were observed by us in experiments with a 6-cavity magnetron, where the resonance properties of the coupling channel of the resonators were due to the relatively small mismatch of the waveguide tee (the common load). However, for a long waveguide channel, these resonance properties become so significant that in addition to the pulling effect (under

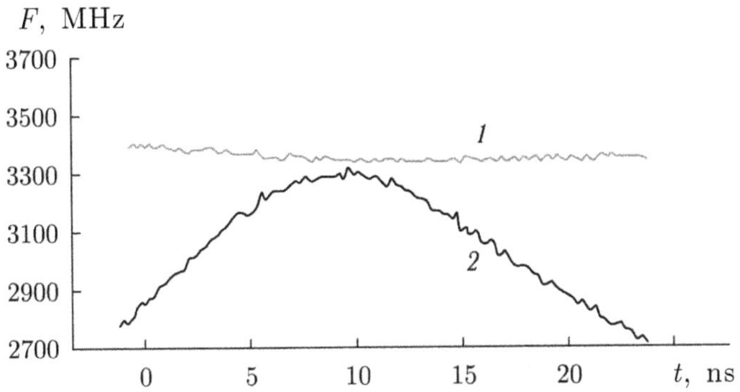

Fig. 6.39. The dynamics of the emission frequency of the 8-cavity RM with a waveguide–slit array (*1*), the shape of the microwave power pulse (2).

optimal tuning), they also cause the appearance of subtypes of the π-mode of oscillations.

Thus, as a result of experiments with the use of a microwave frequency meter, the frequency of the microwave radiation of a relativistic magnetron was measured during the generation pulse. For a magnetron with uncoupled outputs, a small interval of time for maintaining a stable frequency during a pulse is characteristic, beyond which the frequency varies at a rate of 5 MHz/ns. A significant influence on this is due to the excitation of neighbouring modes at the pulse front and plasma phenomena in the interaction space of the device. The generation of microwave radiation of a relativistic magnetron with coupled resonators is characterized by the fact that practically during the entire radiation pulse there is one-mode generation with a frequency deviation of 3 MHz/ns caused by plasma processes and a change in the supply voltage. A relativistic magnetron with a waveguide-slit array in the coupling channel of resonators is characterized by a high stability of the frequency, which is caused by the radiating system.

The results described in the chapter demonstrate the possibility of effectively controlling the energy and spectral characteristics of the microwave radiation of a relativistic magnetron by changing the parameters of the external coupling of the resonators. Controllable strong external coupling provide a selective influence on the specified processes and characteristics and at the optimum organization of the communication channel allow simultaneously to solve problems of stabilization of the spectral composition of the microwave oscillations of the RM and the problem of the output and spatial formation of its radiation. An indicator of the effectiveness of the effects of coupling is not only the narrowing of the spectrum and an increase in the mode stability of the operating mode of oscillations, but also the instability initiated under certain conditions in the form of competing oscillation subtypes. Depending on the configuration (symmetry) of the communication channel, stable modes with summation or subtraction of the resonator powers in the common load-radiator can be realized in the system for the selected type. When a system of load-radiators is installed into the communication channel, the relativistic magnetron acquires the properties of a super-power generating module with the directional radiation of microwave energy.

References

1. Bychkov S.I., Questions of the theory and practical application of devices of the magnetron type. Moscow, Sov. radio, 1967.
2. Polovkov I.P., Stabilization of the frequency of microwave generators by an external volume resonator. Moscow, Sov. radio, 1967.
3. Samsonov D.E., Basics of calculation and design of magnetrons. Moscow, Sov. radio, 1974.
4. Gutsait E.M., Electronics and its application, Itogi Nauki i Tekhniki VINITI Akademii Nauk SSSR. 1976. V. 8. S. 5–42.
5. Migulin V.V., et al., Fundamentals of the theory of oscillations. Textbook, Moscow, Nauka, 1988.
6. Kapranov M.V., Theory of oscillations in radio engineering, Textbook, Moscow, Nauka, 1984.
7. Utkin G.M., Self-oscillating systems and wave amplifiers. Moscow, Sov. radio, 1978.
8. Bogolyubov N.N., Mitropolsky Yu.A., Asymptotic methods in the theory of nonlinear oscillations. Moscow, Nauka, 1974.
9. Vintizenko I.I., et al., Pis‘ma Zh. Teor. Fiz. 1983. Vol. 9. No. 8. P. 482–485.
10. Kitsanov S.A., et al., Zh. Teor. Fiz. 2002. Vol. 72. No. 5. P. 82–90.
11. Korovin S.D., et al., Pis‘ma Zh. Teor. Fiz. 2005. T. 31. No. 10. P. 17–23.
12. Patent for invention No. 2190281 of the Russian Federation, IPC H 01 J 25/50. Relativistic magnetron, I.I. Vintizenko, A.I. Zarevich, S.S. Novikov. No. 2001128794; Declared 25.10.2001, BI. 2002. No. 27.
13. Patent for invention No. 2228560 of the Russian Federation, IPC H 01 J 25/50. Relativistic magnetron, I.I. Vintizenko, A.I. Zarevich, S.S. Novikov. No. 2002124144; Declared. 11.09.2002, BI. 2004. No. 13.
14. Dem'yanchenko A.G., Synchronization of generators of harmonic oscillations. Moscow, Energiya, 1976.
15. Novikov S..S., ZarevichA.I. in: Proc. of 13th Int. Symp. On High Current Electronics. Tomsk, 2004. P. 273–276.
16. Novikov S.S., Izv. Fizika. 2007. No. 10/3. P. 206–214.
17. Novikov S.S., Izv. Fizika. 2007. No. 10/3. P. 215-221.
18. Dvornikov A.A., Utkin G.M., Phased autogenerators of radio transmitting devices.

Moscow, Energiya, 1980.
19. Tsarapkin D.P., Microwave generators on Gunn diodes. Moscow, Radio i svyaz'. 1982.
20. Baskakov S.I., Radio engineering circuits and signals. Moscow, Vysshaya shkola, 1988.
21. Patent for invention No. 2190281 Russian Federation, IPC H 01 J 25/50. Relativistic magnetron, I.I. Vintizenko, AI Zarevich, SS Novikov. No. 2001128794. Declared 25.10.2001. Publ. 2002. BI. No. 27.
22. Zarevich A.I., in :Proc. of 13th International Symposium on High Current Electronics. Tomsk, 2004. P. 300–303.
23. Vintizenko I.I., et al., Izv. Fizika. 2006. No. 9. P. 114–118.
24. Zarevich A.I., et al., Izv. Fizika. 2006. No. 11. Appendix. P. 454–457.
25. Vintizenko I.I., et al., Pis'ma Zh. Teor. Fiz. 2006. Vol. 32. No. 23. P. 40–47.
26. Vintizenko I.I., et al., Pis'ma Zh. Teor. Fiz. 2003. V. 29. No. 7. P. 64–70.
27. Patent for invention No. 2228560 Russian Federation, IPC H 01 J 25/50. Relativistic magnetron, I.I. Vintizenko, AI Zarevich, SS Novikov. No.2002124144. September 11, 2002. Publ. 2004. BI. No. 13.
28. Zarevich A.I., et al., in: Proc. of 13th International Symposium on High Current Electronics. Tomsk, 2004. P. 269–272.
29. Vintizenko I.I., et al., Pis'ma Zh. Teor. Fiz. 2005. V. 31. No. 9. P. 63–68.
30. Vintizenko I.I., Novikov S.S., Izv. Fizika. 2007. No. 10/3. Pp. 115–120.
31. Voskresenskii D.I., et al., Antennas and microwave devices (Design of phased arrays): Moscow. Radio i svyaz'. 1981.
32. Vintizenko I.I., Novikov S.S., Izv. Fizika. 2008. No. 9/2. Pp. 154–156.
33. Vintizenko I.I., Novikov S.S., in: Proc. of 15th International Symposium on High Current Electronics. Tomsk, 2008. P. 429–430.
34. Patent for invention No. 2337426 Russian Federation. Relativistic magnetron with external channels of resonator coupling. Vintizenko I.I. Priority 04.07.07. Date of registration on 27.10.08. Publ. 2008. BI. No. 30.
35. Vintizenko I.I., et al., Izv. Fizika. 2009. No. 11/2. Pp. 271–277.
36. Vintizenko I.I., Melnikov G.V., Pis'ma Zh. Teor. Fiz. 2010. V. 39. No.15. P. 45–52.
37. Babichev D.A., et al., Prib. Tekh. Eksper. 2003. No. 3. P. 93–96.
38. Lebedev I.V., Microwave technology and devices. Moscow, Vysshaya shkola. 1972..
39. Gleizer I.Z., et al., Pis'ma Zh. Teor. Fiz. 1980. Vol. 6. No. 1. P. 44–49.

Index

D

E

F

H

I

L

M

N

P

R

S

T

V

Milton Keynes UK
Ingram Content Group UK Ltd.
UKHW021636071024
449327UK00020BA/1326

9 780367 571115